Linear Feedback Controls

Linear Feedback Controls

The Essentials

Mark A. Haidekker

College of Engineering
Driftmier Engineering Center
University of Georgia
Athens, GA, USA

AMSTERDAM • BOSTON • HEIDELBERG • LONDON • NEW YORK • OXFORD
PARIS • SAN DIEGO • SAN FRANCISCO • SINGAPORE • SYDNEY • TOKYO

Elsevier
225 Wyman Street, Waltham, MA 02451, USA
32 Jamestown Road, London NW1 7BY, UK

First edition 2013

Copyright © 2013 Elsevier Inc. All rights reserved.

No part of this publication may be reproduced or transmitted in any form or by any means, electronic or mechanical, including photocopying, recording, or any information storage and retrieval system, without permission in writing from the publisher. Details on how to seek permission, further information about the Publisher's permissions policies and our arrangements with organizations such as the Copyright Clearance Center and the Copyright Licensing Agency, can be found at our website: www.elsevier.com/permissions.

This book and the individual contributions contained in it are protected under copyright by the Publisher (other than as may be noted herein).

Notices
Knowledge and best practice in this field are constantly changing. As new research and experience broaden our understanding, changes in research methods, professional practices, or medical treatment may become necessary.

Practitioners and researchers must always rely on their own experience and knowledge in evaluating and using any information, methods, compounds, or experiments described herein. In using such information or methods they should be mindful of their own safety and the safety of others, including parties for whom they have a professional responsibility.

To the fullest extent of the law, neither the Publisher nor the authors, contributors, or editors, assume any liability for any injury and/or damage to persons or property as a matter of products liability, negligence or otherwise, or from any use or operation of any methods, products, instructions, or ideas contained in the material herein.

Library of Congress Cataloging-in-Publication Data
A catalog record for this book is available from the Library of Congress

British Library Cataloguing-in-Publication Data
A catalogue record for this book is available from the British Library

ISBN: 978-0-12-405875-0

For information on all Academic Press publications
visit our website at store.elsevier.com

This book has been manufactured using Print On Demand technology. Each copy is produced to order and is limited to black ink. The online version of this book will show color figures where appropriate.

Contents

Preface

Upon looking at this book, the casual reader might ask, why another book on feedback controls? In fact, there are literally dozens of excellent, comprehensive books available on the subject of feedback controls. *Comprehensive* is the important word in this context. The subject of feedback controls is one of the core disciplines of engineering and has been extensively developed over the past two centuries. Analysis of linear systems and design of feedback control systems is based heavily on mathematics. As a consequence, any book that intends to cover this field exhaustively becomes very voluminous. Many books on feedback controls exceed a thousand pages. Even some books that intend to provide mere overviews or outlines readily exceed 500 pages.

This trend gave rise to the present book Linear Feedback Controls—*The Essentials*, with an emphasis on the essentials. The present book strictly focuses on linear systems under the premise that nonlinear control theory is a completely separate subject area. This book covers all of the core areas of classical feedback control systems, including introduction of the mathematical tools needed for control analysis and design. In contrast to many other books, this book presents fewer examples. These examples, however, guide the reader through the entire book, thus allowing a better comparison of the various aspects of different analysis and design methods. The examples in this book are representative for a large number of analogous systems in electrical and mechanical engineering.

This book is built on the philosophy that the individual methods (such as the Laplace-domain treatment of differential equations, interpretation of pole locations, Bode and root locus design methods) comprise the elements of a design engineer's toolbox. Accordingly, this book presents the individual tools and builds the mathematical basis for those tools in the earlier chapters. One important aspect of this book is the seamless integration of time-discrete systems, that is, controls based on digital circuits, microprocessors, or microcontrollers, because digital, time-discrete controls rapidly gain importance and popularity.

The first four chapters of this book provide some important foundations, most importantly, the Laplace transform and its time-discrete counterpart, the z-transform. Moreover, in Chapter 2, a brief review of linear systems is provided, because the analysis of signals and systems is fundamental to understanding feedback controls.

Progressing towards more practical aspects, the following chapters deal with a simple first-order example, the temperature-controlled waterbath. This example system is complemented with a second system, a linear positioning system. The second example invites a focus on the dynamic response of a system, as opposed to the equilibrium response that is the focus of the waterbath example.

In the course of the later chapters, numerous design tools are introduced: formal representation of systems with block diagrams, treatment of nonlinear components, stability analysis, design methods based on a system's frequency response, and the powerful root locus design method.

Chapter 13 (PID-control) and Chapter 14 (design examples) provide the link to the practical realization of the materials in this book. Although the book focuses on mathematical methods, examples for computer-based analysis and simulation are provided. In addition, the book contains many examples of how control elements can be realized with analog circuits (predominantly based on operational amplifiers), and contains algorithms for the realization of time-discrete control elements with micro-controllers.

Overall, I hope that *The Essentials* makes the sometimes daunting subject of feedback controls more accessible and facilitates the entry into this fascinating and important subject. I believe that this book provides a big-picture overview, with examples that relate to the practical application of control systems. In the sense of closing the feedback loop, I also welcome any comments and suggestions of how to improve *The Essentials*.

Athens, April 2013 *Mark A. Haidekker*

Acknowledgments

I owe thanks to many individuals who helped to make this book possible. Among several students of my course *Feedback Control Systems* who provided suggestions, I particularly thank Robert Wainwright, Richard Speir, and Dr. Adnan Mustafic for detailed reviews and Alex Squires for developing Table A.6, which makes the collection of Laplace correspondences one of the most comprehensive correspondence tables available. I also thank the anonymous reviewers for their in-depth review and one not-so-anonymous reviewer, Dr. Matthew Nipper, for insisting to include digital systems—clearly, digital systems add a significant modern element to the contents of this book. I am especially indebted to Dr. Javad Mohammadpour for his particularly extensive review of the completed manuscript and his numerous valuable suggestions for improvement.

Additional credit is due to the millions of skilled programmers and other contributors to Free Software around the world. This book has been entirely produced with Free Software (for the use of the capitalized Free, see http://www.gnu.org/philosophy/free-sw.html): Typesetting was performed with the LaTeX environment. Diagrams were mostly sketched with Inkscape or a combination of Inkscape with Scilab or gnuplot. For circuit schematics, gEDA was used. For simulations, Scilab and its graphical module Xcos were employed. Some control algorithms were developed with Microchip controllers by using the gputils toolchain and the Piklab integrated development environment. All software was run on GNU/Linux-based systems.

Last, but most importantly, the team at Elsevier deserves my special thanks for their support and encouragement throughout the entire publication process: my Acquisition Editors Wayne Yuhasz for his advice in the early stages of the book design and Erin Hill-Parks for her support and promotion of the book, and her guidance through the entire publication process; the Editorial Project Managers, Tracey Miller and Sarah Lay, for their invaluable help with the practical details; and the Production Project Manager, Radhakrishnan Lakshmanan, and his team for putting the book in its final format.

List of Commonly used Symbols

a	General real-valued scalar or coefficient
C	Capacity
D	Spring constant (Hooke's law)
$D(s)$	Disturbance
$\delta(t)$	Kronecker delta impulse function
E	Energy
$E(s)$	Tracking error
ϵ	Control deviation
ϵ	A very small quantity, often used in the context of $\lim_{\epsilon \to 0}$
F	Force
$F(s)$	General Laplace-domain function, often $F(s) = \mathscr{L}\{f(t)\}$
f	Frequency (sometimes also referred to as *linear frequency* to distinguish f from the angular frequency ω)
$f(t)$	General time-dependent function
$\dot{f}(t)$	First derivative of $f(t)$ with respect to time
$\ddot{f}(t)$	Second derivative of $f(t)$ with respect to time
$f(t)^{<n>}$	nth derivative of $f(t)$ with respect to time
$\mathscr{F}$	Fourier transform
$\mathscr{F}^{-1}$	Inverse Fourier transform
$G(s)$, $H(s)$	General Laplace-domain transfer functions
$h(t)$	Time-domain step response of a linear system. Its Laplace transform is the transfer function $H(s)$
I	Electrical current
$\Im(c)$	Imaginary part of c
J	Rotational inertia
j	Imaginary unit, $j := \sqrt{-1}$
k	Proportionality constant
k_p	Gain of a P-controller
L	Magnetic inductivity

$L(s)$	Loop gain
$\mathscr{L}$	Laplace transform
$\mathscr{L}^{-1}$	Inverse Laplace transform
m	Mass
$N(s)$	Noise signal
ω	Angular frequency or angular velocity, $\omega = 2\pi f$
ω_c	Cutoff frequency of an analog filter
P	Electrical power
p	Pole location (may be real-valued or complex)
$p(s)$	Numerator polynomial of a transfer function
$q(s)$	Denominator polynomial of a transfer function
R	Electrical resistance
R_F	Mechanical friction
$R(s)$	Reference signal (setpoint)
$\Re(c)$	Real part of c
s	Complex location in the Laplace-domain with $s = \sigma + j\omega$. Note that s has units of inverse seconds
T	Temperature
T	Time period (often $T = 1/f = 2\pi/\omega$)
T	Sampling period in time-discrete systems
t	Time
τ	Time constant of an exponentially equilibrating process
τ	Torque
$u(t)$	Unit step function, defined as 1 for $t \geq 0$ and 0 otherwise
V	Voltage
w	Complex location in the w-plane with $w = \mu + j\nu$
Z	Digital value
z	Complex variable for the z-transform
z	Zero location in the s-plane (may be real-valued or complex)
$\mathscr{Z}$	z transform
$\mathscr{Z}^{-1}$	Inverse z transform
ζ	Damping factor of a second-order (e.g., spring-mass-damper) system

1 Introduction to Linear Feedback Controls

Abstract

Automation and controls date back thousands of years, and likely begun with the desire to keep water levels for irrigation constant. Much later, the industrial revolution brought a need for methods and systems to regulate machinery, for example the speed of a steam engine. Since about two centuries, engineers have found methods to describe control systems mathematically, with the result that the system behavior could be more accurately predicted and control systems more accurately designed. *Feedback controls* are control systems where a sensor monitors the property of the system to be controlled, such as motor speed, pressure, position, voltage, or temperature. Common to all feedback control systems is the comparison of the sensor signal to a reference signal, and the existence of a *controller* that influences the system to minimize the deviation between the sensor and reference signals. Feedback control systems are *designed* to meet specific goals, such as keeping a temperature or speed constant, or to accurately follow the reference signal. In this chapter, some of the fundamental principles of feedback control systems are introduced, and some common terms defined.

Feedback control systems have a long history. The first engineered feedback control systems, invented nearly 2500 years before our time, were intended to keep fluid levels constant. One application were early water clocks. With constant hydrostatic pressure, the flow rate of water can be kept constant, and the fill time of a container can be used as a time reference. A combination of a floater and a valve served as the control unit to regulate the water level and thus the hydrostatic pressure.

During the early industrial revolution, the wide adoption of steam engines was helped with the invention of the governor. A steam engine suffers from long delays (for example, when coal is added to the furnace), making them hard to control manually. The governor is an automated system to control the rotational speed of a steam engine in an unsupervised fashion. Float regulators were again important, this time to keep the water level in a boiler constant (Figure 1.1).

About 200 years ago, engineers made the transition from intuitively-designed control systems to mathematically-defined systems. Only with the help of mathematics to formally describe feedback control systems it became possible to accurately predict the response of a system. This new paradigm also allowed control systems to grow more complex. The first use of differential equations to describe a feedback control system is attributed to George Biddell Airy, who attempted to compensate a telescope's position for the rotation of the earth with the help of a feedback control system. Airy also made the discovery that improperly designed controls may lead to large oscillatory responses, and thus described the concept of *instability*. The mathematical treatment

Linear Feedback Controls. http://dx.doi.org/10.1016/B978-0-12-405875-0.00001-2
© 2013 Elsevier Inc. All rights reserved.

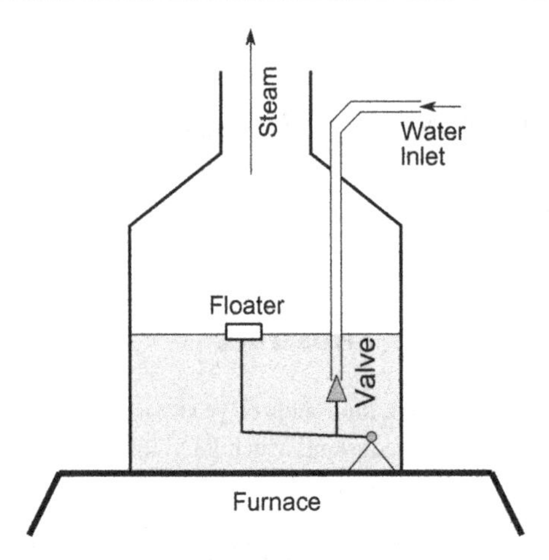

Figure 1.1 Schematic of a water-level regulator with a floater. As steam is produced, the water level sinks, and so does the floater. A valve opens as a consequence, and new water is allowed to enter the boiler, causing the floater to rise and close the valve.

of control systems—notably stability theory—was further advanced by James Clerk Maxwell, who discovered the characteristic equation and found that a system is stable if the characteristic equation has only roots with a negative real part.

Another important step was the frequency-domain description of systems by Joseph Fourier and Pierre-Simon de Laplace. The frequency-response description of systems soon became a mainstream method, driven by progress in the telecommunications. Around 1940, Hendrik Wade Bode introduced double logarithmic plots of the frequency response (today known as Bode-plots) and the notion of phase and gain margins as a metric of relative stability. Concurrently, progress in automation and controls was also driven by military needs, with two examples being artillery aiming and torpedo and missile guidance. The *PID* controller was introduced in 1922 by Nicolas Minorsky to improve the steering of ships.

With the advent of the digital computer came another revolutionary change. The development of digital filter theory found immediate application in control theory by allowing to replace mechanical controls or controls built with analog electronics by digital systems. The algorithmic treatment of control problems allowed a new level of flexibility, specifically for nonlinear systems. Modern controls, that is, both the numerical treatment of control problems and the theory of nonlinear control systems complemented classical control theory, which was limited to linear, time-invariant systems.

In this book, we will attempt to lay the foundations for understanding classical control theory. We will assume all systems to be linear and time-invariant, and make linear approximations for nonlinear systems. Both analog and digital control systems are covered. The book progresses from the theoretical foundations to more and more practical aspects. The remainder of this chapter (Chapter 1) introduces the basic concepts and terminology of feedback controls, and it includes a first example: two-point control systems.

A brief review of linear systems, their differential equations and the treatment in the Laplace domain follows in Chapters 2 and 3. In Chapter 3, several elemental low-order systems are introduced. Chapter 4 complements Chapter 3 for digital systems by introducing sampled signals, finite-difference equations and the z-transform.

A simple first-order system is then introduced in detail in Chapter 5, and the effect of feedback control is extensively analyzed through differential equations. The same analysis is repeated in Chapter 6, but this time the entire analysis takes place in both the Laplace domain and for digital controllers in the z-domain, thus establishing the relationship between time-domain and Laplace-domain treatment. A second-order example introduced in Chapter 9 allows to more closely examine the dynamic response of a system and how it can be influenced with feedback control.

Subsequent chapters can be seen as introducing tools for the design engineer's toolbox: the formal description of linear systems with block diagrams (Chapter 7), the treatment of nonlinear components (Chapter 8), stability analysis and design (Chapter 10), frequency-domain methods (Chapter 11), and finally the very powerful root locus design method (Chapter 12). A separate chapter (Chapter 13) covers the *PID* controller, which is one of the most commonly used control systems.

Chapter 14 is entirely dedicated to providing practical examples of feedback controls, ranging from temperature and motor speed control to more specialized applications, such as oscillators and phase-locked loops. The importance of Chapter 14 lies in the translation of the theoretical concepts to representative practical applications. These applications allow to demonstrate how the mathematical concepts of this book relate to practical design goals.

The book concludes with an appendix that contains a comprehensive set of Laplace- and z-domain correspondences, an introduction to operational amplifiers as control elements, and an overview of key commands for the simulation software *Scilab*. Scilab (www.scilab.org) is free, open-source software, which any reader can freely download and install. Furthermore, Scilab is very similar to MATLAB, and readers can easily translate their knowledge to MATLAB if needed.

1.1 What are Feedback Control Systems?

A feedback control system continuously monitors a process and influences the process in such a manner that one or more process parameters (the *output variables*) stay within a prescribed range. Let us illustrate this definition with a simple example. Assume an incubator for cell culture as sketched in Figure 1.2. Its interior temperature needs to be kept at 37 °C. To heat the interior, an electric heating coil is provided. The incubator box with the heating coil can be seen as the *process* for which feedback control will be necessary, as will become evident soon. For now, let us connect the heating coil to a rheostat that allows us to control the heat dissipation of the heating coil. We can now turn up the heat and try to relate a position of the rheostat to the interior temperature of the incubator. After some experimentation, we'll likely find a position where the interior temperature is approximately 37 °C. Unfortunately, after each change of the rheostat, we have to wait some time for the incubator temperature

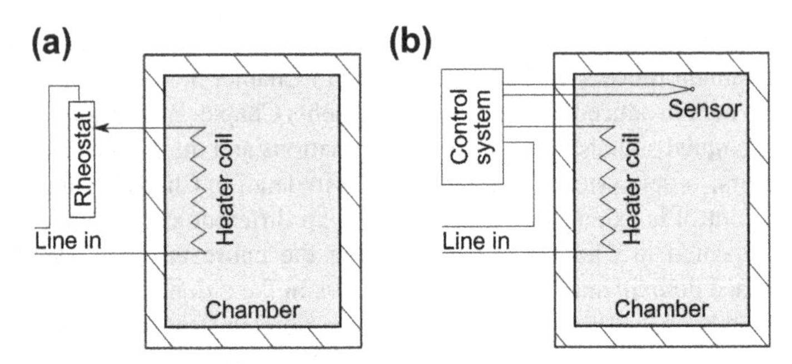

Figure 1.2 Schematic representation of an incubator. The interior of the chamber is supposed to be kept at a constant temperature. A rheostat can be used to adjust the heater power (a) and thus influence the temperature inside the chamber. The temperature inside the chamber equilibrates when energy introduced by the heater balances the energy losses to the environment. However, the energy losses change when the outside temperature changes, or when the door to the incubator is opened. This may require readjustment of the rheostat. The system in (a) is an open-loop control system. To keep the temperature inside the chamber within tighter tolerances, a sensor can be provided (b). By measuring the actual temperature and comparing it to a desired temperature, adjustment of the rheostat can be automated. The system in (b) is a feedback control system with feedback from the sensor to the heater.

to equilibrate, and the adjustment process is quite tedious. Even worse, equilibrium depends on two factors: (1) the heat dissipation of the heater coil and (2) the heat losses to the environment. Therefore, a change in the room temperature will also change the incubator's temperature unless we compensate by again adjusting the rheostat. If somebody opens the incubator's door, some heat escapes and the temperature drops. Once again, it will take some time until equilibrium near 37 °C is reached.

Although the rheostat allows us to control the temperature, it is not *feedback* control. Feedback control implies that the controlled variable is continuously monitored and compared to the desired value (called the *setpoint*). From the difference between the controlled variable and the setpoint, a *corrective action* can be computed that drives the controlled variable rapidly toward the value predetermined with the setpoint. Therefore, a feedback control system requires at least the following components:

- **A process.** The process is responsible for the output variable. Furthermore, the process provides a means to influence the output variable. In the example in Figure 1.2, the process is the chamber together with the heating coil. The output variable is the temperature inside the chamber, and the heating coil provides the means to influence the output variable.
- **A sensor.** The sensor continuously measures the output variable and converts the value of the output variable into a signal that can be further processed, such as a voltage (in electric control systems), a position (in mechanical systems), or a pressure (in pneumatic systems).
- **A setpoint**, or more precisely, a means to adjust a setpoint. The setpoint is related to the output variable, but it has the same units as the output of the sensor. For example,

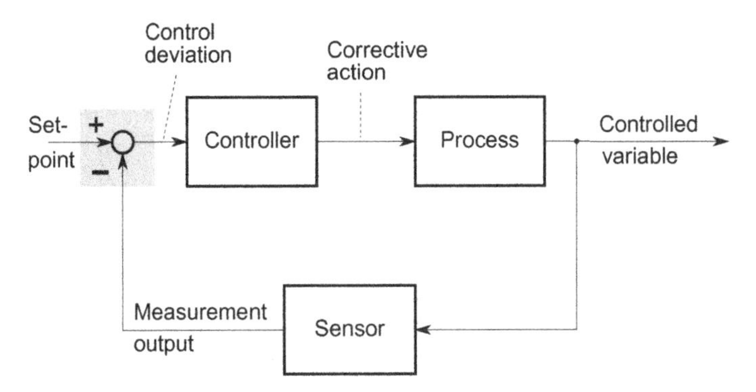

Figure 1.3 Block diagram schematic of a closed-loop feedback system. In the example of the incubator, the process would be the incubator itself with the heater coil, the sensor would be a temperature sensor that provides a voltage that is proportional to the temperature, and the controller is some sort of power driver for the heating coil that provides heating current when the temperature inside the incubator drops below the setpoint. In almost all cases, the measurement output is subtracted from the setpoint to provide the control deviation. This error signal is then used by the actual controller to generate the corrective action. The gray shaded rectangle highlights the subtraction operation.

if the controlled variable is a temperature, and the sensor provides a voltage that is proportional to the temperature, then the setpoint will be a voltage as well.

- **A controller.** The controller measures the deviation of the controlled variable from the setpoint and creates a corrective action. The corrective action is coupled to the input of the process and used to drive the output variable toward the setpoint.

The act of computing a corrective action and feeding it back into the process is known as *closing the loop* and establishes the closed-loop feedback control system. A closed-loop feedback control system that follows our example is shown schematically in Figure 1.3. Most feedback control systems follow the example in Figure 1.3, and it is important to note that almost without exception the control action is determined by the controller from the *difference* between the setpoint and the measured output variable. This difference is referred to as *control deviation*.

Not included in Figure 1.3 are disturbances. A disturbance is any influence other than the control input that causes the process to change its output variable. In the example of the incubator, opening the door constitutes a disturbance (transient heat energy loss through the open door), and a change of the room temperature also constitutes a disturbance, because it changes the heat loss from the incubator to the environment.

1.2 Some Terminology

We introduced the concept of closed-loop feedback control in Figure 1.3. We now need to define some terms. Figure 1.4 illustrates the relationship of these terms to a closed-loop feedback system.

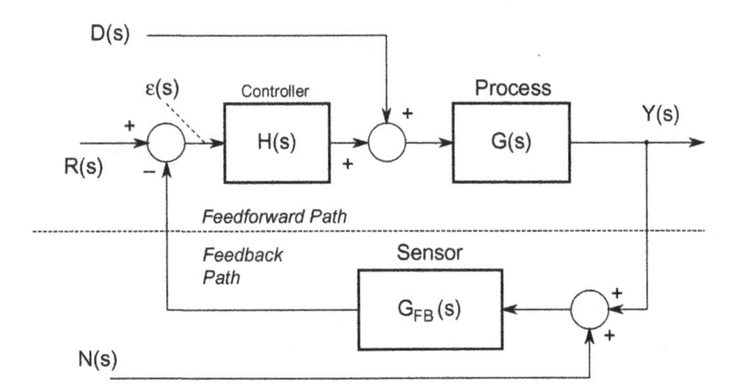

Figure 1.4 General block diagram for a feedback control system. The controlled variable is $Y(s)$, and the setpoint or reference variable is $R(s)$. A disturbance $D(s)$ can be modeled as an additive input to the process, and sensor noise $N(s)$ is often modeled as an additive input to the sensor. Note that we have nonchalantly used Laplace-domain functions, $H(s)$, $G(s)$, and $G_{FB}(s)$ as well as signals $R(s)$, $D(s)$, $N(s)$, and $Y(s)$ in place of their respective time-dependent functions. Laplace-domain treatment of differential equations is covered in Chapter 3, and the majority of the examples in this book are treated in the Laplace domain.

- **Process:** Also referred to as *plant*—the process is a system that has the controlled variable as its property. The process has some means to influence the controlled variable. Therefore, the process can be interpreted as a linear system with one output and one (or more) inputs.
- **Sensor:** The sensor is an apparatus to measure the controlled variable and make the measurement result available to the controller. The sensor itself may have its own transfer function, such as a gain or delay function.
- **Controller:** The controller is a device that evaluates the control deviation and computes an appropriate control action. In many cases, the controller is the only part of the feedback control system that can be freely *designed* by the design engineer to meet the design goals of the closed-loop system.
- **Disturbance:** Any influence other than the control action that influences the controlled variable. Examples are variable load conditions in mechanical systems, electromagnetic interference, shear winds for airplanes etc. A disturbance $d(t)$ can often be modeled as an additive input to the process.
- **Noise:** Noise is a broadband random signal that may be introduced in many places of a system. Amplifiers typically create noise; dirt or rust in a mechanical system would also create noise-like random influences. It is often possible to model noise as a single signal $n(t)$ as an additive input to the sensor.
- **Setpoint:** Also referred to as *reference signal.* This signal determines the operating point of the system, and, under consideration of the sensor output signal, directly influences the controlled variable.
- **Control deviation:** The control deviation $\epsilon(t)$ is the time-dependent difference between setpoint and sensor output. The control deviation is often also referred to as the *error variable* or *error signal.*

- **Control action:** Also termed *corrective action*. This is the output signal of the controller and serves to actuate the process and therefore to move the controlled variable toward the desired value.
- **Controlled variable:** This is the output of the process. The design engineer specifies the controlled variable in the initial stages of the design.
- **Dependent variable:** Any signal that is influenced by other signals anywhere in the entire system (such as the control deviation) is a dependent variable.
- **Independent variable:** Any external signal that is not influenced by other signals is an independent variable. Examples are the setpoint $r(t)$ and the disturbance $d(t)$.
- **Feedforward path:** The feedforward path includes all elements that directly influence the controlled variable.
- **Feedback path:** The feedback path carries the information about the controlled variable back to the controller. Note that the definition of the feedforward and feedback paths is not rigid and merely serves as orientation.

When the feedback path is not connected to the feedforward path at the first summation point (and thus $\epsilon(t) = r(t)$), the system is called *open-loop* system. The open-loop system is useful—often necessary—to characterize the individual components and to predict the closed-loop response. Chapters 11 and 12 describe methods how the closed-loop response can be deduced from the open-loop system. Once the feedback path is connected to the feedforward path, the system becomes a *closed-loop* system, and feedback control takes place.

1.3 Design of Feedback Control Systems

Feedback control systems must be *designed* to suit a predetermined purpose. Normally, only the controller can be appropriately designed, whereas the process and the sensor are predetermined or constrained. Feedback control systems can be designed to achieve specific behavior of the output variable, for example

- To keep the output variable within a tightly constrained range, irrespective of changes in the environment. The incubator introduced above is a good example.
- To rapidly and accurately follow a change in the reference signal. A good example is a disk drive head, where a signal requests positioning of the head over a selected track. The drive head is supposed to reach the target track as quickly as possible and then accurately follow the track for a read/write operation.
- To suppress a sharp, transient disturbance. One example is an active car suspension. When the car rolls over a pothole, the suspension control minimizes the movement of the cabin.
- To reduce the influence of process changes. The process itself may suffer from variations (for example, thermal expansion, accumulation of dirt and dust, mechanical degradation, changes in fluid levels or pressure, etc.). A feedback control system can compensate for these influences and keep the behavior of the controlled system within specifications.

- To linearize nonlinear systems. Hi-fi audio equipment makes extensive use of feedback control to produce high-quality audio signals with low distortion in spite of its components (transistors), which have highly nonlinear characteristics.

The above examples illustrate some possible design goals. Often, the design goals conflict, and a compromise needs to be found. Rational controller design consists of several steps:

1. Definition of the controlled variable(s) and control goals.
2. Characterization of the process. The response of the process to any change of the input must be known. Ideally, the process can be described by an equation (usually, a differential equation). If the process equation is not known, it can be deduced by measuring the response to specific input signals, such as a step change of the input.
3. Characterization of any nonlinear response of the process. If a process is not linear, a feedback control system cannot be designed with the methods of classical linear feedback control theory. However, when a process exhibits nonlinear characteristics, it can often be approximated by a linear system near the operating point.
4. Design of the sensor. The sensor needs to provide a signal that can be subtracted from the setpoint signal and used by the controller.
5. Design of the controller. This step allow the greatest flexibility for the design engineer. Key design criteria are:
   - The stability, i.e., whether a setpoint change or a disturbance can cause the closed-loop system to run out of control.
   - The steady-state response of the closed-loop system, i.e., how close the output variable follows the setpoint, provided that the closed-loop system had enough time to equilibrate.
   - The steady-state disturbance rejection, i.e., how well the influence of a disturbance (in the incubator example, the disturbance could be a change of the room temperature) is suppressed to have minimal long-term impact on the output variable.
   - The dynamic response, i.e., how fast the closed-loop system responds to a change of the setpoint and how fast it recovers from a disturbance.
   - The integrated tracking error, i.e., how well the system follows a time-varying setpoint.
6. Test of the closed-loop system. Since the mathematical description used to characterize the process and design the controller is often an approximation, the closed-loop system needs to be tested under conditions that can actually occur in practice. In many cases, simulation tools can help in the design and initial testing process. However, even sophisticated simulation tools are based on assumptions and approximations that may not accurately reflect the real system. Moreover, rounding errors or even numerical instabilities (*cf.* chaotic system, turbulence) may cause simulation results to deviate significantly from the actual system.

Figure 1.5 provides some insight as to what the design of a feedback control system can involve. Robots, such as the pictured bottling robot that feeds a filling machine, need to be fast and accurate. Feedback control systems ensure that the robot segments move between time-varying setpoints, such as the pickup point and the conveyor belt where

Figure 1.5 Example for the application of feedback control systems in robotics. The robot in this picture feeds empty bottles onto the conveyor belt of a filling machine. The robot moves rapidly between several setpoints to pick up and then deposit the bottles. To achieve high throughput, the robot needs to reach the setpoints as fast as possible. However, a deviation from its path, for example when the robot overshoots its setpoint, may break some bottles or even damage the conveyor.

the bottles are deposited. It is desirable that the robot reaches its setpoints as fast as possible to ensure high throughput. However, the robot cannot be allowed to overshoot its target, even though a defined overshoot may actually speed up the process. In the first example (Figure 1.3), we only considered the steady-state behavior, that is, the incubator temperature at equilibrium. The robot example shows us that the dynamic behavior (also called the transient response) can be equally important in the design considerations. Simplified, we can characterize the core of feedback control design as follows:

Feedback control design allows to influence a process with an undesirable transfer function by means of a controller such that the combined (i.e., controlled or closed-loop) system has a desirable transfer function.

1.4 Two-Point Control

Two-point control is a nonlinear feedback control method that is briefly covered here because of its ubiquity. Room thermostats, ovens, refrigerators, and many other everyday items contain two-point control systems. Two-point control implies that a corrective action is either turned on or off. A typical example is a room thermostat that controls a heater furnace. It is more cost-efficient to design a furnace that is either on or off, as compared to a furnace with controlled heating power (such as variable flame height). In this example, the closed-loop system with two-point control has therefore a temperature sensor and a two-point controller, i.e., a controller that turns the corrective action—the furnace—either on or off.

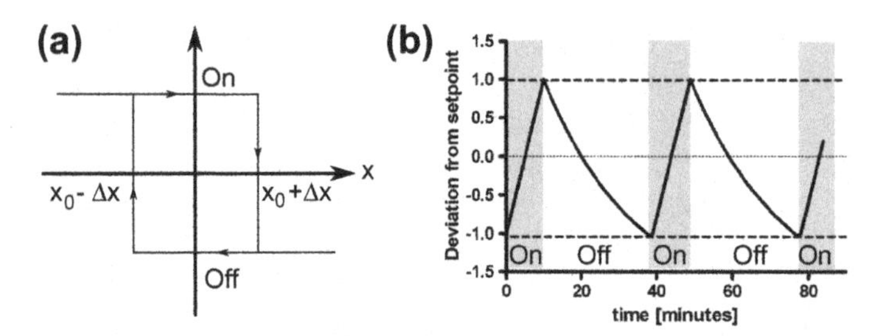

Figure 1.6 Control curve of a two-point controller (a) and a possible timecourse of the controlled variable (b). The controller exhibits hysteresis, that is, it turns the corrective action (heating) on if the input variable x is more than Δx below the setpoint x_0 (here placed in the origin) and keeps it on until x exceeds x_0 by the offset Δx, whereupon it turns the corrective action off. The corrective action stays off until the input variable again drops below $x_0 - \Delta x$. This switching behavior is indicated by the arrows in (a) that mark the direction of the path. If such a controller is used to control a process, for example a room with a heater, the temperature (controlled variable) fluctuates around the setpoint with an amplitude of $2\Delta x$.

A two-point control system needs to be designed with different on- and off-points to prevent undefined rapid switching at the setpoint. The difference in trip points, depending on whether the trip point is approached from above or below, is referred to as *hysteresis*. Figure 1.6a shows the characteristic curve of a two-point switch. The value of the input variable where the switch turns off is lower than the value where it turns on. The center of the hysteresis curve is adjustable and serves as setpoint. When used in the example of a furnace, the controlled variable, i.e., the temperature, fluctuates between the two trip points of the two-point control. A possible timecourse of the temperature deviation from the setpoint is shown in Figure 1.6b.

In two-point control systems, the width of the hysteresis determines the balance between setpoint accuracy and process concerns. The narrower the hysteresis curve, the more often the process will switch, causing start-up stress to the process. If the start phase of the process is associated with cost (stress, wear, higher energy consumption), the cost needs to balanced with accuracy needs. To strike the optimum balance is the duty of the design engineer.

Extensions of the two-point principle are straightforward. For example, the direction of a heat pump can be reversed with an electrical control signal. With this additional signal, the heat pump can therefore be set to alternatively heat or cool. To use this feature, the controller needs to be extended into a three- or four-point switch with different hysteresis curves for heating and cooling (Figure 1.7).

Two-point control systems obey the same mathematical models that apply for linear systems, but the on-phase and the off-phase of the process need to be handled separately. Let us examine the example of a room thermostat with a 1° hysteresis. We assume that the room temperature increases linearly when the heater is on, and decreases linearly when the heater is off, albeit with different rates:

$$\Delta T(t) = \begin{cases} k_1 \cdot t & \text{when heater is on} \\ -k_2 \cdot t & \text{when heater is off} \end{cases} \tag{1.1}$$

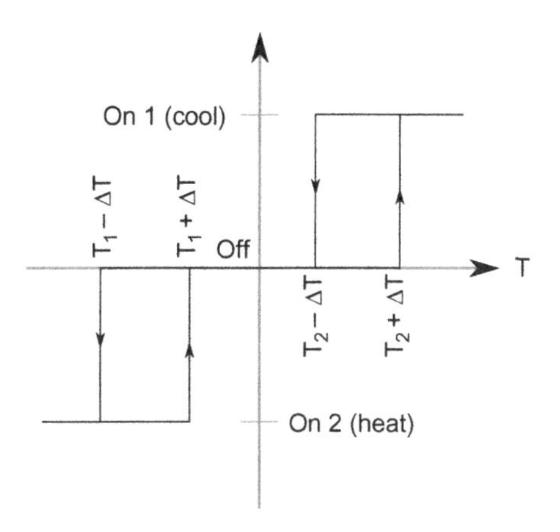

Figure 1.7 Control curve of a four-point controller in the example of a temperature control where a heat pump can provide both cooling and heating. Different trip points for cooling and heating are used, and each subsystem cycles with its own hysteresis. The offsets from the center temperature (denoted ΔT) do not necessarily have to be the same for all four trip points.

Here, $\Delta T(t)$ is the deviation of the temperature from the setpoint (we can move the hysteresis curve in Figure 1.6 to an arbitrarily chosen origin), k_1 depends primarily on heater power and room size, and k_2 depends on room size, quality of the insulation, and the temperature difference to the outside environment. The temperature timecourse is generally nonlinear, but within tight tolerances near the setpoint, the approximation in Eq. (1.1) is sufficiently accurate.

It is common practice to obtain process constants, such as k_1 and k_2, that define the behavior of the process. Frequently, those constants combine multiple factors that do not need to be explicitly known. Often, the individual factors are not accessible or are subject to variations. We need to keep in mind that feedback control can keep the process operating within its constraints in the presence of variable disturbances or even variations within the process. In our example, it would be sufficient to monitor a thermometer while running the heater for a few minutes (thus obtaining k_1 as the temperature rise divided by the observation time period), then continue monitoring when the heater is turned off. This yields k_2 in a similar manner.

We can now relate the time period τ of the on-off cycle to the hysteresis temperature $T_H = 2\,\Delta T$:

$$\tau = \tau_1 + \tau_2 = \frac{T_H}{k_1} + \frac{T_H}{k_2} \tag{1.2}$$

where τ_1 is the time during which the heater is turned on and τ_2 is the time during which the heater is turned off. Eq. (1.2) is sufficient to balance temperature accuracy (low T_H) against the period τ of the heating cycle.

Technical realization of two-point controllers is relatively simple. The principle of a popular room thermostat in older buildings is shown in Figure 1.8. The weight of the mercury adds a positive feedback component and thus provides the hysteresis.

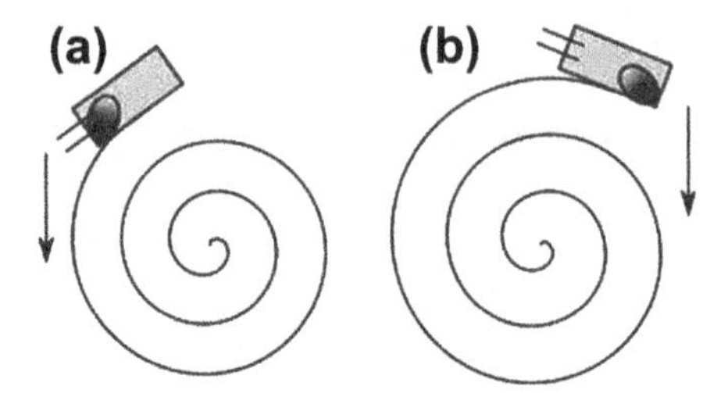

Figure 1.8 Mechanical room thermostat found in older buildings. A bi-metal spiral bends under the influence of temperature changes. At the end of the bi-metal spiral, a small mercury switch is attached. When the spiral expands, the mercury switch turns to the left (a), and the mercury will eventually slide to the left side of the switch. The weight of the mercury pulls the spiral further toward the left. As the spiral contracts, the gravitational force of the mercury (arrows) needs to be overcome. Once the spiral moves past the top point, the mercury slides to the right (b) and further bends the spiral in the new direction.

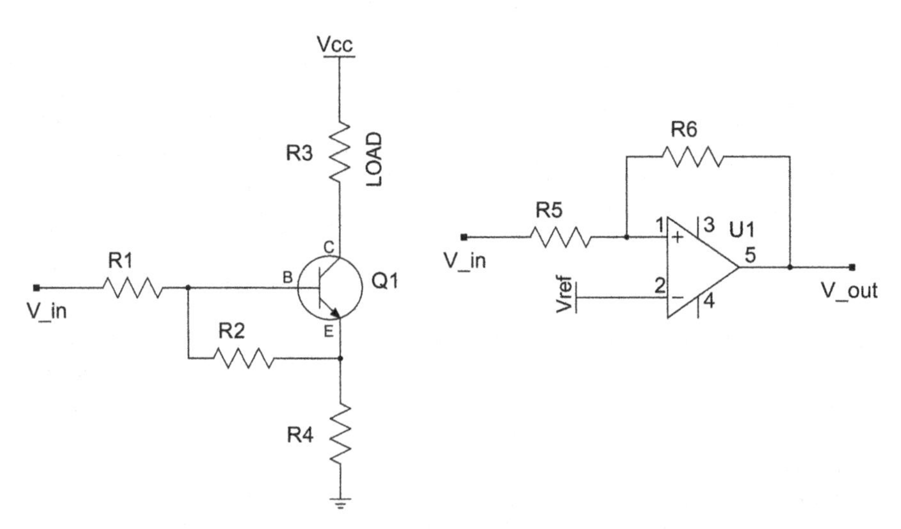

Figure 1.9 Hysteresis generation with semiconductor circuits. Transistor Q1 is configured as a voltage follower. R_2 creates a positive feedback from the emitter to the base and is therefore responsible for the hysteresis. In a similar way, R_6 causes positive feedback from the output of U1 to its positive input. The output voltage swing combined with the voltage divider R_5, R_6 determine the width of the hysteresis.

Figure 1.9 shows two realizations of electronic switches with hysteresis. The transistor needs to be driven to saturation in order to use it as a switch. The op-amp circuit allows easier computation of the hysteresis and easier adjustment of the setpoint (through V_{ref}). If we assume U1 to be a comparator (i.e., its output voltage is either 0 V or the supply voltage, for example, 5 V), the hysteresis is determined by the voltage divider with R5 and R6. If the op-amp is turned off (output voltage at 0 V), the input voltage needs to exceed V_{plus} to switch the output to +5 V:

$$V_{plus} = V_{ref} \frac{R_5 + R_6}{R_6} = V_{ref} \left(1 + \frac{R_5}{R_6}\right) \tag{1.3}$$

Once the op-amp switches to +5 V, R_6 pulls the positive input higher. To bring the op-amp back into the off-state, the input needs to be lowered below V_{minus}:

$$V_{minus} = V_{ref} \left(1 + \frac{R_5}{R_6}\right) - 5\text{V} \cdot \frac{R_5}{R_6} \tag{1.4}$$

The voltage hysteresis V_H is the difference between the two trip points,

$$V_H = V_{plus} - V_{minus} = 5\text{V} \cdot \frac{R_5}{R_6} \tag{1.5}$$

2 Systems and Signals

Abstract

A *system* can be seen as an assembly of components that interact with each other. Examples are a car suspension or a *RC* circuit. A system in this definition has one or more *measurable* properties, referred to as the output signals. These properties (for example speed, temperature, or voltage) are typically functions of time. Moreover, a system allows to influence the output signals, for example, by changing a motor drive voltage or heating power. An input to the system that has the purpose of influencing the output signals is referred to as *input signal*. Input signals are generally also functions of time. To design a feedback control system, each of its components must have a known and defined response (output signal) to a defined input signal. A mathematical description that predicts the output signal for a given input signal is called the *model*. Frequently, the system follows a set of rules that allows it to be called *linear*. For linear systems, the mathematical model is generally an ordinary differential equation with constant coefficients. For a step input signal (that is, a defined change of the input signal at an arbitrary point $t = 0$ to a new value that remains constant at $t > 0$), solutions of the differential equation are exponential functions or exponentially decaying oscillations. For sinusoidal input signals, the output signal is sinusoidal with the same frequency, but altered amplitude and phase. In this chapter, two example systems (the *RC* circuit and the spring-mass-damper system) are presented and their mathematical models explained.

Any feedback control system in the context of this book is a linear, time-invariant system. It is therefore important to review the principles of those systems. A *system* in this context is any physical process, such a mechanical device or an electronic circuit, that responds to one or more stimuli (inputs) in a defined manner, and that has a measurable and quantifiable response. For the efficient design of feedback controls, it is important to characterize the individual components of the feedback system, notably the process and the sensor. In other words, we need to be able to predict how the process reacts to a predetermined input signal. This prediction takes the form of a mathematical description: the *model* of a linear system.

Let us illustrate the concept of a system with two examples: The combination of a resistor and a capacitor into an *RC* lowpass, and an assembly of a spring, a mass, and a frictional damper into a system that acts similar to a car suspension. These two units (*RC* lowpass and spring-mass-damper system) are systems in our definition.

The system alone is not a useful concept. Rather, we need to define the observable output of the system. In the case of the *RC* lowpass, we define the voltage drop across the capacitor as the output variable. For the spring-mass-damper system, the position of the mass is one possible observable output variable. The output variables can be defined as *signals*, where a signal is any observable quantity that exhibits variation with time.

Linear Feedback Controls. http://dx.doi.org/10.1016/B978-0-12-405875-0.00002-4

© 2013 Elsevier Inc. All rights reserved.

In addition to defining the output of the system, we need to define an input variable or stimulus that influences the output variable. The input variable is also a signal. For the *RC* lowpass, a possible input variable is the voltage applied across the resistor and capacitor in series. For the spring-mass-damper system, one possible input variable is the force applied to the inert mass. We can therefore interpret a system as a "black box" with one input and one output, which is often referred to as SISO system: single-input, single-output system. Therefore, the representation as a block with a signal leading into it and another signal emerging from it, as introduced in Figures 1.3 and 1.4, is intuitive. It is important to realize that we generally cannot change the content of the box for a given system, but we can apply an input signal of our choice and observe the output signal that emerges as response to the input.

Usually, a system may have multiple observable variables that could serve as output variables, and it can have more than one input variable that influences the output variable. For example, the rotational speed of a DC motor is determined by the current through the armature coils, but can also be influenced by an external torque or friction. These considerations lead to the concept of a MIMO system (multiple-input, multiple-output system). In the special case of linear systems, the input variables can be treated as additive components and their influence on the output variable examined independently and superimposed. Therefore, any linear, time-invariant MIMO system can be decomposed into a combination of SISO systems.

Any linear, time-invariant SISO system can be described by linear, ordinary differential equations with constant coefficients,

$$a_n \frac{\mathrm{d}^n y(t)}{\mathrm{d}t^n} + a_{n-1} \frac{\mathrm{d}^{n-1} y(t)}{\mathrm{d}t^{n-1}} + \cdots + a_1 \frac{\mathrm{d}y(t)}{\mathrm{d}t} + a_0 y(t) = f(x(t)) \tag{2.1}$$

where a_n through a_0 are real-valued scalars and f is a linear function of the input variable $x(t)$. To obtain a differential equation in the form of Eq. (2.1), all dependent variables of the system have to be removed by substitution, and all terms with the output variable need to be collected on the left-hand side. This mathematical description of a SISO system contains the three elements,

- The *input variable*, also called the independent variable, $x(t)$. We need to be able to influence this variable in order to observe the system response, for example, by applying a voltage to a circuit, or by pushing at a mass.
- The *output variable*, $y(t)$. We need to be able to observe or measure the variable as it changes over time, specifically, after a defined input variable $x(t)$ has been applied.
- The *system* itself, which is defined by the coefficients of the differential equation $a_n, \ldots, a_0$. Unlike the signals $x(t)$ and $y(t)$, the coefficients (i.e., the system itself) cannot be influenced.

The fact that a linear system can be described by a form of Eq. (2.1) allows the elegant treatment in the Laplace domain that is presented in the next chapter. Furthermore, linear, time-invariant systems meet the following criteria:

1. Scaling principle: If $y(t)$ is the response to $x(t)$, then the same input signal, with a linearly scaled magnitude (i.e., $a \cdot x(t)$) causes a response scaled by the same factor, that is, $a \cdot y(t)$ where a is a scalar.

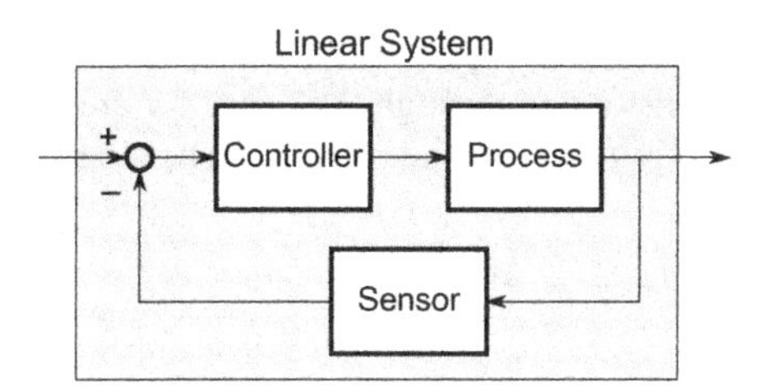

Figure 2.1 A closed-loop feedback control system as a whole (inside the shaded box) behaves like a linear, time-invariant system when all components are linear and time-invariant.

2. Superposition principle: If $y_1(t)$ is the response to an input signal $x_1(t)$, and $y_2(t)$ is the response to $x_2(t)$, then the response to a signal $x_1(t) + x_2(t)$ is $y_1(t) + y_2(t)$.
3. Time invariance: If $y(t)$ is the response to $x(t)$, then a delayed input signal $x(t - \tau)$ causes the same output signal, but equally delayed, i.e., $y(t - \tau)$.

Feedback control systems that are built with linear, time-invariant components behave like a linear, time-invariant system as a whole (Figure 2.1). In practice, some nonlinearity can be expected, usually in the process. With suitable feedback control, however, the nonlinear behavior of a single component becomes less pronounced, and the overall closed-loop system can be approximately linear even when some components show nonlinear behavior.

2.1 Example First-Order System: The *RC* Lowpass

The *RC* lowpass is a voltage divider that contains an energy storage element (the capacitor) as shown in Figure 2.2. The system coefficients are defined by the resistance R and the capacitance C. A measurable output signal is the voltage across the capacitor, and the input signal is the voltage applied across the entire R-C voltage divider.

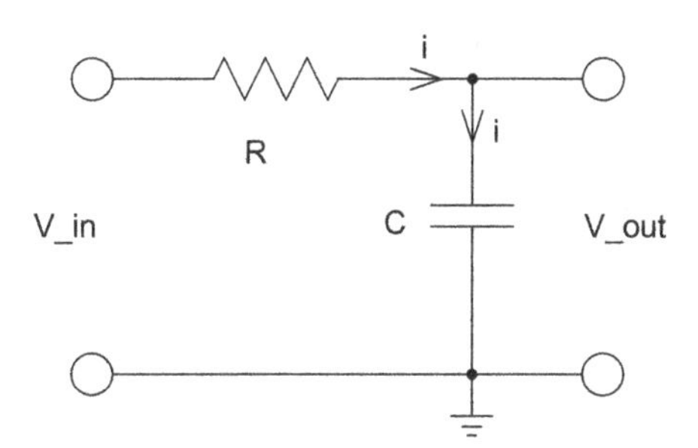

Figure 2.2 *RC* lowpass as an example system. The input signal is a voltage V_{in} applied to the resistor with respect to ground. The output voltage V_{out} is the voltage across the capacitor. The resistance R and capacitance C are process constants that define the system.

We can develop the differential equation of this system by recognizing that the current $i(t)$ through the resistor and through the capacitor are identical, provided that no current flows into the output branch. The current can be related to the voltages across the components as follows:

$$i(t) = \frac{V_{in}(t) - V_{out}(t)}{R} = C \cdot \frac{\mathrm{d}V_{out}(t)}{\mathrm{d}t} \tag{2.2}$$

To obtain the differential equation in the form of Eq. (2.1), we arrange the independent variables on the right-hand side and the dependent variables on the left-hand side. The current is a dependent variable and dropped from the equation; the equation is then multiplied by R to isolate the independent variable $V_{in}(t)$:

$$R \cdot C \cdot \frac{\mathrm{d}V_{out}(t)}{\mathrm{d}t} + V_{out}(t) = V_{in}(t) \tag{2.3}$$

Since the first derivative of the output variable $V_{out}(t)$ is the highest derivative, such a system is referred to as *first-order system*. The order is always the highest derivative of the output variable. Derivatives are introduced by any form of energy storage, such as a capacitor, an inductor, a spring, or a mass.

Once we choose a signal $V_{in}(t)$ to apply to the RC lowpass, we can solve Eq. (2.3) to predict the output signal $V_{out}(t)$. Standard approaches to solve ordinary differential equations apply, and a simple example will suffice. At $t = 0$, we apply a constant voltage V_1 to the energy-free system. Initially, the current is limited by the resistor to $i = V_1/R$, and the voltage across the capacitor increases at a rate $\dot{V}_{out} = V_1/RC$. As the capacitor accumulates a charge, the output voltage rises and the current decays. The system reaches equilibrium when $i = 0$ and $V_{out} = V_{in}$. Solutions of ordinary differential equations are combinations of exponential terms, and the special solution in our example is

$$V_{out}(t) = V_1 \cdot \left(1 - e^{-t/RC}\right) \tag{2.4}$$

A more comprehensive treatment of a first-order system that includes the determination of the system coefficients, is provided in Chapter 5.

2.2 Example Second-Order System: The Spring-Mass-Damper System

Spring-mass-damper systems can be found in many mechanical systems. One common example is the suspension of a car. The car itself is the mass; it is suspended by an elastic spring. A damper (the actual shock absorber) prevents oscillations. The basic schematic of a spring-mass-damper system is shown in Figure 2.3. The equilibrium of forces requires us to consider four force components: the external force F, the force needed to linearly extend the spring in the x-direction $D{\cdot}x$, the force required to accelerate the mass $m \cdot \ddot{x}$, and the force to overcome the velocity-dependent viscous friction $R_F{\cdot}\dot{x}$.

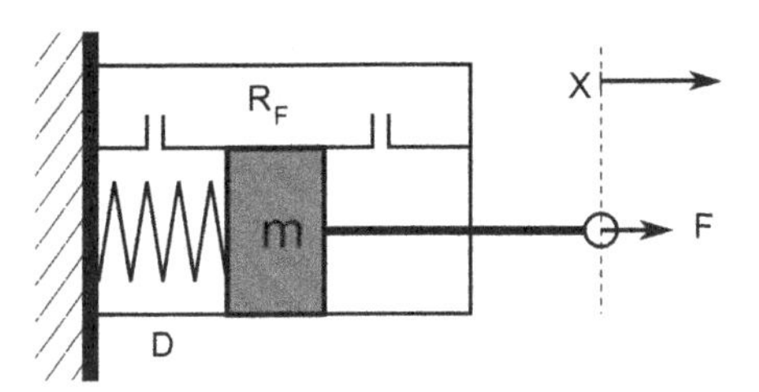

Figure 2.3 A spring-mass-damper system. The inert mass m moves in the direction of the x axis inside a sealed cylinder. At the dotted line, we define $x = 0$. A spring with Hooke's constant D attempts to move the mass toward $x = 0$. The mass is embedded in a viscous fluid—very much like inside a car's shock absorber—that creates a friction R_F. An external force F is applied to the handle that is rigidly connected to the mass.

To determine how the mass moves under some time-variable force $F(t)$, we need to solve the balance-of-forces equation (Eq. (2.5)). Furthermore, there are initial conditions: the position of the mass at $t = 0$ is x_0, and the initial velocity is $v_0 = \dot{x}(t = 0)$.

$$\begin{aligned} & m\,\ddot{x}(t) + R_F\dot{x}(t) + D\,x(t) = F(t) \\ & x(t = 0) = x_0 \\ & \dot{x}(t = 0) = v_0 \end{aligned} \tag{2.5}$$

The solution approach for differential equations of the type given in Eq. (2.1) relies on the fact that an additive combination of solutions for any ordinary differential equation is itself a solution. A good starting point is to solve the differential equation for a zero-input signal, in this case, $F(t) = 0$. It can be seen that the solution is again composed of exponential terms of the form e^{kt}, and we can substitute $x(t)$ in the homogeneous differential equation to obtain

$$\begin{aligned} & m\,k^2 e^{kt} + R_F k\, e^{kt} + D\, e^{kt} \\ & = e^{kt}(m\,k^2 + R_F k + D) \end{aligned} \tag{2.6}$$

It is interesting to realize that a special solution $x(t) = e^{kt}$ isolates a polynomial of k. For the unforced differential equation ($F(t) = 0$), we can find k through

$$m\,k^2 + R_F k + D = 0 \tag{2.7}$$

which provides two solutions for k,

$$k_{1,2} = -\frac{R_F}{2m} \pm \sqrt{\frac{R_F^2}{4m^2} - \frac{D}{m}} \tag{2.8}$$

The polynomial of k is referred to as the *characteristic polynomial* of the differential equation, and the solutions of Eq. (2.7) as its *roots*. The solutions $k_{1,2}$ can be real-valued (with $k_{1,2} \leq 0$), or they become complex when $R_F^2 < 4mD$. Complex roots have a negative real component and occur in complex conjugate pairs. For large values of R_F and real-valued exponents k, the motion of the mass exhibits an exponential characteristic. For small values of R_F and complex exponents k, expressed as $k = \sigma + j\omega$, the exponential solution can be rewritten as

$$e^{-(\sigma+j\omega)t} = e^{-\sigma t} \cdot e^{-j\omega t} = e^{-\sigma t}\left(\cos(\omega t) - j\sin(\omega t)\right) \tag{2.9}$$

Equation (2.9) describes an exponentially attenuated harmonic oscillation and is related to resonance effects. A closer examination of the solutions of Eq. (2.5) is simplified with the use of the Laplace transform and is covered in Chapter 3.

2.3 Obtaining the System Response from a Step Input

Frequently, the response of a system to a step change at the input is examined. The step response reveals the nature of the system with good accuracy. In Chapter 5, the relationship of the step response to the differential equation and its coefficients is provided in detail. In addition, if the step response is known, the response of the system to any arbitrary input signal $x(t)$ can be predicted. Let us define a general step function $v(t)$ as

$$v(t) = \begin{cases} 0 & \text{for } t < 0 \\ V & \text{for } t \geq 0 \end{cases} \tag{2.10}$$

where V is a constant for $t \geq 0$. Because of the time invariance, the system response to a delayed step input, $v(t - \tau)$, is the same as the response to $v(t)$, but delayed by τ. We can approximate any arbitrary signal by a sequence of delayed and scaled step functions. At this point, it is more intuitive to use the delta function, which is the first derivative of the step function, that is, at $t = 0$ it assumes the value V but is zero otherwise. The delta function is defined more rigorously in the next chapter. Here, we can imagine one delta function $\delta(t - \tau)$ as an infinitesimally thin stick of height 1, placed at $t = \tau$. We can scale this stick to our function at τ by multiplying with $x(\tau)$ as shown in Figure 2.4. The resulting product, $x(\tau) \cdot \delta(t - \tau)$, is the infinitesimal contribution of the scaled δ-"stick" at time $t = \tau$. By adding (i.e., integrating) all of these contributions, the original signal is obtained:

$$x(t) = \int_{-\infty}^{\infty} x(\tau) \cdot \delta(t - \tau) d\tau \tag{2.11}$$

If $h(t)$ is the step response of the system, then $\dot{h}(t)$ is the impulse response. We can therefore replace the δ function in Eq. (2.11) by the equally delayed impulse response $\dot{h}(t - \tau)$, and the integral yields the system response $y(t)$ to the input signal $x(t)$:

$$y(t) = \int_{-\infty}^{\infty} x(\tau) \cdot \dot{h}(t - \tau) d\tau = x(t) \otimes \dot{h}(t) \tag{2.12}$$

Equation (2.12) describes the *convolution integral* of the function $x(t)$ with another function, in this case $\dot{h}(t)$. The symbol $\otimes$ denotes the convolution operation. We can summarize that for any system of known step response $h(t)$, we can predict the response

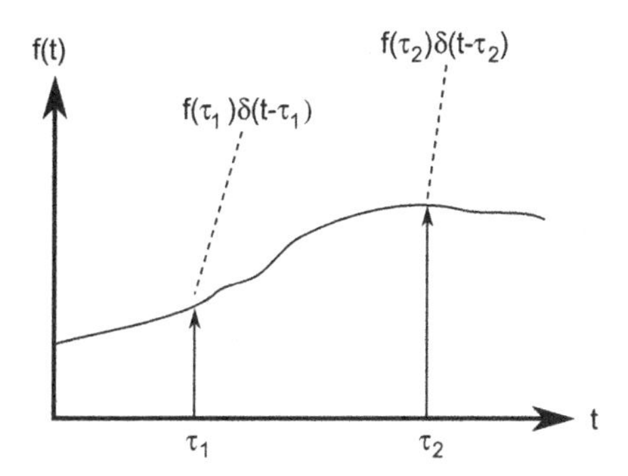

Figure 2.4 A function $f(t)$ and two function values selected at $t = \tau_1$ and $t = \tau_2$. If the function is represented at these time points by a delta function, the scaled and shifted delta functions are $f(\tau_1)\delta(t - \tau_1)$ and $f(\tau_2)\delta(t - \tau_2)$, respectively. Note that $f(\tau_1)$ and $f(\tau_2)$ are scaling constants with respect to the delta function.

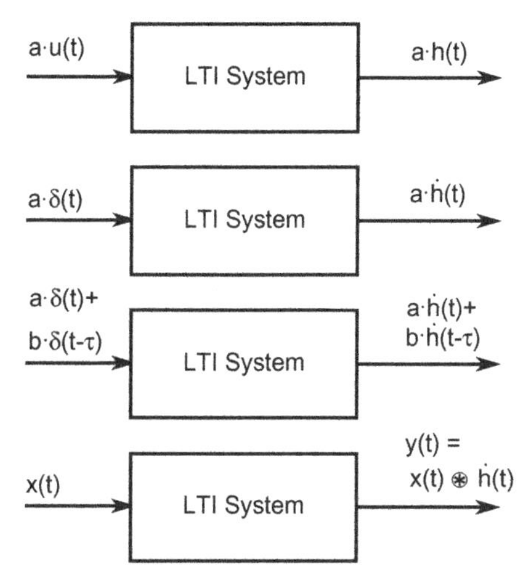

Figure 2.5 Convolution property of LTI systems. Let us assume that a LTI system has a step response of $h(t)$. We know (a) that its response to a scaled step function $a \cdot u(t)$ is an equally scaled step response $a \cdot h(t)$. Its impulse response is $\dot{h}(t)$ because of the linearity of the derivative, and because $\delta(t) = \dot{u}(t)$. The scaling principle still holds, and the response to a scaled delta function $a\delta(t)$ is $a\dot{h}(t)$. Time-invariance stipulates that a delayed delta pulse $\delta(t - \tau)$ elicits a delayed impulse response, that is, $\dot{h}(t - \tau)$. Two non-simultaneous and scaled delta pulses $a\delta(t) + b\delta(t - \tau)$ produce two time-shifted impulse responses $a\dot{h}(t) + b\dot{h}(t - \tau)$. Finally, the infinitely dense sequence of delayed and scaled delta pulses that constitute the function $x(t)$ causes an infinite sequence of scaled and delayed impulse responses, which, added together, yield $y(t)$ as defined in the convolution integral Eq. (2.12).

to an arbitrary signal $x(t)$ by convolving $x(t)$ with the first derivative of $h(t)$ as illustrated in Figure 2.5.

In summary, the mathematical model (i.e., the differential equation that describes how the components of a system interact) allows us to predict the output variable for a given input signal by (a) determining either the impulse response or the step response, and (b) by applying the convolution integral for the given input signal. This concept leads to the notion of the *transfer function* in the frequency domain, because a convolution in the time domain corresponds to a simple multiplication in the frequency (Laplace or Fourier) domain.

2.4 State-Space Models

In many control problems, the state-space representation of a system is used. State-space models allow an elegant and compact matrix representation of control systems. The key difference to the description as a high-order differential equation is the introduction of a *state vector* that represents the rate of change of the dynamic system. As such, the state-space representation uses only first-order differential equations. Within the scope of this book, there is no benefit in using state-space models, and apart from the short introduction given in this section for reasons of completeness, space-state models are not further covered.

The general form of a state-space model is

$$\begin{aligned} \dot{\mathbf{x}} &= \mathbf{A}\mathbf{x} + \mathbf{B}\mathbf{u} \\ \mathbf{y} &= \mathbf{C}\mathbf{x} + \mathbf{D}\mathbf{u} \end{aligned} \tag{2.13}$$

where $\mathbf{x}$ is the state vector, $\mathbf{u}$ is the vector of input signals, and $\mathbf{y}$ is the vector of output signals. The matrices $\mathbf{A}$ through $\mathbf{D}$ contain the system coefficients, and for linear, time-invariant systems, these matrices contain only constant coefficients. The first part of Eq. (2.13) is referred to as the *state differential equation*, and the second part is known as the *output equation*.

To examine the individual components, let us assume that a SISO third-order system can be described with the differential equation

$$a_3 \frac{\mathrm{d}^3 y(t)}{\mathrm{d}t^3} + a_2 \frac{\mathrm{d}^2 y(t)}{\mathrm{d}t^2} + a_1 \frac{\mathrm{d}y(t)}{\mathrm{d}t} + a_0 y(t) = u(t) \tag{2.14}$$

where $u(t)$ denotes the input signal in state-space convention (elsewhere in this book, the general input signal is $x(t)$). Since this is a SISO system, $u(t)$ and $y(t)$ are one-dimensional. We now introduce the elements of the state vector as follows:

$$x_1(t) = y(t); \quad x_2(t) = \dot{x}_1(t); \quad x_3(t) = \dot{x}_2(t) \tag{2.15}$$

With this definition, we can rewrite Eq. (2.14) in terms of the state vector elements $x_1(t)$ through $x_3(t)$ as

$$\frac{\mathrm{d}x_3(t)}{\mathrm{d}t} = -\frac{a_2}{a_3} x_3(t) - \frac{a_1}{a_3} x_2(t) - \frac{a_0}{a_3} x_1(t) + \frac{1}{a_3} u(t) \tag{2.16}$$

The definition of the state vector is straightforward, and the output equation readily emerges:

$$\mathbf{x} = \begin{bmatrix} x_1(t) \\ x_2(t) \\ x_3(t) \end{bmatrix} = \begin{bmatrix} y(t) \\ \dot{x}_1(t) \\ \dot{x}_2(t) \end{bmatrix}; \quad y(t) = \begin{bmatrix} 1 & 0 & 0 \end{bmatrix} \mathbf{x} \tag{2.17}$$

For this system, the state differential equation is

$$\frac{\mathrm{d}}{\mathrm{d}t} \begin{bmatrix} x_1(t) \\ x_2(t) \\ x_3(t) \end{bmatrix} = \begin{bmatrix} 0 & 1 & 0 \\ 0 & 0 & 1 \\ -\frac{a_0}{a_3} & -\frac{a_1}{a_3} & -\frac{a_2}{a_3} \end{bmatrix} \cdot \begin{bmatrix} x_1(t) \\ x_2(t) \\ x_3(t) \end{bmatrix} + \begin{bmatrix} 0 \\ 0 \\ \frac{1}{a_3} \end{bmatrix} \cdot u(t) \tag{2.18}$$

We can see that Eq. (2.18), with substitutions from Eq. (2.17), yields Eq. (2.16). Furthermore, the first 3×3 matrix in Eq. (2.18) corresponds to $\mathbf{A}$ in Eq. (2.13), and the 3×1 matrix before $u(t)$ is $\mathbf{B}$. Moreover, $\mathbf{C} = [1\,0\,0]$, and all elements of $\mathbf{D}$ are zero. This example demonstrates how a conventional model can be converted into a state-space model and *vice versa*.

2.5 Systems and Signals in Scilab

Throughout this book, examples are provided how the Scilab software package can be used to assist with the analysis and design of linear systems. Scilab can be used in two distinctly different ways:

- To perform numerical calculations, such as complex arithmetic, division of polynomials, or obtaining the roots of polynomials.
- To perform simulations of linear and nonlinear systems. For this purpose, the toolbox `xcos` is used, which provides a canvas on which a system can be described with basic building blocks in the style of a block diagram. Various input signals can be provided, and the evolution of the output signals over time can be simulated and monitored.

Scilab provides a number of predefined variables. In the context of linear systems, `%i` represents the imaginary unit ($\sqrt{-1}$), `%s` is the position variable s of the Laplace transform (Chapter 3), and `%z` is the position variable of the z-transform (Chapter 4). Most operations in the context of this book require the use of either s or z. For example, to determine the roots of the spring-mass-damper characteristic polynomial (Eq. (2.7)), we use s as the variable for which to solve the equation

$$m\,s^2 + R_F s + D = 0 \tag{2.19}$$

With given values of $m = 10$, $R_F = 8$, and $D = 800$ (we ignore the units, because they would not correspond to s), the roots can be found in Scilab as follows:

```
s = %s
q = 10*s^2 + 8*s + 800
roots(q)
```

In the first line, a convenient shortcut is created by assigning the value of the Laplace variable to the previously unused variable `s`. In the second line, a polynomial $q(s)$ is defined. The third line returns the solution of $q(s) = 0$, that is,

```
ans =
  - 0.04 + 8.9353232i
  - 0.04 - 8.9353232i
```

The answer to the question `roots(q)` is a vector with two complex elements, for which the notation in this book would be $s_{1,2} = -0.04 \pm 8.9j$.

A second example deals with complex arithmetic. Let us define a function $H(s)$ as a fraction of two polynomials,

$$H(s) = \frac{s+3}{s^2 + 2s + 10} \tag{2.20}$$

and we want to evaluate $H(s)$ for $s = -0.04 + 8.9j$. In this case, we assign the specific value of the variable s to a new variable r so as to not overwrite s. Next, we enter the polynomial:

```
r = -0.04 + 8.9* %i
(r+3)/(r^2 + 2*r + 10)
```

This time, the answer provided is $-0.0172 - 0.129j$. In these basic steps, Scilab is very similar to Matlab and Octave, and the reader is encouraged to use either online documentation or one of the numerous Matlab books. The Scilab online documentation (`http://www.scilab.org/resources/documentation`) covers similarities and differences to Matlab.

Scilab comes with a comprehensive package for simulating linear systems. Most operations that involve linear systems require the definition of a compound variable by using `syslin`. We can, for example, simulate the step response of the spring-mass-damper system introduced in Eq. (2.5). For this purpose, we need to define the spring-mass-damper system as a linear system with a transfer function (see Chapter 3). Using the same values as in the example above, the first three lines define the linear system (we call its variable `smd`):

```
s = %s
q = 10*s^2 + 0.8*s + 800
smd = syslin ('c', 1/q)
t = linspace (0,15, 301);
resp = csim ('step', t, smd);
plot (t, resp)
```

In the fourth line, a vector of discrete time values is created for which the simulation will calculate the response. The fifth line performs the actual simulation of the linear system `smd` for a unit step input and the time vector t. The last line finally plots the step response over the time vector.

A more sophisticated simulation tool exists in the `xcos` toolbox. Xcos (formerly called Scicos) can be started from the Applications menu or simply by typing "xcos" in

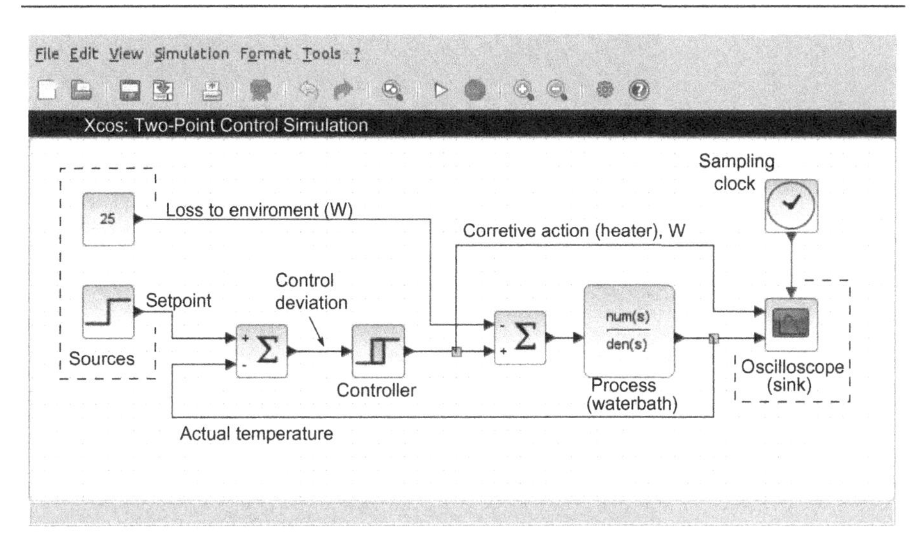

Figure 2.6 Example block diagram in xcos. The process is a heated waterbath, which is defined by its Laplace-domain transfer function (see Chapter 6). Heat loss is modeled as a constant loss to the environment of 25 W, and heat is introduced from the two-point controller (indicated by the hysteresis symbol). The two-point controller switches at ±0.5 °C and is therefore provided with the control deviation, i.e., the difference between setpoint and actual temperature (first summation symbol). A simulation shows the rising temperature from the initial condition and subsequent oscillations around the setpoint as shown in Figure 1.6b.

the console. This command opens two new windows: A canvas for the system elements, and a browser for the component palettes. The component palettes can be divided into three main categories: Functions, sources, and sinks. The functions contain mathematical operations, such as summation, multiplication, or complete transfer functions. Sources provide signals, including step functions and sinusoids. Sinks are utilities to view signals or to save a signal to a file. An example block diagram that can be used for simulation is shown in Figure 2.6. A simulation run opens a new window in which the signals connected to the oscilloscope (i.e., the heater power and the waterbath temperature) are shown as functions of time.

3 Solving Differential Equations in the Laplace Domain

Abstract

The *Laplace transform* is a particularly elegant way to solve linear differential equations with constant coefficients. The Laplace transform describes signals and systems not as functions of time, but as functions of a complex variable s. When transformed into the Laplace domain, differential equations become polynomials of s. Solving a differential equation in the time domain becomes a simple polynomial multiplication and division in the Laplace domain. However, the input and output signals are also in the Laplace domain, and any system response must undergo an inverse Laplace transform to become a meaningful time-dependent signal. In this chapter, the Laplace transform is introduced, and the manipulation of signals and systems in the Laplace domain explained. Tools to find time-domain and Laplace-domain correspondences are presented.

In this chapter, we will briefly introduce the Laplace transform and the related Fourier transform, some of their properties, and the most pertinent applications for linear control systems. This chapter is not meant to provide a comprehensive coverage of the Laplace and Fourier transforms, nor is it even remotely exhaustive. Rather, readers of this book should be reasonably familiar with the mathematical foundations of the Laplace transform. The family of integral transforms, including the Laplace and Fourier transforms, are a topic of considerable depth, and exhaustive treatment of integral transforms could easily fill a book twice as thick as this one.

3.1 The Laplace Transform

The Laplace transform $\mathscr{L}$ of a given signal $f(t)$ is defined through[1]

$$F(s) = \mathscr{L}\{f(t)\} = \int_{0^+}^{\infty} f(t)e^{-st}\,\mathrm{d}t \tag{3.1}$$

where s is a complex number with $s = \sigma + j\omega$. A specific value of s defines a single point in the complex plane. This plane is often referred to as the s-plane. To obtain the complete Laplace transform of a signal $f(t)$, we would have to evaluate the integral in Eq. (3.1) for every possible value of s. Often, we can find a closed-term solution for $F(s)$. The lower integration bound, denoted as 0^+, indicates that we are evaluating the integral an infinitesimally short moment after the start of our experiment at a time arbitrarily defined as $t = 0$.

[1] This book tries to follow the convention that time-domain signals, such as $f(t)$, are denoted by lower-case letters and frequency-domain correspondences by capital letters, such as $F(s)$. Some variables, such as temperature T or voltage V, have upper-case letters, and the only distinction is the argument, such as $T(t)$ *versus* $T(s)$.

Linear Feedback Controls. http://dx.doi.org/10.1016/B978-0-12-405875-0.00003-6
© 2013 Elsevier Inc. All rights reserved.

The definition of a *transform* implies that an inverse transform exists to restore $f(t)$ from the transformed function $F(s)$, and the Laplace transform is no exception. The inverse Laplace transform is

$$f(t) = \mathscr{L}^{-1}\{F(s)\} = \frac{1}{2\pi j} \int_{\sigma-j\infty}^{\sigma+j\infty} F(s)e^{+st}\mathrm{d}t \tag{3.2}$$

To demonstrate the use of the Laplace transform, let us use the normalized step response of the RC lowpass that was introduced in Eq. (2.4), $x(t) = (1 - e^{-kt})$. We substitute $x(t)$ for the general function $f(t)$ in Eq. (3.1) and split the integral in two additive parts, at the same time joining the exponents in the second integral:

$$X(s) = \int_{0^+}^{\infty} \left(1 - e^{-kt}\right) e^{-st}\,\mathrm{d}t = \int_{0^+}^{\infty} e^{-st}\,\mathrm{d}t - \int_{0^+}^{\infty} e^{-kt-st}\,\mathrm{d}t \tag{3.3}$$

The integrals can be solved and evaluated at their integration bounds,

$$X(s) = -\frac{1}{s}e^{-st}\bigg|_0^{\infty} + \frac{1}{k+s}e^{-(k+s)t}\bigg|_0^{\infty} = \frac{k}{s(s+k)} \tag{3.4}$$

We define $X(s)$ as the *Laplace correspondence* of $x(t)$, indicated by the correspondence symbol ○—●:

$$\left(1 - e^{-kt}\right) \;\circ\!\!-\!\!\bullet\; \frac{k}{s(s+k)} \tag{3.5}$$

Many of these correspondences can be found in Appendix A (see Table A.3 for this specific example and Tables A.1 through A.6 for a comprehensive list of correspondences). The Laplace-domain function $X(s)$ can be plotted as a two-dimensional function of the real and imaginary components of s. The functions given in Eq. (3.5) are shown in Figure 3.1. Most of the information contained in the s-plane plot is redundant and not needed for the specific purpose of describing linear systems. However, there are locations where a Laplace-domain function diverges, namely, when its denominator becomes zero. These locations are called *poles*. In more complex functions, the numerator can also become zero, and those locations are called the *zeros* of the Laplace-domain function. We are predominantly interested in the poles of a Laplace-domain function, because each of these poles corresponds to an exponential term in the solution of the differential equation. In fact, with a bit of experience it is possible to approximately predict the dynamic response of a system just by knowing the poles and zeros in the Laplace domain. This idea leads to the concept of pole-zero plots. Pole-zero plots are plots of the locations of poles and zeros in the s-plane. By convention, poles are indicated by the symbol $\times$, and zeros are indicated by the symbol $\circ$. The pole-zero plot is useful for a quick assessment of the dynamic behavior of a system.

In the previous chapter, we mentioned that any linear, time-invariant system can be described by linear, ordinary differential equations with constant coefficients,

$$a_n\frac{\mathrm{d}^n y(t)}{\mathrm{d}t^n} + a_{n-1}\frac{\mathrm{d}^{n-1} y(t)}{\mathrm{d}t^{n-1}} + \cdots + a_1\frac{\mathrm{d}y(t)}{\mathrm{d}t} + a_0 y(t) = f(x(t)) \tag{3.6}$$

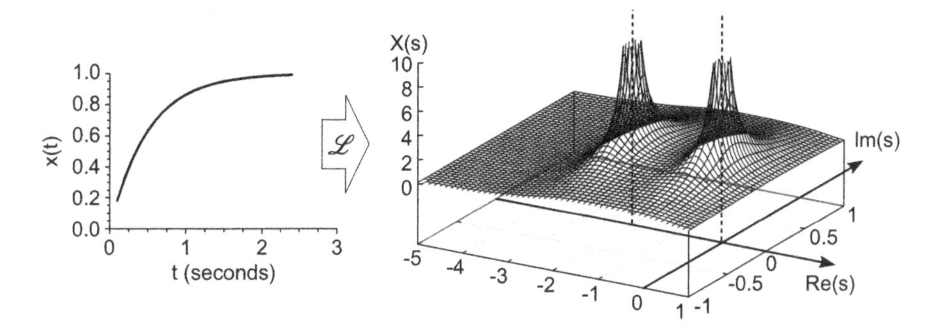

Figure 3.1 Plot of the time-domain function $x(t) = (1 - e^{-kt})$ and its Laplace correspondence $X(s)$ (Eq. (3.5)) for $k = 2$. The Laplace transform of $x(t)$ is rather abstract, but it shows two conspicuous points where $X(s)$ diverges (dotted lines) as the denominator becomes zero, namely at $s = 0$ and $s = -2$.

where a_n through a_0 are real-valued scalars and f is a linear function of the input variable $x(t)$. Note that we collected the dependent (i.e., output) variable on the left-hand side of the equation and left all independent variables on the right-hand side. The *derivative rule* (see also Table A.7) of the Laplace transform is

$$\frac{\mathrm{d}}{\mathrm{d}t} y(t) \quad \circ\!\!-\!\!\bullet \quad sY(s) - y(0^+) \tag{3.7}$$

where $y(0^+)$ is the initial condition of $y(t)$ for $t \to 0$. With this rule, we can now perform the Laplace transform of Eq. (3.6). For now, we assume that all initial conditions are zero (we will later examine cases with nonzero initial conditions). We are primarily interested in the left-hand side of the equation and simply denote $\mathscr{L}\{f(x(t))\}$ as RHS(s). The Laplace transform of Eq. (3.6) becomes

$$a_n s^n Y(s) + a_{n-1} s^{n-1} Y(s) + \cdots + a_1 s Y(s) + a_0 Y(s) = \mathrm{RHS}(s) \tag{3.8}$$

We can factor out $Y(s)$ and obtain the left-hand side of the equation as the product of the output variable with a polynomial of s:[2]

$$Y(s) \cdot \left(a_n s^n + a_{n-1} s^{n-1} + \cdots + a_1 s + a_0\right) = \mathrm{RHS}(s) \tag{3.9}$$

For a SISO (Single-Input, Single-Output) system, the right-hand side can be treated analogously and yields the input variable, $X(s)$, multiplied with a different polynomial, $b_m s^m + b_{m-1} s^{m-1} + \cdots + b_1 s + b_0$, where generally $m < n$. If we denote the polynomial on the left-hand side with $q(s)$ and the one on the right-hand side with $p(s)$, Eq. (3.9) can be elegantly rewritten as

$$Y(s) \cdot q(s) = X(s) \cdot p(s) \iff Y(s) = X(s) \frac{p(s)}{q(s)} \tag{3.10}$$

[2]We first introduced this idea in a nonsystematic fashion in Eq. (2.6).

The *transfer function* $H(s)$ of the system is defined as the function with which an input signal is multiplied (in the Laplace domain!) to obtain the output signal, that is,

$$H(s) = \frac{Y(s)}{X(s)} = \frac{p(s)}{q(s)} \tag{3.11}$$

It is crucial to recognize that $q(s)$ describes the dynamic system, whereas $p(s)$ merely describes the input signal to the system (in fact, often $p(s) = 1$). Because $q(s)$ contains the information on the dynamic behavior of the system, it is called the *characteristic polynomial* of the system. Let us specifically note:

- Each root of the characteristic polynomial (i.e., those s for which $q(s) = 0$) is a pole of the transfer function $p(s)/q(s)$, that is, a location in the s-plane where the transfer function becomes infinity.
- Each root of the characteristic polynomial corresponds to a solution of the differential equation. The set of all functions that correspond to the roots is called the *fundamental set*, and a linear combination of the fundamental set functions is the special solution for the differential equation.
- Roots always are either real-valued or occur as complex conjugate pairs of the form $\sigma \pm j\omega$. A solitary complex root without its conjugate counterpart *cannot* occur.
- The distribution of the poles in the complex s-plane tells us how the system will react to changes of the input signal. Real-valued roots cause an exponential response, and complex conjugate roots cause an oscillatory response.
- Roots with a positive real part correspond to a solution of the differential equation with exponentially increasing magnitude. A system that has roots of the characteristic polynomial with positive real part (i.e., poles in the right half-plane) is *unstable*.
- The elegance of the Laplace-domain treatment of a linear system comes from the fact that the output signal can be computed from the transfer function $H(s)$ and a known input signal $X(s)$ by simple multiplication. However, the response is obtained in its Laplace-domain form as $Y(s)$, and the inverse transform must be determined to obtain $y(t)$. For this purpose, correspondence tables can be used. If a function is not found in the tables, partial fraction expansion (Section 3.5) is usually an option.

We now return to the Laplace transform of the derivatives with full consideration of the initial conditions. Higher-order derivatives require the repeated application of Eq. (3.7) and have multiple initial conditions. The general form of Eq. (3.7) for the nth derivative becomes

$$\frac{\mathrm{d}^n}{\mathrm{d}t^n} y(t) \quad \circ\!\!-\!\!\bullet \quad s^n Y(s) - \sum_{k=0}^{n-1} s^{(n-k-1)} y(0^+)^{\langle k \rangle} \tag{3.12}$$

where the superscript $\langle k \rangle$ indicates the k-th derivative with respect to time. The Laplace transform of Eq. (3.6) with initial conditions, best written as a sum, is

$$\sum_{i=0}^{n} a_i \left(s^i Y(s) - \sum_{k=0}^{i-1} s^{(i-k-1)} y(0^+)^{\langle k \rangle} \right) = \mathrm{RHS}(s) \tag{3.13}$$

Let us examine Eq. (3.13) for the special case of a first- and second-order differential equation. In a first-order system, only the initial condition of the output variable exists:

$$a_1 \left(sY(s) - y(0^+) \right) + a_0 Y(s) = \text{RHS}(s) \tag{3.14}$$

A second-order system can have two nonzero initial conditions: the value of the output variable at $t = 0$ and its rate of change, that is, $\dot{y}(t = 0)$. The Laplace transform of the second-order differential equation is

$$a_2 \left(s^2 Y(s) - sy(0^+) - \dot{y}(0^+) \right) + a_1 \left(sY(s) - y(0^+) \right) + a_0 Y(s) = \text{RHS}(s) \tag{3.15}$$

Since the initial conditions are constant, they are usually moved to the right-hand side and treated like input signals to the system—but with the crucial distinction that they are not functions of time, unlike the function $x(t)$ in Eq. (3.6). If we rearrange Eq. (3.15) in this fashion, we obtain

$$Y(s) \left(a_2 s^2 + a_1 s + a_0 \right) = \text{RHS}(s) + a_2 \left(sy(0^+) + \dot{y}(0^+) \right) + a_1 y(0^+) \tag{3.16}$$

It is important to realize that the characteristic polynomial has not changed. As a side note, let us recall that the variable s has units of inverse seconds. It is therefore consistent that the $(n-1)$th derivative is multiplied with s of one order higher than the nth derivative, and the units still match in the summation. In fact, the physical units of constants and variables can be treated in the Laplace domain with the same rigor as in the time domain.

3.2 Fourier Series and the Fourier Transform

Joseph Fourier discovered that any 2π-periodic signal can be described as an infinite sum of harmonic oscillations. Given any signal $f(t)$ that repeats after $t = 2\pi$ (i.e., $f(t) = f(t + 2k\pi)$), a summation can be formulated where

$$f(t) = \frac{a_0}{2} + \sum_{k=1}^{\infty} a_k \cos(kt) + b_k \sin(kt) \tag{3.17}$$

Equation (3.17) is referred to as *Fourier synthesis*. The signal $f(t)$ depends only on the Fourier coefficients a_k and b_k. To synthesize a specific given signal $f(t)$ through Eq. (3.17), these coefficients need to be determined through Fourier analysis:

$$\begin{aligned} a_k &= \frac{2}{\pi} \int_{-\pi/2}^{\pi/2} f(t) \cos(kt)\, \mathrm{d}t \\ b_k &= \frac{2}{\pi} \int_{-\pi/2}^{\pi/2} f(t) \sin(kt)\, \mathrm{d}t \end{aligned} \tag{3.18}$$

Fourier synthesis is relevant, because the superposition principle stipulates that the response of a system to a linear sum of signals is the same as the sum of the response to the individual signals. Consequently, if we can determine the system response to the individual sinusoidal components in a Fourier series, we know the system response to the signal $f(t)$ that the Fourier series represents.

It is more convenient to use the complex notation of the Fourier transform, which builds on the Euler relationship

$$e^{j\phi} = \cos\phi + j\sin\phi \tag{3.19}$$

and which allows us to rewrite Eq. (3.17) as

$$f(t) = \frac{C_0}{2} + \sum_{k=1}^{\infty} C_k e^{jkt} \tag{3.20}$$

with the complex Fourier coefficients C_k obtained through

$$C_k = \frac{2}{\pi}\int_{-\pi/2}^{\pi/2} f(t)e^{jkt}\,\mathrm{d}t \tag{3.21}$$

The Fourier coefficients, either the pair of a_k and b_k or the complex C_k, form the *Fourier spectrum*, which tells us how strong each harmonic component is represented in the signal $f(t)$. The assumed periodicity $f(t) = f(t+2k\pi)$ is inherent in the periodicity of the basis functions in Eq. (3.17).

Simplified, we can interpret the Fourier transform as casting a time-varying signal $f(t)$ into an amplitude-frequency space. A sine wave $f(t) = A\cdot \sin\omega t$, for example, can be described as a single point in an amplitude-frequency coordinate system. Similarly, a superposition of two sine waves, $f(t) = A_1\sin(\omega_1 t) + A_2\sin(\omega_2 t)$, is represented by the two points (A_1, ω_1) and (A_2, ω_2) in amplitude-frequency space.

This simplified interpretation does not allow for the possibility of a phase shift. For example, a simple amplitude-frequency space cannot account for the signal $f(t) = A_1\sin(\omega_1 t) + A_2\sin(\omega_2 t + \varphi)$: a third axis is required. One possibility is to define a three-dimensional space of amplitude, frequency, and phase. Alternatively, we can use the relationship

$$A\sin(\omega t + \varphi) = (A\cos\varphi)\sin\omega t + (A\sin\varphi)\cos\omega t \tag{3.22}$$

where two basis functions (sin and cos) with the same frequency, but different amplitude are combined. In this case, we have two orthogonal amplitude components with the same frequency. This interpretation relates directly to Eq. (3.17) and explains the necessity to use both sine and cosine functions as basis functions for the Fourier synthesis.

The notion of decomposing a signal into its harmonic content leads to several formulations that are similar in spirit, but different in detail. The above introduction (specifically, Eqs. (3.20) and (3.21)) relates a continuous signal $f(t)$ to a discrete spectrum with integer multiples ($k = 0, 1, 2, \ldots$) of a fundamental frequency ω. This specific form of harmonic synthesis is usually referred to as *Fourier series*.

Often, it is desirable to obtain a continuous spectrum. For this case, an alternative formulation of the Fourier analysis exists that is often simply referred to as the *continuous Fourier transform*:

$$\mathcal{F}\{f(t)\} = F(\omega) = \int_{-\infty}^{+\infty} f(t)e^{-j\omega t}\,\mathrm{d}t \tag{3.23}$$

where the integral needs to be evaluated for all possible values of ω. The inverse Fourier transform is defined as

$$f(t) = \mathcal{F}^{-1}\{F(\omega)\} = \frac{1}{2\pi}\int_{-\infty}^{+\infty} F(\omega)e^{+j\omega t}\,\mathrm{d}\omega \tag{3.24}$$

There are numerous cases in which the Fourier integral in Eq. (3.23) does not exist, yet the function $f(t)$ still has a frequency spectrum. For example, Eq. (3.23) diverges for a constant, nonzero $f(t)$. We will not cover those cases as they are not needed for the frequency-domain description of linear systems. On the other hand, it is interesting to relate the Fourier- to the Laplace transform. We obtain from the Laplace correspondence tables that a sinusoidal oscillation corresponds to a conjugate pole pair on the imaginary axis of the s-plane,

$$\sin\omega t \quad \circ\!\!-\!\!\bullet \quad \frac{\omega}{s^2+\omega^2} \tag{3.25}$$

where the pole pair resides at $\pm j\omega$. In other words, periodic, sinusoidal, unattenuated oscillations are locations in the s-plane where the real part σ of s is zero. In a somewhat unscientific fashion, setting $s = j\omega$ in the Laplace transform,

$$F(j\omega) = \mathcal{L}\{f(t)\}|_{\sigma=0} = \int_{0}^{\infty} f(t)e^{-j\omega t}\,\mathrm{d}t \tag{3.26}$$

creates a transform highly reminiscent of the Fourier transform in Eq. (3.23). Intuitively, the Fourier transform of a periodic signal can be found on the imaginary axis of its Laplace transform. Two distinctions emerge from the assumption that a system is subject to the sinusoidal input signal for a sufficiently long time that transients have decayed to zero: first, the one-dimensional frequency on the Fourier transform contrasted with the two-dimensional variable s of the Laplace transform; second, the lower integration bound that starts at $-\infty$ for the Fourier transform.

The Fourier transform has a number of important properties, which almost all (except for the symmetry) correspond to properties of the Laplace transform:

- *Linearity:* The Fourier transform is a linear operation for which both superposition and scaling principles hold.
- *Symmetry:* The time domain function $f(t)$ is generally real-valued. If $f(t)$ is an odd function, $F(\omega)$ is odd and imaginary. If $f(t)$ is an even function, $F(\omega)$ is even and real-valued. Any function $f(t)$ can be split into an even and an odd component:

$$f_{even}(t) = \frac{1}{2}\left(f(t) + f(-t)\right); \quad f_{odd}(t) = \frac{1}{2}\left(f(t) - f(-t)\right) \tag{3.27}$$

Due to the linearity, the components $f_{even}(t)$ and $f_{odd}(t)$ can be subjected to the Fourier transform separately and recombined. For any $f(t)$, the Fourier transform shows a conjugate symmetry with $F(\omega) = F^*(-\omega)$ where F^* denotes the complex conjugate of F.

- *Time shift:* A time delay τ corresponds to a multiplication with an exponential:

$$\mathcal{F}\{f(t-\tau)\} = F(\omega) \cdot e^{-j\omega\tau} \tag{3.28}$$

- *Time scaling:* When the signal is compressed on the time axis by a factor of a, the spectrum is expanded by the same factor:

$$\mathcal{F}\{f(at)\} = \frac{1}{|a|} F\left(\frac{\omega}{a}\right) \tag{3.29}$$

- *Convolution:* A convolution in the time domain corresponds to a multiplication in the frequency domain and *vice versa.*
- *First derivative:*

$$\mathcal{F}\{\dot{f}(t)\} = j\omega F(\omega) \tag{3.30}$$

Let us examine three examples of the continuous Fourier transform. First, we examine the Fourier transform of $f(t) = A\cos(\nu t)$ where ν is the frequency of the cosine (contrasted with ω, which is the frequency variable of the Fourier transform):

$$\mathcal{F}\{A\cos(\nu t)\} = \int_{-\infty}^{+\infty} A\cos(\nu t) e^{-j\omega t}\, \mathrm{d}t \tag{3.31}$$

To solve the Fourier integral, we move the constant A out of the integral and use Euler's relationship to convert the cosine into a sum of exponentials:

$$\begin{aligned} \cdots &= \frac{A}{2}\int_{-\infty}^{+\infty} \left(e^{j\nu t} + e^{-j\nu t}\right) e^{-j\omega t}\, \mathrm{d}t \\ &= \frac{A}{2}\int_{-\infty}^{+\infty} \left(e^{-j(\omega-\nu)t} + e^{-j(\omega+\nu)t}\right) \mathrm{d}t \end{aligned} \tag{3.32}$$

The infinite integral of a harmonic oscillation is always zero unless the argument of the exponential is zero. Therefore, the result of the integration is nonzero only at the frequency of the cosine:

$$F(\omega) = \mathcal{F}\{A\cos(\nu t)\} = \begin{cases} A/2 & \text{for } \omega = -\nu \\ A/2 & \text{for } \omega = +\nu \\ 0 & \text{otherwise} \end{cases} \tag{3.33}$$

It is interesting to note that a spectral component exists for $\omega = -\nu$, that is, a *negative* frequency. Although negative frequencies may not seem intuitive at first, they are mathematically straightforward. For example, if we substitute ν for $-\nu$ in the argument of the cosine, we obtain a function $g(t) = \cos(-\nu t)$. Due to the even symmetry of the cosine function, $\cos(-\nu t) = \cos(\nu t)$. The amplitude "energy" of the signal is split evenly across the positive and negative component, and each has the amplitude of $A/2$.

As a second example, we examine a nonperiodic signal $f(t) = e^{-at}u(t)$, that is, an exponential decay. The Fourier integral is

$$\mathscr{F}\left\{e^{-at}u(t)\right\} = \int_0^{\infty} e^{-(a+j\omega)t}\,\mathrm{d}t \tag{3.34}$$

The lower integration bound of zero is a result of the step function $u(t)$, which ensures that $f(t) = 0$ for $t < 0$. The integration is straightforward and yields

$$\mathscr{F}\left\{e^{-at}u(t)\right\} = \frac{1}{j\omega + a} \tag{3.35}$$

The function $f(t)$ apparently has an infinite number of Fourier coefficients. Moreover, consider that e^{-at} is the impulse response of the system $1/(s+a)$. In Eq. (3.26), we related Fourier and Laplace transforms by setting the real component σ to zero and obtained $s = j\omega$. The term $j\omega$ in the denominator of Eq. (3.35) clearly corresponds to s in the Laplace transform of the function.

In the third example, we examine the Fourier transform of the *comb* function. The comb function is a train of equidistant delta pulses and can be described as

$$\text{comb}(t, T_0) = \sum_{k=-\infty}^{\infty} \delta(t - kT_0) \tag{3.36}$$

where T_0 is the distance in time of the delta pulses. It is easy to prove that the delta function and the constant 1 are Fourier transform pairs. Moreover, a time-shifted delta function has a Fourier transform of

$$\mathscr{F}\left\{\delta(t - kT_0)\right\} = e^{-j\omega k T_0} \tag{3.37}$$

Since the comb function meets the periodicity condition for the Fourier series, we can use Eq. (3.20) to express the comb function as an infinite sum of harmonic oscillations with the Fourier coefficients C_k (Eq. (3.21)):

$$C_k = \frac{2\pi}{T_0}\int_{-T_0/2}^{T_0/2} \delta(t) e^{jk\frac{2\pi}{T_0}t}\,\mathrm{d}t \tag{3.38}$$

The periodicity is reflected in the substitution of the integration bounds from $\pm\pi$ to $\pm T_0/2$ with the appropriate scaling factor in the exponent. Evaluation of the integral gives $C_k = 2\pi/T_0$ for all k. The pulse repetition frequency is

$$\omega_0 = \frac{2\pi}{T_0} \tag{3.39}$$

and we find that the Fourier transform of a comb function with the pulse distance T_0 is a frequency-domain comb function with the pulse distance ω_0 (Figure 3.2). We can

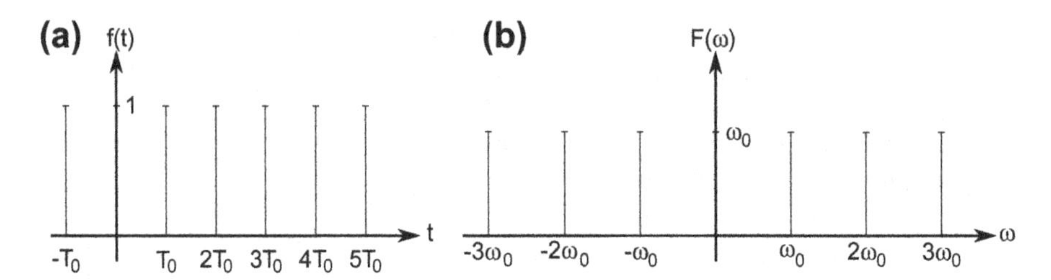

Figure 3.2 Graphical representation of the comb function (a) and its Fourier transform (b).

use Eq. (3.21) to explicitly write the spectrum of the comb function:

$$\mathrm{Comb}(\omega, \omega_0) = \omega_0 \sum_{k=-\infty}^{\infty} \delta(\omega - k\omega_0) \tag{3.40}$$

This result has important implications for discretely sampled signals. If we sample a signal $f(t)$ in intervals of T_0, the resulting discrete sequence of samples, $f_s(t)$, can be seen as $f(t)$ multiplied with the comb function (*cf.* Section 4.1, specifically Eq. (4.1)):

$$f_s(t) = f(t) \cdot \sum_{k=-\infty}^{\infty} \delta(t - kT_0) = f(t) \cdot \mathrm{comb}(t, T_0) \tag{3.41}$$

The convolution property now tells us that a multiplication in the time domain corresponds to a convolution in the frequency domain. In this case, the spectrum of the sampled signal $f_s(t)$ becomes

$$\begin{aligned} \mathscr{F}\{f_s(t)\} &= F(\omega) \otimes \mathrm{Comb}(\omega, \omega_0) \\ &= F(\omega) \otimes \omega_0 \sum_{k=-\infty}^{\infty} \delta\left(\omega - k\omega_0\right) \\ &= \omega_0 \sum_{k=-\infty}^{\infty} F\left(\omega - k\omega_0\right) \end{aligned} \tag{3.42}$$

Equation (3.42) tells us that the spectrum of a discretely sampled signal is an infinite replicate of the spectrum of the continuous signal at equidistant frequencies of ω_0. A problem emerges if the spectrum of $f(t)$ contains frequency components higher than $\omega_0/2$, because those frequency components overlap with the neighboring replicate images of the spectrum. This effect is known as *aliasing* and is discussed in more detail in Section 4.1.

3.3 Representation of the RC Lowpass and Spring-Mass-Damper Systems in the Laplace Domain

We will now use the principles introduced in this chapter to determine the system response of the RC lowpass (Figure 2.2) and the spring-mass-damper system (Figure 2.3). Recall that the differential equation of the RC lowpass is

$$R \cdot C \cdot \dot{V}_{out}(t) + V_{out}(t) = V_{in}(t) \tag{3.43}$$

Assume that we have chosen an input signal $V_{in}(t)$ and that we know its Laplace correspondence $V_{in}(s)$. In addition, the capacitor carries a charge at the start of the experiment, and the initial condition is $V_{out}(t = 0) = V_0$. By using the derivative rule (Eq. (3.7)), the Laplace correspondence of Eq. (3.43) can be found:

$$R \cdot C \left(s \cdot V_{out}(s) - V_0\right) + V_{out}(s) = V_{in}(s) \tag{3.44}$$

The initial condition V_0 can be moved to the right-hand side, because it can be treated similar to an independent variable. However, the initial condition is *fixed* once the experiment starts at $t = 0$. Therefore, any initial condition is not a variable (such as $x(t)$ or $X(s)$), but rather a constant. As such, it remains constant in the Laplace domain unlike a variable, for which a Laplace correspondence must always be found. By factoring out $V_{out}(s)$ and dividing by the characteristic polynomial $q(s) = RCs+1$, the output variable can be isolated:

$$V_{out}(s) = \frac{1}{RCs+1} \cdot V_{in}(s) + \frac{RC}{RCs+1} \cdot V_0 \tag{3.45}$$

The characteristic polynomial has one root at $s = -1/RC$, and the system response is single-exponential with a time constant $\tau = RC$. The two summation terms in Eq. (3.45) can be handled separately and superimposed. A direct correspondence exists for the term with the initial condition (Table A.3), and the contribution of the initial condition is a voltage that decays with $V_0 \cdot e^{-t/RC}$. To stay with the example in Section 2.1, we choose for the input voltage $V_{in}(t)$ a step function that jumps from 0 to V_1 at $t = 0$. Although the voltage is constant for $t > 0$, the input voltage must be treated as a *variable*, and the Laplace correspondence for the step function, in this case $V_{in}(s) = V_1/s$, needs to be used (Table A.1). Equation (3.45), written out explicitly for the step input voltage, becomes

$$V_{out}(s) = \frac{1}{RCs+1} \cdot \frac{V_1}{s} + \frac{RC}{RCs+1} \cdot V_0 \tag{3.46}$$

With the correspondences in Table A.3, the time-domain function $V_{out}(t)$ can be found:

$$V_{out}(t) = V_1 \cdot \left(1 - e^{-t/RC}\right) + V_0 \cdot e^{-t/RC} \tag{3.47}$$

For the energy-free system, that is, $V_0 = 0$, the solution is the same as the one in Eq. (2.4).

The next example is the spring-mass-damper system (Section 2.2, *cf.* Figure 2.3). By using the balance-of-forces approach, the differential equation can be found as

$$m\ddot{x}(t) + R_F\dot{x}(t) + Dx(t) = F(t) \tag{3.48}$$

where $x(t)$ is the position of the mass, $\ddot{x}(t)$ is the acceleration, and $\dot{x}(t)$ is the velocity. Two possible initial conditions exist, the position of the mass at $t = 0$, defined as x_0, and the initial velocity defined as $v_0 = \dot{x}(t = 0)$. The general form of this equation was given in Eq. (3.15), and the Laplace-domain equation of the spring-mass-damper system is

$$m \cdot \left(s^2X(s) - sx_0 - v_0\right) + R_F \cdot \left(sX(s) - x_0\right) + D \cdot X(s) = F(s) \tag{3.49}$$

As in the example of the RC lowpass, the constant terms (initial conditions) are moved to the right-hand side, and the dependent variable $X(s)$ is factored out on the left-hand side:

$$X(s)\left(ms^2 + R_Fs + D\right) = F(s) + x_0\left(ms + R_F\right) + mv_0 \tag{3.50}$$

By dividing by the characteristic polynomial, $q(s) = ms^2+R_Fs+D$, the movement of the mass $X(s)$ can be determined for any given force $F(t)$ and initial condition,

$$X(s) = \frac{F(s) + m\left(sx_0 + v_0\right) + R_Fx_0}{ms^2 + R_Fs + D} \tag{3.51}$$

and can be simplified for the energy-free system when $v_0 = 0$ and $x_0 = 0$,

$$X(s) = \frac{1}{ms^2 + R_Fs + D} \cdot F(s) \tag{3.52}$$

With the help of correspondence tables or partial fraction expansion, we can now obtain the inverse Laplace transform of $X(s)$ and thus the time-dependent motion $x(t)$. The spring-mass-damper system is a second-order system and therefore has two poles at the locations where $q(s) = ms^2 + R_Fs + D = 0$. The *system* is defined by the three components, mass m, spring (Hooke's constant D) and damper (viscous friction R_F). The combination of these components determines the location of the poles. Different responses can be obtained with different combinations of these components. Furthermore, a system can change over time, much like a car's shock absorber that slowly wears out.[3]

We can examine the behavior of a shock absorber with wear by varying the coefficient R_F. However, the behavior of a frictional element depends on the friction itself, but also on the energy stored in both the mass and the spring, which needs to be dissipated by the friction element. A common normalization involves the definition of a natural frequency ω_n and a damping coefficient ζ,

$$\omega_n = \sqrt{\frac{D}{m}}; \quad \zeta = \frac{R_F}{2\sqrt{D \cdot m}} \tag{3.53}$$

[3]Strictly, such a system is no longer time-invariant. However, if the degradation occurs very slowly, it can be interpreted as a series of different time-invariant systems.

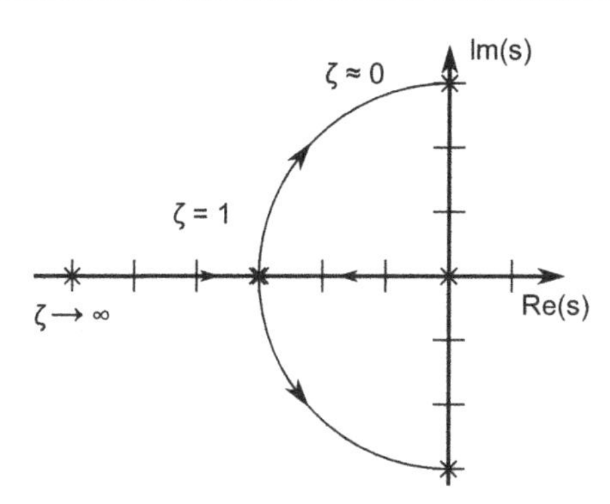

Figure 3.3 Location of the roots of a spring-mass-damper system as the damping factor ζ goes from ∞ to 0. For high values of ζ, both poles lie on the real axis. As ζ is lowered (lower friction), the poles move toward each other along the real axis. With $\zeta = 1$, the two poles meet in one location and form a double pole. This case is the critically damped case. When ζ is further decreased, a complex conjugate pole pair moves along a quarter circle towards the imaginary axis. When $\zeta = 0$, a pole pair forms on the imaginary axis that indicates a periodic, unattenuated oscillation.

which allows us to divide $q(s)$ by m and rewrite the equation $q(s) = 0$ as

$$s^2 + 2\zeta\omega_n + \omega_n^2 = 0 \tag{3.54}$$

The two solutions for this equation are the poles p_1 and p_2:

$$p_{1,2} = -\omega_n \left(\zeta \pm \sqrt{\zeta^2 - 1}\right) \tag{3.55}$$

Starting from a very high friction, $R_F \to \infty$, and moving toward very low friction values, $R_F \to 0$, the normalized friction, i.e., the damping coefficient ζ likewise goes from ∞ to 0. The location of the poles as ζ goes from ∞ to 0 is shown in Figure 3.3. There are five important cases to consider:

1. For extremely large values of R_F, one pole is near the origin and one pole is at $-R_F/m$. The friction is so high (and the spring in comparison so weak) that we just linearly move the mass as we push against it with constant force.
2. For comparatively large R_F (to be precise, for $\zeta > 1$), the poles $p_{1,2}$ in Eq. (3.55) are both real-valued. The transfer function in Eq. (3.52) can be rewritten in product form:

$$H(s) = \frac{X(s)}{F(s)} = \frac{1}{ms^2 + R_F s + D} = \frac{1}{m(s + p_1)(s + p_2)} \tag{3.56}$$

   The system response contains two exponential terms. The time constant of the two exponential functions becomes more similar as $\zeta \to 1$.
3. When $\zeta = 1$, the expression under the square root of Eq. (3.55) vanishes. The two poles are now at the same location $p_1 = p_2 = -\zeta\omega_n = -R_F/2m$. The system response is still exponential, but both exponential terms have the same time constant. A second-order system with $\zeta = 1$ is called *critically damped*.

4. For $\zeta < 1$, the expression under the square root in Eq. (3.55) turns negative. We can deal with this situation by factoring out $j = \sqrt{-1}$, whereupon Eq. (3.55) turns into

$$p_{1,2} = -\omega_n \left(\zeta \pm j\sqrt{1-\zeta^2} \right) \tag{3.57}$$

and $p_{1,2}$ form a complex conjugate pole pair. We can see that each pole of the pair follows a quarter circle as ζ decreases from 1 to 0. As shown in Eq. (2.9), a complex conjugate pole pair indicates that the output signal contains an attenuated oscillatory component.
5. The last case occurs when $R_F = 0$ (frictionless case) and correspondingly $\zeta = 0$. The first-order term of the characteristic polynomial vanishes, and the pole pair is now on the imaginary axis at $p_{1,2} = \pm j\omega_n$. The system response is a periodic, unattenuated oscillation at the natural frequency ω_n.

It is interesting to examine the step response of such a system. The step response in this example means that a sudden and constant force is applied at $t = 0$ (somebody pulls at the handle with a constant force), and the response is the resulting motion of the moving parts of the system. The step responses for some different values of ζ are shown in Figure 3.4. With very high friction values, the system responds slowly, because one pole is near the origin. Such a pole p is associated with an exponential response e^{-pt}, and because of the low value of p, the term vanishes slowly with time. The critically damped case is of particular interest, because it has the fastest step response

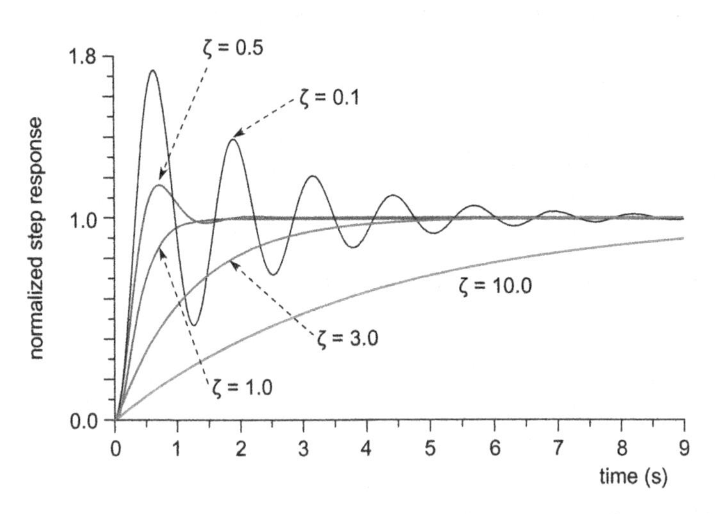

Figure 3.4 Step responses of a second-order (spring-mass-damper) system for some of the cases listed in the text. With a high damping coefficient, in this case $\zeta = 10$, the dynamic response is very slow as it is dominated by the real-valued pole near the origin. With lowered damping ($\zeta = 3$), the poles move closer to the location of the double pole (see Figure 3.3), and the step response becomes faster. The special critically damped case with $\zeta = 1$, where the two poles meet on the real axis, has the fastest step response without overshoot. When ζ is further decreased, overshoot occurs, and with very low values of ζ, in this case, $\zeta = 0.1$, distinct multiple oscillations can be observed.

without overshoot. In a car suspension, for example, this is desirable, because the wheels should return to contact with the pavement as quickly as possible—without bouncing. Bouncing, better described as repeated oscillations, occurs with very low values of ζ. Overshoot with repeated oscillations is associated with pronounced resonance effects that are usually detrimental to a system's behavior. We will see later that feedback control can attenuate resonance effects and therefore reduce the susceptibility of a system to resonance build-up.

3.4 Transient and Steady-State Response

We have seen repeatedly that a linear system responds to any input signal with a linear combination of exponential functions. In a stable system, the exponentials decay with time; however, the exponential responses may be complex and contain oscillatory components. The *transient response* refers to the system response near $t = 0$ when the exponential components dominate the output signal. Conversely, the *steady-state response* refers to the system response when $t \to \infty$ and the exponential components have sufficiently decayed. In this context, *sufficiently decayed* is a relative term: a decision must be made how much influence from the exponential terms is allowable. Clearly, if any part of the decaying functions falls below the measurement limit, it can be neglected. Often, a threshold is defined. For example, we can define that the steady state has been reached when the system response falls within $\pm 2\%$ of its final value. A system with a time constant τ reaches the 2% tolerance band after approximately 4τ.

The response of a system depends on both the transfer function and the input signal. This is intuitive when we look at Eq. (3.51). When no forces exist ($F(s) = 0$), the response is determined by the initial conditions alone. Conversely, the motion of the mass depends on whether we briefly yank at the handle, apply a constant force, or even introduce an oscillatory force. Only when both the transfer function $H(s)$ and the system input $X(s)$ are known, it becomes possible to compute $Y(s) = H(s) \cdot X(s)$. There are several typical input functions that are frequently used to describe a system's response. The four most important functions are the impulse function, the step function, the ramp function, and the harmonic oscillation.

An *impulse function* (also known as delta impulse function or Kronecker delta pulse) has an infinite amplitude of zero duration, but its integral is unity. We can define the delta impulse as

$$\delta(t) = \lim_{\tau \to 0} \begin{cases} 0 & \text{for } t < 0 \\ 1/\tau & \text{for } 0 \leq t < \tau \\ 0 & \text{for } t \geq \tau \end{cases} \tag{3.58}$$

The delta impulse is the only function in this series of functions that introduces a finite amount of energy. The Laplace transform of the delta pulse is $\mathcal{L}\{\delta(t)\} = 1$.

The *unit step function* (also known as Heaviside step function), here denoted as $u(t)$, is the integral of the delta impulse over time and is defined as:

$$u(t) = \lim_{\tau \to 0} \begin{cases} 0 & \text{for } t < 0 \\ t/\tau & \text{for } 0 \le t < \tau \\ 1 & \text{for } t \ge \tau \end{cases} \tag{3.59}$$

Usually, the simpler definition in Eq. (3.60) is found:

$$u(t) = \begin{cases} 0 & \text{for } t \le 0 \\ 1 & \text{for } t > 0 \end{cases} \tag{3.60}$$

When we consider that the first derivative operation corresponds to a multiplication with s in the Laplace domain, and conversely the integration operation corresponds to a multiplication with $1/s$, it makes intuitive sense that $\mathcal{L}\{u(t)\} = 1/s$. Direct evaluation of the Laplace integral yields the same result. If we integrate this signal again, we obtain the unit ramp, defined as

$$r(t) = \begin{cases} 0 & \text{for } t < 0 \\ t & \text{for } t \ge 0 \end{cases} \tag{3.61}$$

There is no uniform symbol for the ramp, so we arbitrarily denote it as $r(t)$. By using the integration operator in the Laplace domain again, we obtain $\mathcal{L}\{r(t)\} = 1/s^2$. Note that the ramp function is not bounded. The energy introduced by the unit step function is also not bounded, although the step function reaches a finite value.

The unit sinusoid with frequency ω is different from the above. The time-domain function and its Laplace transform are

$$f(t) = \sin(\omega t) \quad \circ\!\!-\!\!\bullet \quad F(s) = \frac{\omega}{s^2 + \omega^2} \tag{3.62}$$

We can now examine the response of a process to some input signal. For example, we apply a unit step function to a process with the transfer function $\alpha/(s + b)$. The Laplace-domain response is therefore

$$Y(s) = \frac{1}{s} \cdot \frac{\alpha}{s + b} \quad \bullet\!\!-\!\!\circ \quad y(t) = \frac{\alpha}{b}\left(1 - e^{-bt}\right) \tag{3.63}$$

How do we know $y(t)$? If we are lucky, we can find $Y(s)$ in Laplace correspondence tables (see Appendix A). If the response cannot be found in the available correspondence tables, the next best option is partial fraction expansion, which is described in the next section.

Often it is sufficient to know the value of the output signal after the system has equilibrated. This value is known as the *steady-state response*. It is possible to determine the time-domain response as in the example above and let $t \to \infty$. However, the *final value theorem* allows to arrive at this goal much faster. If the response $Y(s)$ is known, the final value theorem states that

$$y(t \to \infty) = \lim_{s \to 0} s \cdot Y(s) \tag{3.64}$$

A counterpart for $t \rightarrow 0^+$ exists, called the *initial value theorem*:

$$y(t \rightarrow 0) = \lim_{s \rightarrow \infty} s \cdot Y(s) \tag{3.65}$$

Let us apply the final value theorem in one example. The transfer function of a system is given as

$$H(s) = \frac{s}{s+a} \tag{3.66}$$

and we want to examine if the output diverges when a ramp input is applied, or if the output reaches a finite equilibrium value. The ramp function is $X(s) = 1/s^2$. We apply the final value theorem to the output signal (i.e., the product of transfer function and input signal):

$$y(t \rightarrow \infty) = \lim_{s \rightarrow 0} s \cdot \frac{1}{s^2} \cdot \frac{s}{s+a} = \frac{1}{a} \tag{3.67}$$

Note that we found the time-domain equilibrium value without formally performing an inverse Laplace transform.

A continuous oscillation also counts as a steady-state response. One example is the RC lowpass with a sinusoidal signal at its input. Implicitly, in the Laplace domain, the sinusoidal signal begins at $t = 0$. More precisely, we would define the input signal as $x(t) = \sin(\omega t) \cdot u(t)$. In the Laplace domain, we find the response function

$$Y(s) = \frac{a}{s+a} \cdot \frac{\omega}{s^2 + \omega^2} \tag{3.68}$$

where the first fraction is the transfer function of the RC lowpass with $a = 1/RC$ and the second fraction is the Laplace transform of the sinusoidal input signal, $X(s)$. The final value theorem is not applicable, because no *constant* final value exists. Without covering the details (a detailed example is presented at the end of Section 3.5.1), partial fraction expansion yields:

$$Y(s) = \left(\frac{a\omega}{a^2 + \omega^2}\right) \cdot \frac{1}{s+a} + \left(\frac{a^2}{a^2 + \omega^2}\right) \cdot \frac{\omega}{s^2 + \omega^2} - \left(\frac{a\omega}{a^2 + \omega^2}\right) \cdot \frac{s}{s^2 + \omega^2} \tag{3.69}$$

The expressions in large parentheses are real-valued, constant scalars. The second part of the product in each summation term is a function of s, which we can find in the correspondence tables. The superposition principle allows us to add the inverse Laplace transform of each summation term to obtain $y(t)$:

$$y(t) = \frac{a\omega}{a^2 + \omega^2} e^{-at} + \frac{a^2}{a^2 + \omega^2} \sin \omega t - \frac{a\omega}{a^2 + \omega^2} \cos \omega t \tag{3.70}$$

We can see from Eq. (3.70) that the response of the RC lowpass to an input signal $x(t) = \sin(\omega t) \cdot u(t)$ consists of a transient exponential and a continuous oscillation. The output oscillation has the same frequency as the input oscillation, but experiences a phase shift (see Chapter 11). Since $x(t)$ is assumed to be zero for $t < 0$ and the RC lowpass energy-free (meaning, $y(t) = 0$ for $t < 0$), the output signal $y(t)$ must be continuous at $t = 0$. We therefore require $y(t = 0^+) = 0$. Equation (3.70) reveals that the exponential transient ensures continuity at $t = 0$, because Eq. (3.70) becomes for $t = 0$

$$y(t = 0) = \frac{a\omega}{a^2 + \omega^2} e^0 - \frac{a\omega}{a^2 + \omega^2} \cos(0) = 0 \tag{3.71}$$

Equation (3.70) also reveals that the output signal becomes a continuous sinusoidal oscillation when the exponential term has decayed for $t \gg 0$:

$$y(t \gg 0) = \frac{a}{a^2 + \omega^2} \left(a \sin \omega t - \omega \cos \omega t \right) \tag{3.72}$$

3.5 Partial Fraction Expansion

There are cases where we need to obtain the inverse Laplace transform of a Laplace-domain function, even if it cannot be found in any correspondence table. Usually, a system response in the Laplace domain exists as the fraction of two polynomials,

$$F(s) = \frac{p(s)}{q(s)} = \frac{\sum_{j=0}^{M} b_j s^j}{\sum_{k=0}^{N} a_k s^k}, \quad a_N = 1, N \geq M \tag{3.73}$$

This system has N poles at $s = p_k$ (i.e., roots of the denominator polynomial). Note that the value of p_k is generally negative. The corresponding term for a pole, $(s - p_k)$, looks more familiar if we substitute a numerical value, such as $p_k = -3$, where we obtain $(s + 3)$.

With the known poles p_k, we can rewrite the denominator polynomial $q(s)$ as

$$q(s) = \sum_{k=0}^{N} a_k s^k = \prod_{k=1}^{n} (s - p_k)^{r_k} \tag{3.74}$$

where we normalized $a_N = 1$ (as given in Eq. (3.73)). The term r_k is used to consider multiple roots such that $\sum_{k=1}^{n} r_k = N$. Complex poles follow the same scheme, but the multiplicative terms occur in conjugate pairs $(s - p_k)(s - p_{k+1})$ where $p_k = p_{k+1}^*$.

For any $F(s)$ described by Eq. (3.73), a partial fraction expansion according to Eq. (3.75) exists:

$$F(s) = b_N + \sum_{k=1}^{n} \sum_{l=1}^{r_k} \frac{c_{kl}}{(s - p_k)^l} \tag{3.75}$$

where $b_N = 0$ when $M < N$ and $b_N = b_M$ when $M = N$. The c_{kl} are the *residues* (often also termed *residuals*) of $F(s)$ at p_k, and these need to be determined to complete the expansion. Once we have found the additive form of $F(s)$ in Eq. (3.75), the individual additive components can be subjected to the inverse Laplace transform one-by-one, and their corresponding time-domain functions can be added due to the superposition principle of linear systems. The inverse Laplace transform of $F(s)$ is therefore

$$\mathscr{L}^{-1}\{F(s)\} = b_N\delta(t) + \sum_{k=1}^{n}\sum_{l=1}^{r_k} \frac{c_{kl}}{(l-1)!} t^{l-1} e^{p_k t} \tag{3.76}$$

Once again we note that the p_k are generally negative (in stable systems), and thus the exponential term vanishes with time. If p_k is part of a complex conjugate pair, the exponential term describes an exponentially decaying oscillation.

Two options exist to obtain the unknown c_{kl}. We can multiply Eq. (3.75) with the smallest common denominator and obtain a lengthy expression in the numerator with the c_{kl} multiplied by some of the pole terms. Multiplying these out and arranging them by powers of s leads to a matrix equation that can be solved by matrix inversion or with Cramer's rule. Alternatively, the residues can be determined with Eq. (3.77):

$$c_{kl} = \frac{1}{(r_k - l)!} \cdot \frac{\partial^{r_k - l}}{\partial s^{r_k - l}} \left((s - p_k)^{r_k} \cdot F(s)\right)\Big|_{s=p_k} \tag{3.77}$$

where the index k refers to the root, and the index l counts the repeated roots.

Let us consider several special cases:

- **No repeated roots:** If no repeated roots exist, Eqs. (3.75) and (3.77) can be simplified to

$$F(s) = b_N + \sum_{k=1}^{N} \frac{c_k}{s - p_k} \tag{3.78}$$

$$c_k = \left((s - p_k) \cdot F(s)\right)\big|_{s=p_k} \tag{3.79}$$

- **Double pole:** For the special case of a double root of the denominator polynomial, two values of c_{kl} need to be determined to satisfy the inner sum of Eq. (3.75):

$$\begin{aligned} c_{k1} &= \frac{\partial}{\partial s} \left((s - p_k)^2 \cdot F(s)\right)\Big|_{s=p_k} \\ c_{k2} &= \left((s - p_k)^2 \cdot F(s)\right)\big|_{s=p_k} \end{aligned} \tag{3.80}$$

- **Complex roots:** Complex roots can be treated in exactly the same way as real-valued roots. However, since each complex root always has a complex conjugate counterpart, the two corresponding residuals can be determined in one term. Let us assume that one complex conjugate root pair in Eq. (3.75) leads to

$$F(s) = \frac{c_1}{s - p_1} + \frac{c_2}{s - p_2} + \frac{c_3}{s - p_3} + \cdots \tag{3.81}$$

where $p_1 = p_2^*$ and notably $c_1 = c_2^*$. Let us define $p_1 = a + jb$ and consequently $p_2 = a - jb$. We can cross-multiply the first two additive terms in Eq. (3.81) and obtain

$$F(s) = \frac{C_1 s + C_2}{s^2 - 2as + a^2 + b^2} + \frac{c_3}{s - p_3} + \cdots \tag{3.82}$$

For the first-order residual term $C_1 s + C_2$ we now use

$$C_1 s + C_2|_{s=p1} = \left((s - p_1)(s - p_2) \cdot F(s)\right)\Big|_{s=p_1} \tag{3.83}$$

Since p_1 is complex, we obtain two equations (real and imaginary parts of Eq. (3.83)) to determine C_1 and C_2. The choice of p_1 is arbitrary and the same values for C_1 and C_2 are obtained when p_2 is used instead. Combining the complex conjugate root pairs into one second-order term also facilitates the inverse Laplace transform, because we can conveniently use the sine and cosine correspondences in Table A.5.

3.5.1 Partial Fraction Expansion Examples

To illustrate partial fraction expansion, let us turn back to the spring-mass-damper system in Eq. (3.51) with an energy-free system (i.e., $x_0 = 0$ and $v_0 = 0$). Furthermore, we use $m = 1$ kg, $D = 8$ N/m, and $R_F = 6$ N s/m. We want to determine the response $x(t)$ when a constant force is applied at $t = 0$, that is, we use the unit step function $F(s) = 1/s$ for the force. The response is therefore

$$X(s) = \frac{1}{s(ms^2 + R_F s + D)} \tag{3.84}$$

We can see that the denominator polynomial has three roots,

$$p_{1/2} = -\frac{1}{2m}\left(R_F \pm \sqrt{R_F^2 - 4mD}\right); \quad p_3 = 0 \tag{3.85}$$

with the numerical values, $p_1 = -2$, $p_2 = -4$, and $p_3 = 0$. As a side note, we can see from Eq. (3.85) that the poles carry units of inverse seconds, which is consistent with the fact that the poles are locations in the s-plane.

No repeated roots: In the example above, we do not have repeated roots, and we can use the simplified form in Eq. (3.78), which provides us with the general solution (note $N = n = 3$)

$$X(s) = \frac{c_1}{s - p_1} + \frac{c_2}{s - p_2} + \frac{c_3}{s} = \frac{c_1}{s + 2} + \frac{c_2}{s + 4} + \frac{c_3}{s} \tag{3.86}$$

We now need to find the three residues c_k. From Eq. (3.79), we obtain

$$c_k = \left((s - p_k) \cdot X(s)\right)\Big|_{s=p_k} = \left.\frac{(s - p_k)}{s(s - p_1)(s - p_2)}\right|_{s=p_k} \tag{3.87}$$

By examining Eq. (3.87) for p_1, p_2, and p_3, we obtain the c_k:

$$\begin{aligned} c_1 &= \left.\frac{1}{s(s-p_2)}\right|_{s=p_1} = \frac{1}{p_1^2 - p_1 p_2} = -\frac{1}{4} \\ c_2 &= \left.\frac{1}{s(s-p_1)}\right|_{s=p_2} = \frac{1}{p_2^2 - p_1 p_2} = \frac{1}{8} \\ c_3 &= \left.\frac{s}{s(s-p_1)(s-p_2)}\right|_{s=0} = \frac{1}{p_1 p_2} = \frac{1}{8} \end{aligned} \tag{3.88}$$

Finally, we replace the c_k in Eq. (3.86) with the specific values found above and get

$$\begin{aligned} X(s) &= \frac{1}{(p_1^2 - p_1 p_2)(s-p_1)} + \frac{1}{(p_2^2 - p_1 p_2)(s-p_2)} + \frac{1}{p_1 p_2 s} \\ &= -\frac{1}{4(s+2)} + \frac{1}{8(s+4)} + \frac{1}{8s} \end{aligned} \tag{3.89}$$

All three additive components are simple first-order terms for which the inverse Laplace transform can be found in Table A.3, and we obtain the inverse Laplace transform $x(t) = \mathscr{L}^{-1}\{X(s)\}$ as

$$x(t) = -\frac{1}{4}e^{-2t} + \frac{1}{8}e^{-4t} + \frac{1}{8} \tag{3.90}$$

We can verify the result in this simple case, because the Laplace-domain equation of the form

$$X(s) = \frac{1}{s(s+2)(s+4)} \tag{3.91}$$

can be found in the correspondence tables (Table A.4), and we obtain the same time-domain correspondence as in Eq. (3.90).[4]

Solving the residues with a linear equation system: As an alternative approach, let us try to use a linear equation system instead of Eq. (3.79) for the residues. The starting point is Eq. (3.86), and we expand all three summation terms to get the common denominator:

$$X(s) = \frac{c_1}{s+2} + \frac{c_2}{s+4} + \frac{c_3}{s} = \frac{c_1 s(s+4) + c_2 s(s+2) + c_3(s+2)(s+4)}{s(s+2)(s+4)} \tag{3.92}$$

[4]In its practical application, partial fraction expansion is a magnet for small oversights and minor mistakes, but even minor mistakes can affect the final outcome in a major way. In addition, mistakes can easily lead to zero-valued denominators, divisions of zero by zero or other unfortunate obstacles. If you can find any way to avoid partial fraction expansion, avoid it.

The numerator is equal to 1. By multiplying out the numerator and rearranging by powers of s, we obtain the equation

$$(c_1 + c_2 + c_3)s^2 + (4c_1 + 2c_2 + 6c_3)s + 8c_3 = 1 \tag{3.93}$$

In matrix notation, Eq. (3.93) becomes

$$\begin{bmatrix} 1 & 1 & 1 \\ 4 & 2 & 6 \\ 0 & 0 & 8 \end{bmatrix} \cdot \begin{bmatrix} c_1 \\ c_2 \\ c_3 \end{bmatrix} = \begin{bmatrix} 0 \\ 0 \\ 1 \end{bmatrix} \tag{3.94}$$

Matrix inversion or simple elimination ($c_3 = 1/8$ is immediately obvious) yields the same values $c_1 = -1/4$ and $c_2 = 1/8$.

Repeated roots: If $p_1 = p_2$, the simplification in Eq. (3.78) cannot be used, because we have repeated roots. In this example, we use $p = p_1 = p_2 = -3$ with $p_3 = 0$ and obtain $X(s)$ by expanding the summations in Eq. (3.75),

$$X(s) = \frac{c_{11}}{(s - p_3)} + \frac{c_{21}}{(s - p)} + \frac{c_{22}}{(s - p)^2} = \frac{c_{11}}{s} + \frac{c_{21}}{(s + 3)} + \frac{c_{22}}{(s + 3)^2} \tag{3.95}$$

and computing the residues with

$$\begin{aligned} c_{11} &= \left. \frac{(s - p_3)}{s(s - p)^2} \right|_{s=p_3=0} = \frac{1}{p^2} = \frac{1}{9} \\ c_{21} &= \left. \frac{\partial}{\partial s} \frac{(s - p)^2}{s(s - p)^2} \right|_{s=p} = -\frac{1}{p^2} = -\frac{1}{9} \\ c_{22} &= \left. \frac{(s - p)^2}{s(s - p)^2} \right|_{s=p} = \frac{1}{p} = -\frac{1}{3} \end{aligned} \tag{3.96}$$

Substituting c_{11}, c_{21}, and c_{22} into Eq. (3.95) yields the Laplace-domain solution:

$$X(s) = \frac{1}{p^2 s} - \frac{1}{p^2(s - p)} + \frac{1}{p(s - p)^2} = \frac{1}{9s} - \frac{1}{9(s + 3)} - \frac{1}{3(s + 3)^2} \tag{3.97}$$

By using the correspondences for each summation term, the time-domain solution emerges as

$$x(t) = \frac{1}{9} - \frac{1}{9}e^{-3t} - \frac{1}{3}te^{-3t} \tag{3.98}$$

Once again, we find the function $X(s) = 1/(s(s + 3)^2)$ directly in Table A.4, and we thus can verify our result.

Complex roots: Now, let us reduce the damping coefficient to $R_F = 4$ N s/m. We still have the pole from the step function in the origin, that is, $p_3 = 0$. However, now $p_{1,2} = -2 \pm 2j$. Since $p_1 = p_2^*$, these do not count as repeated roots, and we could use the same case for nonrepeating roots above. However, it is simpler to allow a second-order term in the additive form ($p_{1,2} = a \pm jb$):

$$X(s) = \frac{C_1 s + C_2}{s^2 - 2as + a^2 + b^2} + \frac{c_3}{s} = \frac{C_1 s + C_2}{s^2 + 4s + 8} + \frac{c_3}{s} \tag{3.99}$$

We use Eq. (3.83) for the residues C_1 and C_2 and choose arbitrarily $s = p_1$:

$$C_1 p_1 + C_2 = \frac{1}{p_1} = \frac{p_2}{p_1 p_2} \tag{3.100}$$

where $p_2 = p_1^*$. By using $p_1 = a + jb$ and $p_2 = a - jb$, and by separating the real and imaginary components, Eq. (3.100) becomes

$$C_1 a + C_2 + jC_1 b = \frac{a}{a^2 + b^2} - j\frac{b}{a^2 + b^2} \tag{3.101}$$

from where we immediately obtain $C_1 = -1/(a^2 + b^2)$ by comparing the imaginary parts. For the real part, we have

$$C_1 a + C_2 = \frac{a}{a^2 + b^2} \quad \Rightarrow \quad C_2 = \frac{2a}{a^2 + b^2} \tag{3.102}$$

For the third residue, we evaluate $1/(s^2 + 2as + a^2 + b^2)$ at $s = 0$ and obtain $c_3 = 1/(a^2 + b^2)$. In summary, the residues are

$$C_1 = -\frac{1}{a^2 + b^2} = -\frac{1}{8}; \quad C_2 = \frac{2a}{a^2 + b^2} = -\frac{1}{2}; \quad c_3 = \frac{1}{a^2 + b^2} = \frac{1}{8} \tag{3.103}$$

The partial fraction expansion is therefore

$$\begin{aligned} X(s) &= \frac{1}{8}\left(\frac{-s-4}{s^2 + 4s + 8} + \frac{1}{s}\right) \\ &= \frac{1}{8}\left(-\frac{s+2}{(s+2)^2 + 4} - \frac{2}{(s+2)^2 + 4} + \frac{1}{s}\right) \end{aligned} \tag{3.104}$$

In the second line of Eq. (3.104), the terms have been rearranged to match those in the correspondence tables. We now find all three additive terms in the correspondence tables and finally obtain for $x(t)$

$$x(t) = \frac{1}{8}\left(1 - e^{-2t}(\cos(2t) + \sin(2t))\right) \tag{3.105}$$

Continuous oscillation: Lastly, let us examine the output signal of a RC lowpass when the input signal is a continuous sinusoid. In this example, the frequency of the sinusoid is twice the filter's cutoff frequency: $\omega = 2\pi 60$ Hz and $\omega_c = 1/(RC) = 2\pi 30$ Hz. We recall the transfer function of the RC lowpass from Eq. (3.45) and multiply it with the Laplace transform of the signal $\sin \omega t$:

$$Y(s) = \frac{1}{RCs + 1} \cdot \frac{\omega}{s^2 + \omega^2} = \frac{\omega_c}{s + \omega_c} \cdot \frac{\omega}{s^2 + \omega^2} \tag{3.106}$$

The approach to the solution with the unknown residues c_1, C_2, and C_3 is

$$Y(s) = \frac{c_1}{s+\omega_c} + \frac{C_2 s + C_3}{s^2+\omega^2} \tag{3.107}$$

and the most straightforward path appears to be the setting up of a 3×3 matrix by cross-multiplying the additive terms in Eq. (3.107),

$$\begin{bmatrix} 1 & 1 & 0 \\ 0 & \omega_c & 1 \\ \omega^2 & 0 & \omega_c \end{bmatrix} \cdot \begin{bmatrix} c_1 \\ C_2 \\ C_3 \end{bmatrix} = \begin{bmatrix} 0 \\ 0 \\ \omega\omega_c \end{bmatrix} \tag{3.108}$$

which yields the residues

$$c_1 = \frac{\omega\omega_c}{\omega^2+\omega_c^2} = 0.4; \quad C_2 = -\frac{\omega\omega_c}{\omega^2+\omega_c^2} = -0.4; \quad C_3 = \frac{\omega\omega_c^2}{\omega^2+\omega_c^2} = 75.4 \tag{3.109}$$

and the final partial fraction expansion

$$Y(s) = \frac{1}{\omega^2+\omega_c^2}\left(\frac{\omega\omega_c}{s+\omega_c} + \frac{\omega\omega_c^2 - \omega\omega_c s}{s^2+\omega^2}\right) = \frac{0.4}{s+188.5} + \frac{75.4-0.4s}{s^2+377^2} \tag{3.110}$$

The time-domain solution is the sum of a nonoscillatory decay and a phase-shifted sinusoid:

$$y(t) = \frac{2}{5}e^{-188.5t} + \frac{1}{5}\sin(\omega t) - \frac{2}{5}\cos(\omega t) \tag{3.111}$$

The exponentially-decaying term is of interest, because frequency-domain analysis methods would not provide this term. The difference between the Fourier-domain and Laplace-domain solutions comes from the assumption that the oscillation starts at $t = -\infty$ (Fourier transform and frequency-domain methods) contrasted with an oscillation that starts at $t = 0$ (Laplace-domain methods). The time-domain response needs to be continuous at $t = 0$, which the rapidly decaying term $e^{-188.5t}$ ensures in light of the phase shift.

3.5.2 Partial Fraction Expansion in Scilab

The Scilab command `pfss` performs numerical partial fraction expansion. To remain with the example of the underdamped system, we first define a linear system with the given transfer function and apply `pfss` to the linear system:

```
s=poly ('0',s);
h = 1/(s*(s^2 + 4*s + 8));
H = syslin ('c', h);
pfss (H)
```

The result is a vector with two elements, $0.125/s$ and $(-0.5-0.125s)/(8+4s+s^2)$, which perfectly matches our result in Eq. (3.104).

3.6 Building Blocks of Linear Systems

We now introduce several basic low-order SISO systems, their transfer functions, and their dynamic behavior. More complex systems can usually be represented by multiplicative or additive combinations of first- and second-order SISO systems. In fact, through partial fraction expansion, any rational function $F(s)$ can be transformed into a sum of first- and second-order systems (see Section 3.5). Furthermore, some low-order SISO systems can be used as compensators. The term *compensator* generally refers to a part of a control system that is introduced to improve stability or dynamic response.

3.6.1 Gain Blocks

The gain block has a real-valued, scalar transfer function. The input signal is multiplied with a constant factor a. Examples are levers or voltage dividers, but also dissipative elements, such as a resistance or friction. Gain blocks have no energy storage. The transfer function of a gain block has no zeros and no poles.

- Time-domain function:

$$y(t) = a \cdot x(t) \tag{3.112}$$

- Laplace-domain transfer function:

$$H(s) = \frac{Y(s)}{X(s)} = a \tag{3.113}$$

- Unit step response:

$$y_{step}(t) = a \cdot u(t) \tag{3.114}$$

Note that a gain block can also include attenuation when $0 < a < 1$, and that a may carry physical units. One example is the insulation-dependent loss of energy to the environment k_e in the waterbath example in Chapter 5. k_e has units of Watts per Kelvin. By convention, a is normally a positive number. Sign changes are often realized at summation points. Sign changes with $a < 0$ are allowable, but frowned upon.

3.6.2 Differentiators

The output signal of a differentiator is the first derivative of the input signal, generally multiplied with a constant factor. Once again, the constant factor may carry physical units, and the first derivative operation additionally introduces inverse seconds to the units. Examples are springs, electric inductors, and fluid inertia: For example, the voltage across an inductor is proportional to the first derivative of the current through the inductor, and the pressure in a tube changes with the first derivative of the flow velocity. Differentiators are energy storage elements: The inductor stores energy in the magnetic field, and the hydraulic pipe stores energy in the momentum of the fluid. The transfer function of the differentiator has one zero in the origin.

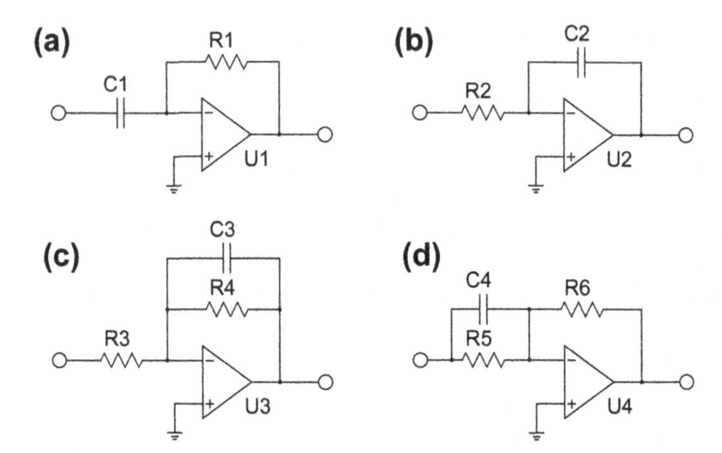

Figure 3.5 Example circuits for low-order operations build with operational amplifiers. (a) differentiator with the transfer function $H(s) = -R_1C_1s$; (b) integrator with the transfer function $H(s) = -1/(R_2C_2s)$; (c) first-order phase-lag system (lowpass) with the transfer function $H(s) = -(R_4/R_3)/(C_3R_4s + 1)$. If R3 is replaced by a capacitor, the circuit turns into a first-order highpass. (d) First-order phase-lead system with the transfer function $H(s) = -(R_6/R_5) \cdot (C_4R_5s + 1)$. All functions have a negative sign, and an additional inverter is necessary if a positive transfer function is required.

- Time-domain function:

$$y(t) = a \cdot \dot{x}(t) \tag{3.115}$$

- Laplace-domain transfer function:

$$H(s) = \frac{Y(s)}{X(s)} = a \cdot s \tag{3.116}$$

- Unit step response:

$$y_{step}(t) = a \cdot \delta(t) \tag{3.117}$$

In the example of the ideal inductor, the proportionality constant is the inductance L, and the transfer function is $V(s) = L \cdot s \cdot I(s)$, where $V(s)$ is the voltage across the inductor and $I(s)$ is the current through the inductor. A differentiator based on an operational amplifier is shown in Figure 3.5a.

3.6.3 Integrators

The integrator is the complementary element to the differentiator. Its output is the integral of the input signal over time, multiplied with a proportionality constant. Typical examples are the capacitor, which accumulates charges, or a water tank, which accumulates fluid. The inert mass is also an integrator as its velocity is proportional to the force acting on the mass, integrated over time. The energy storage property of the integrator is particularly obvious in the inert mass example. The transfer function of the integrator has one pole in the origin.

- Time-domain function:

$$y(t) = a \cdot \int_0^t x(\tau)\mathrm{d}\tau \tag{3.118}$$

- Laplace-domain transfer function:

$$H(s) = \frac{Y(s)}{X(s)} = \frac{a}{s} \tag{3.119}$$

- Unit step response:

$$y_{step}(t) = a \cdot t \tag{3.120}$$

An operational amplifier circuit that realizes the integrator transfer function is shown in Figure 3.5b. Integrators are important control elements, because the output can only reach a steady state when the input is zero. An integrating element in a controller can have the ability to drive the steady-state control deviation to zero, a property that is used in the *PID* controller.

3.6.4 Phase-Lag System, First-Order Lowpass

Phase-lag systems are very common. These systems occur when an energy storage unit and an energy dissipator are combined. One example is the RC lowpass in Chapter 2, and an additional example is the waterbath that is covered in detail in Chapters 5 and 6. The name *phase-lag system* comes from the fact that a sinusoidal input causes a sinusoidal output, but with a phase lag. The reason why such a system is alternatively called a *lowpass* is that higher-frequency sinusoidal input signals are attenuated with increasing frequency. The transfer function of the first-order lowpass has one pole on the negative real axis at $-1/\tau$. An example realization with an operational amplifier is shown in Figure 3.5c, where the pole location and the overall gain can be adjusted independently.

- Time-domain function:

$$\tau\dot{y}(t) + y(t) = x(t) \tag{3.121}$$

- Laplace-domain transfer function:

$$H(s) = \frac{1}{\tau s + 1} \tag{3.122}$$

- Unit step response:

$$y_{step}(t) = 1 - e^{-t/\tau} \tag{3.123}$$

3.6.5 First-Order Highpass

The first-order highpass is complementary to the first-order lowpass, that is, $H_{HP}(s) = 1 - H_{LP}(s)$. Its output signal is the first derivative of the complementary lowpass filter, and a sinusoidal output has a phase lead over the sinusoidal input. One example is the RC highpass. The transfer function of the first-order highpass has one pole on the negative real axis at $-1/\tau$ and one zero in the origin, and it can be seen that the first-order highpass is equivalent to the first-order lowpass followed by a differentiator. In a circuit realization, a highpass circuit can be derived from a lowpass circuit by exchanging the resistors for capacitors and *vice versa*.

- Time-domain function:

$$\tau \dot{y}(t) + y(t) = \dot{x}(t) \tag{3.124}$$

- Laplace-domain transfer function:

$$H(s) = \frac{\tau s}{\tau s + 1} \tag{3.125}$$

- Unit step response:

$$y_{step}(t) = e^{-t/\tau} \tag{3.126}$$

A practical example with operational amplifiers builds on Figure 3.5c, but with a capacitor at the input replacing R_3. When we designate this capacitor C_6, the highpass transfer function emerges as $H(s) = -(R_4 C_6)s/(R_4 C_3 s + 1)$.

3.6.6 PD System or Phase-Lead Compensator

The *PD* system can be interpreted as the additive combination of a gain block and a differentiator. The *PD* system is a very popular phase-lead compensator, and we will see later that the D-component improves the relative stability of closed-loop systems with a *PD* controller. The transfer function of the *PD* system has one zero on the negative real axis at $-1/\tau$. An example realization with an operational amplifier is shown in Figure 3.5d, where the location of the zero and the overall gain can be adjusted independently.

- Time-domain function:

$$y(t) = x(t) + \tau_D \dot{x}(t) \tag{3.127}$$

- Laplace-domain transfer function:

$$H(s) = 1 + \tau_D s \tag{3.128}$$

- Unit step response:

$$y_{step}(t) = 1 + \tau_D \delta(t) \tag{3.129}$$

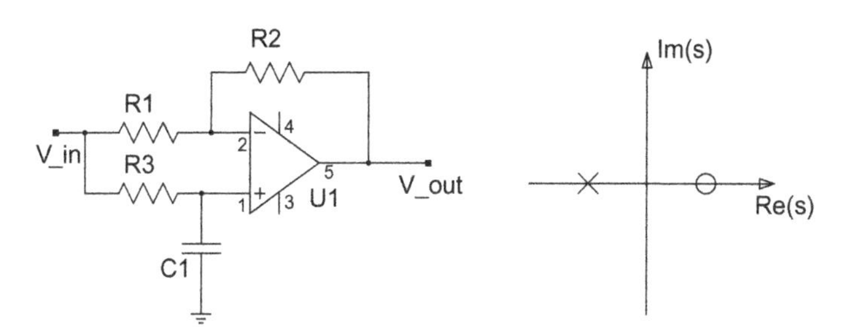

Figure 3.6 Realization of a first-order allpass with an operational amplifier and its corresponding pole-zero plot. Unit gain is achieved when $R_1 = R_2$, and the characteristic time constant of this allpass is $\tau = R_3 \cdot C_1$. The transfer function has one stable pole at $s = -1/\tau$ and a zero at $s = +1/\tau$. The frequency-dependent phase shift is $\varphi = -2 \arctan \omega\tau$.

3.6.7 Allpass Compensator

The first-order allpass is a special electronic circuit (Figure 3.6), which is best described in terms of its frequency response (*cf.* Chapter 11). The magnitude of its transfer function is unity for all frequencies, and the phase shift changes from 0° at low frequencies to −180° at high frequencies.

- Time-domain function:

$$\tau \dot{y}(t) + y(t) = x(t) - \tau \dot{x}(t) \tag{3.130}$$

- Laplace-domain transfer function:

$$H(s) = \frac{1 - \tau s}{1 + \tau s} \tag{3.131}$$

- Unit step response:

$$y_{step}(t) = 1 - 2e^{-t/\tau} \tag{3.132}$$

3.6.8 Second-Order System

The second-order system is unique in this context, because its characteristic equation may have complex conjugate roots. The second-order system is the lowest-order system capable of an oscillatory response to a step input. Typical examples are the spring-mass-damper system and the electronic RLC circuit. Second-order systems with potential oscillatory responses require two different and independent types of energy storage, such as the inductor and the capacitor in RLC filters, or a spring and an inert mass. The transfer function of the general second-order system has two poles in one of three configurations: both poles can be real-valued and on the negative real axis, they can form a double-pole on the negative real axis, or they can form a complex conjugate pole pair. The significance of the pole position of a second-order system is examined in Section 3.3 and in more detail in Chapter 9.

- Time-domain function:

$$\frac{1}{\omega^2}\ddot{y}(t) + \frac{2\zeta}{\omega}\dot{y}(t) + y(t) = x(t) \tag{3.133}$$

- Laplace-domain transfer function:

$$H(s) = \frac{\omega^2}{s^2 + 2\zeta\omega s + \omega^2} \tag{3.134}$$

The unit step response depends on the roots of the characteristic equation. If both roots are real-valued, the second-order system behaves like a chain of two first-order systems, and the step response has two exponential components. If the roots are complex, the step response is a harmonic oscillation with an exponentially decaying amplitude.

3.6.9 Dead-Time System (Time-Delay System)

Dead-time systems, also called time-delay systems, occur when a signal reaches the output of the system after a delay τ. Examples are transmission lines or conveyor belts. Another example is a power amplifier system in conference rooms or halls, where the sound from the loudspeakers may reach the microphone after a delay $\tau = d/c$, with d the distance between microphone and loudspeaker and c the speed of sound.

The dead-time system creates a frequency-dependent phase lag that often reduces the stability margin of a control system. A dead-time system cannot be treated in the Laplace domain in a straightforward manner, because the transfer function is no longer a rational polynomial.

- Time-domain function:

$$y(t) = x(t - \tau) \tag{3.135}$$

- Laplace-domain transfer function:

$$H(s) = e^{-s\tau} \tag{3.136}$$

- Unit step response:

$$y_{step}(t) = u(t - \tau) \tag{3.137}$$

When the time delay τ is short compared to other system components, the Padé approximation in the Laplace domain becomes applicable. The Padé approximation emerges from the series expansion of the transcendental function $e^{-s\tau}$, and the first-order Padé approximation of a time delay τ is

$$e^{-s\tau} \approx \frac{1 - \frac{\tau}{2}s}{1 + \frac{\tau}{2}s} \tag{3.138}$$

This approximation resembles a first-order allpass with a pole at $-2/\tau$ and an opposing zero at $+2/\tau$. Phase-lead compensators can often be used to improve the stability of feedback control systems with time delays.

4 Time-Discrete Systems

Abstract

With more and more powerful digital processing devices (computers and microcontrollers), digital feedback controls have become ubiquitous. However, a fundamental difference to time-continuous systems exists: data processing systems take samples of continuous signals at constant time intervals T. The signal is *discretized*, and the true value of the signal is not known until the next sample is taken. The resulting time lag T introduces an irrational term e^{-sT} in the Laplace domain and thus makes it impossible to accurately compute a system response with Laplace-domain methods. For time-discrete systems, an equivalent transform exists, the z-transform, which allows the mathematical treatment of discrete systems in the z-domain in a similar fashion to the treatment of continuous systems in the Laplace domain. In this chapter, the z-transform and its justification though a discrete sampler model are introduced. The relationship between Laplace- and z-domains is explored, and methods are provided to convert z-transformed signals into time-domain sequences of signals.

Digital control systems are time-discrete systems. The fundamental difference between continuous and time-discrete systems comes from the need to convert analog signals into digital numbers, and from the time a computer system needs to compute the corrective action and apply it to the output. A typical digital controller is sketched in Figure 4.1. An analog signal, either a sensor signal or the control deviation, is fed into an analog-to-digital converter (ADC). The ADC samples the analog signal periodically with a sampling period T. The sampling process can be interpreted as a sample-and-hold circuit followed by digital read-out of the converted digital value Z. The microcontroller reads Z and computes the corrective action A, which is applied as a digital value at its output. Since the computation takes a certain amount of time, A becomes available with a delay T_D after Z has been read. The digital value A is converted into an analog signal through a digital-to-analog converter (DAC). The analog signal still has the stair-step characteristic of digital systems, and a lowpass filter smoothes the steps. Both the sampling interval T and the processing delay T_D introduce phase shift terms in the Laplace domain that fundamentally change the behavior of time-discrete systems when compared to continuous systems. Time-delay systems with a delay τ have a Laplace transfer function $H(s) = \exp(-s\tau)$ (see Section 3.6.9). We can see that $H(s) \rightarrow 1$ when $\tau \rightarrow 0$. In other words, when the sampling frequency is very high compared to the dynamic response of the system, treatment as a continuous system is possible. As a rule of thumb, the approximated Laplace-domain pole created at $-2/T$ should be at least ten to twenty times further to the left than the fastest system pole. If this is not the case, the methods introduced in this chapter need to be applied.

Linear Feedback Controls. http://dx.doi.org/10.1016/B978-0-12-405875-0.00004-8
© 2013 Elsevier Inc. All rights reserved.

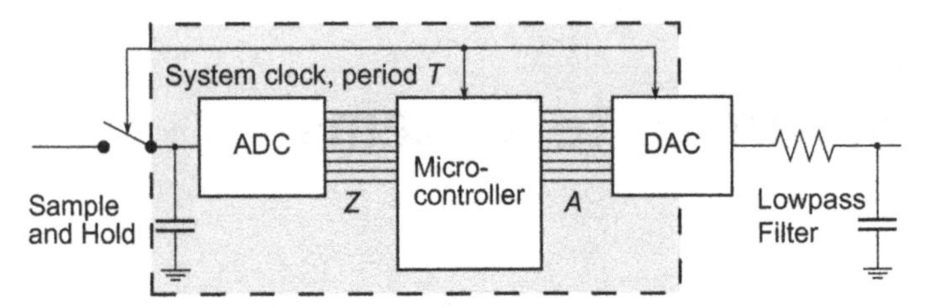

Figure 4.1 Basic schematic of a digital control system. The analog-to-digital converter (ADC) samples the analog input signal in discrete time intervals T. The sampling process can be interpreted as a sample-and-hold circuit with subsequent digital read-out. The microcontroller receives a digital value Z as its input, from which it computes the corrective action A. A, itself a digital value, is applied to a digital-to-analog converter (DAC), and its analog output signal passes through a lowpass filter to remove frequency components related to the sampling frequency. The gray-shaded part of the system represents the time-discrete domain.

For the considerations in this chapter, we need to make two assumptions. First, the sampling interval is regular, that is, the time from one discrete sample to the next is the constant T. Second, we either assume that the processing time T_D is small compared to T, and the corrective action is available immediately after a sample has been taken, or we assume that the processing time T_D is constant and can be included in the definition of T.

4.1 Analog-to-Digital Conversion and the Zero-Order Hold

Many integrated ADC have a digital input signal that initiates a conversion. At the exact instant the conversion is started, the ADC takes a sample of the analog signal $f(t)$ and, after completion of the conversion, provides the proportional digital value Z. Strictly, the sampled values of $f(t)$ are known on the digital side only for integer multiples of the sampling interval T. When a sample is taken at $t = k \cdot T$, the corresponding digital value is $Z_k(t) = f(kT) \cdot \delta(t - kT)$. The discretized (sampled) signal $f_s(t)$ can therefore be interpreted as the set of all Z_k, superimposed:

$$f_s(t) = f(t) \cdot \sum_{k=-\infty}^{\infty} \delta(t - kT) = \sum_{k=-\infty}^{\infty} f(kT) \cdot \delta(t - kT) \tag{4.1}$$

Equation (4.1) describes a discrete convolution of the original signal $f(t)$ with a sequence of equidistant delta-pulses. The interpretation of the sampled signal as a sequence of scaled delta-pulses in shown in Figure 4.2b. An alternative interpretation is possible where any discretely sampled value Z_k remains valid until the next sampling takes place. This interpretation leads to a stair-step-like function as depicted in Figure 4.2c.

We can argue that the stair-step function in Figure 4.2c is a sequence of scaled step functions followed by the delayed step function with a negative sign. For example, at $t = 0$ (corresponding to $k = 0$) we obtain the value Z_0. For $t = T$ (corresponding

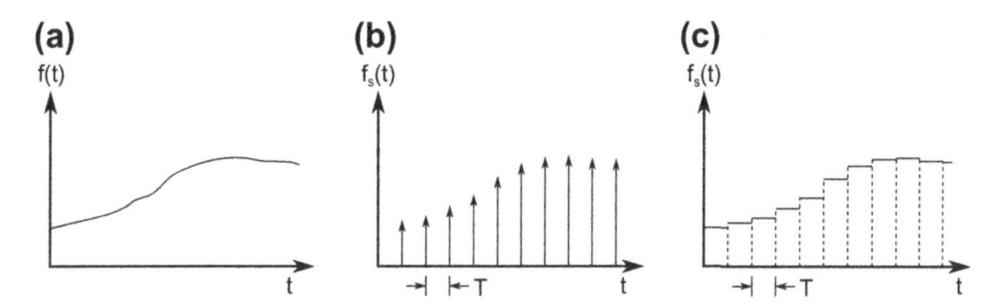

Figure 4.2 Continuous signal $f(t)$ and two interpretations of the discretely-sampled signal $f_s(t)$. The center diagram (b) shows the individual samples as delta pulses with a rate of T (Eq. (4.1)), and the right diagram (c) shows the interpretation of the signal where the discrete sample remains valid and unchanged until the next sampling occurs.

to $k = 1$) we obtain the value Z_1. To make this work, we compose the first stair-step from $Z_0 u(t) - Z_0 u(t - T)$. The second stair-step follows immediately as $Z_1 u(t - T) - Z_1 u(t - 2T)$, and so on. The Laplace transform of the sample-and-hold unit (also referred to as *zero-order hold*) emerges as:

$$u(t) - u(t - T) \quad \circ\!\!-\!\!\bullet \quad \frac{1}{s} - \frac{1}{s} e^{-sT} = \frac{1 - e^{-sT}}{s} \tag{4.2}$$

Once again, we can see that any zero-order hold element (that is, any digital, time-discrete processing element) introduces an irrational term, namely, e^{-sT} into the Laplace transform. It is therefore no longer possible to determine the system response by computing the roots of the characteristic polynomial. However, we can examine the Laplace transform of the discretely sampled signal by transforming the convolution in Eq. (4.1):

$$\mathcal{L}\{f_s(t)\} = \int_0^{\infty} \left[\sum_{k=-\infty}^{\infty} f(kT) \cdot \delta(t - kT) \right] e^{-st}\, \mathrm{d}t = \sum_{k=0}^{\infty} f(kT) \cdot e^{-skT} \tag{4.3}$$

Once again, the time-delay term e^{-sT} is part of the transform. The Laplace transform in Eq. (4.3) becomes particularly interesting when we define a new complex variable, z, as

$$z = e^{sT} \tag{4.4}$$

and rewrite Eq. (4.3) as

$$\mathcal{L}\{f_s(t)\} = \sum_{k=0}^{\infty} f(kT) \cdot z^{-k} \tag{4.5}$$

The Laplace transform of a discretely sampled signal has now become an infinite sum, and a new variable z has been introduced that depends on the sampling period T. This infinite sum leads directly to the definition of the z-transform (Section 4.2).

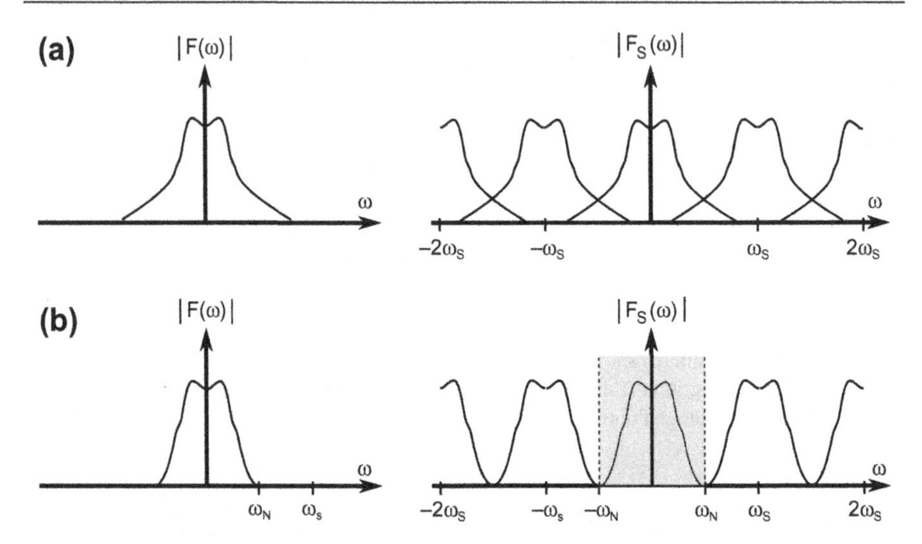

Figure 4.3 (a) Sketch of the frequency spectrum $|F(\omega)|$ of $f(t)$ and the corresponding spectrum of the sampled signal $f_s(t)$. The spectrum of the sampled signal contains replicated image spectra of the baseband spectrum at intervals of ω_s. If the signal $f(t)$ is not bandlimited, spectral components extend beyond $\omega_s/2$ and overlap with the image spectra. Original and overlapping spectral components cannot be separated. (b) Spectrum of $f(t)$ after limitation of the bandwidth (lowpass filtering). Since no spectral components extend beyond the Nyquist frequency ω_N, an ideal lowpass filter (shaded area, dashed lines) can be used to separate the baseband spectrum from the image spectra and thus restore $f(t)$.

At this point, the sampling theorem needs to be briefly introduced. The *comb-function*, that is, the sequence of shifted delta-pulses that is used to sample the signal, has its own frequency spectrum. The multiplication of a signal with a comb function (shown in Eq. (4.1)) therefore corresponds to a convolution of the frequency spectra of $f(t)$ and the comb function. The Fourier spectrum of the comb function was derived in Section 3.2. In simple terms, frequency spectra are replicated along multiples of the sampling frequency $\omega_s = 2\pi/T$ as shown in Figure 4.3. Any frequency components that extend above ω_s overlap with frequency components of the replicates. This ambiguity of the spectra in the frequency domain is known as *aliasing*. Reconstruction of the original signal $f(t)$ is not possible, because the original spectrum and the aliased components cannot be separated.

In practice, any signal to be sampled needs to be band-limited to half of the sampling frequency. We define this frequency $\omega_N = 2\pi f_N = \omega_s/2$ as *Nyquist frequency*. Band-limiting can be achieved by using an *analog* lowpass filter before the ADC. Often, a first-order lowpass is sufficient when only high-frequency noise needs to be attenuated and when the sampling frequency is sufficiently high. On the other hand, higher-order filters are needed when fast transients of the system play a major role in the controller design. These filters need to be considered in the overall transfer function of the system.

Here, the importance of the lowpass filter at the output of the digital system in Figure 4.1 becomes evident. The discrete, sampled signal at the output of the DAC consists of the *baseband spectrum*, that is, the original spectrum centered at $\omega = 0$

and the *image spectra* (i.e., the replicated spectra at $\pm k\omega_s$). A lowpass filter can isolate the baseband spectrum and thus restore the original signal. The boxcar filter indicated in Figure 4.3b does not exist in practice. Rather, a compromise needs to be found between the effort for designing a steep filter—both at the input and the output—and for simply raising the sampling rate: the more separation exists between the bandlimit of the baseband spectrum and the Nyquist frequency, the less effort is needed to filter the unwanted frequency components.

An example for this challenge can be found in the early designs of the compact disc (CD) player. Music on a CD is digitized at 44.1 kHz, and a bandlimit of $f_N = 22.05$ kHz is necessary. The maximum frequency was chosen for physiological reasons. When the CD was first introduced, digital signal processors were just barely able to perform reconstruction and error correction in real time, and high-order analog Tchebyshev filters were used at the output to remove frequency components from the sampler. Many users complained about a "cold" or "technical" aspect of the sound, which may be attributed to the poor step response and the passband ripple of the filters. As digital signal processors became more powerful, the problem was solved by digitally interpolating new samples between existing ones and thus raising the apparent sampling frequency to 88.2 kHz or even higher. The use of less steep filters with better step response now became feasible. Not unusual for high-end audio, however, other listeners now complained about perceived deficiencies in the sound quality …

4.2 The z-Transform

Given a continuous signal $f(t)$ and a sequence of samples taken at time points $t_k = kT$, we can interpret the sequence $f_k = f(kT)$ as the discretely sampled version $f_s(t)$ of the signal $f(t)$. The z-transform of a sequence is defined as

$$F(z) = \mathscr{Z}\{f_s(t)\} = \mathscr{Z}\{f(t)\} = \sum_{k=0}^{\infty} f(kT) \cdot z^{-k} \tag{4.6}$$

We note that the z-transform of the continuous signal $f(t)$ and the sampled signal $f_s(t)$ are identical, and that the sampled signal f_k has identical z- and Laplace-transforms. Moreover, the z-transform, like the Laplace transform, is a linear operation, that is, the scaling and superposition principles hold (see Table B.6).

Similar to the Laplace transform, the z-transform of discretized signals can be computed and interpreted as two-dimensional functions in the complex z-plane. Correspondence tables are available to provide the z-transform for many common signals (Tables B.1 through B.5). Let us, for example, determine the z-transform of the unit step $u(t)$. By using Eq. (4.6), we obtain

$$\mathscr{Z}\{u(t)\} = \sum_{k=0}^{\infty} z^{-k} = \frac{1}{1 - z^{-1}} \tag{4.7}$$

Convergence to the closed-term solution on the right-hand side of Eq. (4.7) is only possible when $|z| > 1$, that is, only for z that lie outside the unit circle in the z-plane. This observation will later lead to criteria of stability that differ from the Laplace transform.

As a second example, let us determine the z-transform of an exponential decay, $f(t) = e^{-at}$.

$$\mathscr{Z}\left\{e^{-at}\right\} = \sum_{k=0}^{\infty} e^{-akT} z^{-k} = \sum_{k=0}^{\infty} \left(e^{aT} \cdot z\right)^{-k} = \frac{1}{1-(ze^{aT})^{-1}} = \frac{z}{z-e^{-aT}} \tag{4.8}$$

Note that e^{-aT} is a constant that depends on the exponential decay constant a and the sampling period T. The resulting z-domain function has a zero in the origin and a pole at $z = +e^{-aT}$.

The z-transform is of importance, because the transfer functions of both continuous and time-discrete systems can be expressed in the z-domain. Furthermore, the convolution theorem holds for the z-transform. If a system with the z-domain transfer function $H(z)$ receives an input signal $X(z)$, we can compute its z-domain response through

$$Y(z) = X(z) \cdot H(z) \tag{4.9}$$

The most notable difference between the Laplace- and z-transforms is the discrete nature of the digital output signal. Consistent with time-discrete systems, the output signal is known only at integer multiples of the sampling period, that is, at $t = kT$. This behavior emerges from the digital-to-analog output stage in Figure 4.1, which receives a new value only once per sampling interval. This value is kept valid until it is replaced by the next computed output value (i.e., the behavior indicated in Figure 4.2c). **This behavior also complicates the translation of Laplace-domain problems into z-domain problems, because it is generally not possible to translate a transfer function $H(s)$ into $H(z)$ through direct correspondence.** In the next section, we will discuss how a time-continuous system with a time-discrete control element can be converted into a completely time-discrete model.

If the z-transform of a signal is known, the time-domain samples can be determined through the inverse z-transform. Formally, the inverse z-transform involves evaluating a contour integral for all integer multiples of T:

$$f_k = f(kT) = \mathscr{Z}^{-1}\left\{F(z)\right\} = \frac{1}{2\pi j} \oint_c F(z) z^{k-1}\, \mathrm{d}z \tag{4.10}$$

The contour c encircles the origin of the z-plane and lies inside the region of convergence of $F(z)$. For the contour $|z| = 1$, the contour integral turns into a form of the inverse time-discrete Fourier transform of $F(z)$,

$$f_k = \frac{1}{2\pi j} \int_0^{2\pi} F(e^{j\varphi}) e^{j\varphi(k-1)}\, \mathrm{d}\varphi \tag{4.11}$$

The strict definition of the inverse z-transform does not invite its practical application. Any of the following methods provides a more practical approach to obtain the discretely sampled signal values f_k from a z-domain signal $F(z)$, and each item in the list is followed by an example below:

- Correspondence tables: If $F(z)$ can be expressed as a linear combination of expressions found in the correspondence tables (Appendix B), obtaining $f_k = f(kT)$ is as straightforward as in the Laplace domain.

- Partial fraction expansion: With the exact same method as described in Section 3.5, the fraction of two polynomials $p(z)/q(z)$ can be transformed into a sum of first- and second-order polynomials, whose correspondence can usually be found in Appendix B.
- Polynomial long division: The formal polynomial long division of a fraction $p(z)/q(z)$ leads to an expression in the form $f_0 + f_1 z^{-1} + f_2 z^{-2} + \cdots$ When we consider Eq. (4.6), we can see that the coefficients of the polynomial long division are identical to the sampled values, that is, $f_k = \{f_0, f_1, f_2, \ldots\}$.
- Digital filter response: When the transfer function $p(z)/q(z)$ is known, the present discrete output value can be computed from past output values, and present and past input values. This method advertises itself for a computer-based solution.

Example: The z-transform of a sequence f_k, sampled at $T = 0.1$ s, is known to be

$$F(z) = \frac{z}{z^2 - 0.8296z + 0.1353} \tag{4.12}$$

For partial fraction expansion, it is convenient to substitute $F(z)$ for a shifted series $G(z)/z$. The reason will become clear further below. The roots of the denominator polynomial are $z_1 = 0.6065$ and $z_2 = 0.2231$. We therefore write

$$\begin{aligned} \frac{G(z)}{z} &= \frac{z}{z^2 - 0.8296z + 0.1353} \\ &= \frac{P}{z - 0.6065} + \frac{Q}{z - 0.2331} = \frac{1.582}{z - 0.6065} - \frac{0.582}{z - 0.2331} \end{aligned} \tag{4.13}$$

where the residuals $P = 0.1582$ and $Q = -0.582$ were found with Eq. (3.79). The two additive terms in Eq. (4.13) do not exist in the correspondence tables. However, if we multiply both sides with z, we obtain correspondences for the exponential decay (Table B.2):

$$G(z) = 1.582\frac{z}{z - 0.6065} - 0.582\frac{z}{z - 0.2331} \tag{4.14}$$

We can determine the decay constants a and b with $e^{-aT} = 0.6065$ and $e^{-bT} = 0.2331$ and obtain $a = 5\ \text{s}^{-1}$ and $b = 15\ \text{s}^{-1}$. The sequence $g_k = g(kT)$ is therefore

$$g(kT) = 1.582 \cdot e^{-5kT} - 0.582 \cdot e^{-15kT}; \quad k = 0, 1, 2, \ldots \tag{4.15}$$

Evaluated for $k = 0, 1, 2, \ldots$, we obtain $g_k = \{1, 0.8297, 0.553, 0.3465, 0.2127, \ldots\}$. However, we initially substituted $F(z) = G(z) \cdot z^{-1}$. We need to apply the time shift property (Table B.6), since z^{-1} corresponds to a delay by T:

$$f((k+1)T) \quad \circ\!\!-\!\!\bullet \quad zF(z) - zf(0) \tag{4.16}$$

Clearly, $zF(z) = G(z)$ with the correspondence $f((k+1)T) = g(kT)$. Because of the causality of $g(kT)$, we know $f(0) = g(-T) = 0$. The sequence f_k is therefore obtained directly from g_k by shifting the values one place to the right: $f_k = \{0, 1, 0.8297, 0.553, 0.3465, 0.2127, \ldots\}$.

Table 4.1 Polynomial long division of $z^2 - 0.8296z + 0.1353$ into z (Eq. (4.12)).

		z^{-1}	$+0.8296z^{-2}$	$+0.5529z^{-3}$	$+0.3465z^{-4}$	$+0.2127z^{-5}$ $(\ldots)$
$z^2 - 0.8296z + 0.1353$]	z					
	z	-0.8296	$+0.1353z^{-1}$			
		$+0.8296$	$-0.1353z^{-1}$			
		$+0.8296$	$-0.6882z^{-1}$	$+0.1122z^{-2}$		
			$+0.5529z^{-1}$	$-0.1122z^{-2}$		
			$+0.5529z^{-1}$	$-0.4587z^{-2}$	$+0.0748z^{-3}$	
				$+0.3465z^{-2}$	$-0.0748z^{-3}$	
				$+0.3465z^{-2}$	$-0.2875z^{-3}$	$+0.4688z^{-4}$
					$+0.2127z^{-3}$	$-0.4688z^{-4}$
						$(\ldots)$

The same result can be obtained by polynomial long division, that is, by dividing $z^2-0.8296z+0.1353$ into z as demonstrated in Table 4.1. With the forward z-transform

$$F(z) = \sum_{k=0}^{\infty} f_k \cdot z^{-k} \tag{4.17}$$

we can read the coefficients of z^{-k} from the result of the polynomial division and obtain $f_k = \{0, 1, 0.8296, 0.5529, 0.3465, 0.2127, \ldots\}$ in accordance with the sequence found by partial fraction expansion. Note that the coefficient for z^0 is zero and therefore $f_0 = 0$. Note also that knowledge of the value for T was necessary to determine the decay constants in the z-transform correspondences, but was not needed for the polynomial long division.

To explain the inverse z-transform through implementation of a discrete filter, we re-interpret the output signal in Eq. (4.12) in two ways. First, we divide by z^2 to obtain z^0 as the highest power of z. All z^{-1} are now time delays by T. Second, we interpret $F(z)$ as a transfer function $H(z)$ with a z-domain input of 1 (i.e., a delta pulse). Since the z-transform of the delta pulse is 1, we have $F(z) = H(z)$ when the input is a delta pulse, i.e., $\delta_k = \{1, 0, 0, \ldots\}$. By cross-multiplying with the denominator polynomial, Eq. (4.12) becomes

$$F(z)\left[1 - 0.8296z^{-1} + 0.1353z^{-2}\right] = z^{-1} \cdot 1 \tag{4.18}$$

where the right-hand 1 is the z-transform of the delta pulse, $F(z)$ is the output of the digital filter, and the remaining polynomials of z are the actual filter. A filter that realizes Eq. (4.18) is sketched in Figure 4.4. Each block labeled z^{-1} represents a unit delay, that is, it holds its input value from the previous clock cycle for the duration of T. For example, if the input sequence is $\delta_k = \{1, 0, 0, 0, \ldots\}$, then the output of the first unit delay is $z^{-1}\delta_k = \{0, 1, 0, 0, \ldots\}$. Similarly, when the current output value is f_k,

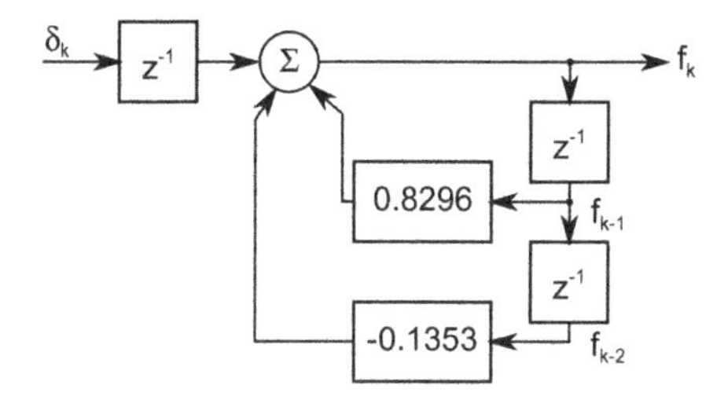

Figure 4.4 Schematic of a discrete filter that realizes the finite-difference equation, Eq. (4.18). Each block with z^{-1} represents a unit delay and can be seen as a memory unit that stores the value from the previous clock cycle. The summation point combines not only input values, but also past output values.

Table 4.2 Finite-difference calculation of the output sequence of Eq. (4.19), which is represented by the digital filter in Figure 4.4. Each line is one finite time step. Note how the output values shift to the right into the past-value columns.

f_k	f_{k-1}	f_{k-2}	x_{k-1}	x_k
0	–	–	0	1
1	0	–	1	0
0.8296	1	0	0	0
0.5529	0.8296	1	0	0
0.3465	0.5529	0.8296	0	0
0.2126	0.3465	0.5529	0	0
0.1295	0.2126	0.3465	0	0
...	...	...	...	...

the term $z^{-1}F(z)$ represents the shifted sequence f_{k-1}, and $z^{-2}F(z)$ corresponds to f_{k-2}. The summation point therefore combines input values with past output values.[1] By using these relationships, Eq. (4.18) turns into a finite-difference equation:

$$f_k = 0.8296 \cdot f_{k-1} - 0.1353 \cdot f_{k-2} + \delta_{k-1} \tag{4.19}$$

From this point, we can compute the output values as shown in Table 4.2.

4.3 The Relationship between Laplace- and *z*-domains

Laplace- and z-domain are strictly related when we consider the definition of the complex variable $z = e^{sT}$. Since $s = \sigma + j\omega$, we can relate any point in the s-plane

[1] Digital filters that feed back past output values are known as *infinite impulse response filters*, as opposed to *finite impulse response filters*, which only output a weighted sum of past and present input values. Digital filter theory goes beyond this book. The interested reader is referred to Richard G. Lyons' book *Understanding Digital Signal Processing*.

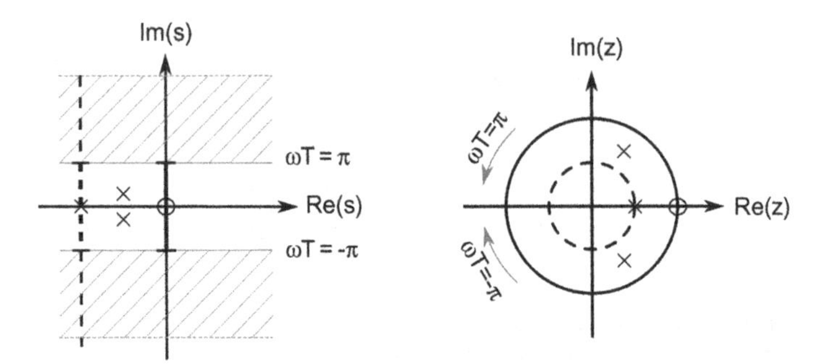

Figure 4.5 Relationship between the s- and z-planes. The imaginary axis of the s-plane is mapped to the unit circle of the z-plane, and the origin of the s-plane maps to $z = 1$ (small circles). A vertical line in the left half-plane near $\sigma = -0.69/T$ (dashed line) maps to a circle inside the unit circle (dashed). Also shown are three stable poles and their approximate corresponding location in the z-plane. The thick section of the imaginary axis maps to the complete unit circle. Increasing the frequency repeats the same circle, and the regions subject to aliasing are hatched in gray. For time-discrete systems, all frequencies need to be kept in the white band between $-\pi < \omega T \leq \pi$.

to its corresponding point in the z-plane through

$$z = e^{(\sigma+j\omega)T} = e^{\sigma T} \cdot e^{j\omega T} \tag{4.20}$$

which describes z in polar coordinates. Vertical lines in the s-plane map to circles in the z-plane as shown in Figure 4.5.

The origin of the s-plane maps to $z = 1$, and the negative branch of the real axis of the s-plane maps to the section of the real axis of the z-plane with $0 < z \leq 1$. Correspondingly, the positive (unstable) branch of the real axis of the s-plane maps to the real axis of the z-plane with $z > 1$. Negative real values of z can only be achieved with complex values of s when $\omega T = \pm\pi$, and the gray horizontal lines in Figure 4.5 map ambiguously to the negative real axis of the z-plane. A hypothetical horizontal line that begins in one of the complex poles in the s-plane and extends to $-\infty$ parallel to the real axis maps to a line in the z-plane connecting the corresponding pole to the origin. The imaginary axis of the s-plane, which separates the stable left half-plane from the unstable right half-plane maps to the unit circle $|z| = 1$. Stable poles of a z-domain transfer function lie inside the unit circle. Moreover, the frequency ambiguity known as aliasing becomes evident as increasing frequencies ω in the s-plane merely map to the same circle over and over. In fact, the harmonic component of z leads to the requirement that $-\pi < \omega T \leq \pi$. With $\omega = 2\pi f$, we arrive again at the Nyquist frequency $f_N = \pm 1/(2T)$.

The unit circle in the z-plane has special significance. From Eq. (4.20) follows $\sigma = 0$ for all $|z| = 1$, which makes intuitive sense from the fact that the imaginary axis of the s-plane maps to the unit circle of the z-plane. For $|z| = 1$ (equivalent to $z = e^{j\omega}$) we

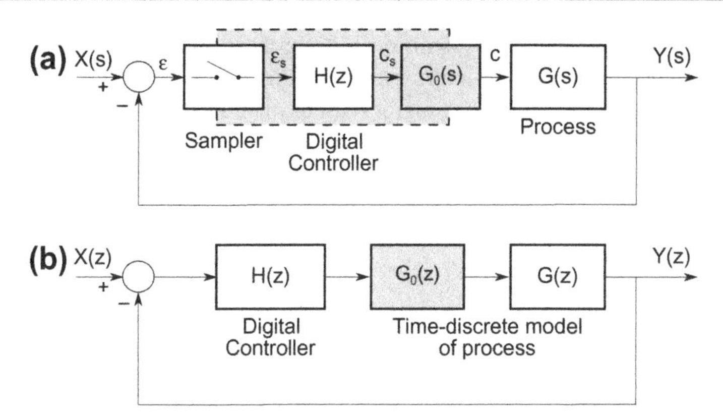

Figure 4.6 Combining time-continuous and time-discrete elements in one feedback loop. (a) The digital controller uses a sampler to sample the input signal ϵ at intervals T. The sampled, time-discrete sequence ϵ_s acts as the input of the controller, and its output is the equally time-discrete sequence c_s. The time-continuous process $G(s)$ interprets the input signal as continuous, and a virtual zero-order hold element $G_0(s)$ (gray shaded box) needs to be inserted in the model between controller and process to correctly consider the transition between discrete and continuous signals. The time-discrete elements are highlighted by the dashed box. (b) With the virtual zero-order hold, the time-discrete correspondences $G(z)$ and $G_0(z)$ of the continuous transfer functions $G(s)$ and $G_0(s)$ can be used, and the entire feedback loop turns into a time-discrete system.

can rewrite the z-transform as

$$F(\omega) = \sum_{k=0}^{\infty} f(kT) \cdot e^{-j\omega k} \tag{4.21}$$

which is actually a form of the discrete Fourier transform of the sequence $f_k = f(kT)$. The Fourier-transform nature of the unit circle becomes important when we discuss the frequency response of time-discrete systems (Chapter 11).

It is important to see that the mapping from the s- to the z-plane always depends on T. A long T (slow controller) reduces the width of the alias-free band in the s-plane, and a long T moves the mapped (stable) poles closer to the origin of the z-plane. Lastly, it is worth noting that s has units of inverse seconds, whereas z carries no units.

Frequently, a process with a time-continuous transfer function $G(s)$ is combined with a time-discrete digital controller. When the controller output is used as input for the process, a *virtual* zero-order hold element $G_0(s)$ needs to be inserted in the mathematical model. The zero-order hold is not a separate physical component, but its model is needed to account for the time-discrete nature of the controller. We can envision the zero-order hold as some form of type-adjusting element that converts a time-discrete input signal mathematically correct into a time-continuous output signal. Only then is it possible to use direct correspondences to convert the Laplace-domain transfer functions into z-domain transfer functions. The model is explained in Figure 4.6.

Let us consider an example system to illustrate these steps. To keep the example simple, we will examine an open-loop system (i.e., the feedback path in Figure 4.6 does not exist). The digital controller—again for reasons of simplification—is a P controller with $H(z) = k_p$. Therefore, $c_s(kT) = k_p \cdot \epsilon_s(kT)$. The process has the transfer function $G(s) = 5/(s(s+5))$. The transfer function of the forward path $L(s)$ in the Laplace domain is therefore

$$L(s) = \frac{Y(s)}{\epsilon(s)} = k_p \cdot G_0(s) \cdot G(s) = k_p \frac{1 - e^{-sT}}{s} \frac{5}{s(s+5)} \tag{4.22}$$

Through partial fraction expansion (note the double root at $s = 0$), we obtain a sum of first- and second-order polynomials:

$$L(s) = k_p \cdot \left(1 - e^{-sT}\right) \left(\frac{1}{s^2} - \frac{1}{5s} + \frac{1}{5(s+5)}\right) \tag{4.23}$$

The three fractions in large parentheses represent a ramp function, a step function, and an exponential decay, respectively. The equivalent z-domain terms can now be substituted:

$$L(z) = k_p \cdot \left(1 - z^{-1}\right) \left(\frac{Tz}{(z-1)^2} - \frac{z}{5(z-1)} + \frac{z}{5(z - e^{-5T})}\right) \tag{4.24}$$

Further arithmetic can be simplified by rearranging the terms and by using numerical values where applicable. If we assume $T = 0.1$ s, we can rewrite $L(z)$ as

$$L(z) = k_p \cdot \left(\frac{0.1065z + 0.0902}{5(z-1)(z-0.6065)}\right) \tag{4.25}$$

where it now could be used to compute the closed-loop impulse or step response, or to optimize the k_p parameter. It can be seen, however, that z-transform arithmetic usually suffers from considerably greater complexity than time-continuous Laplace-domain arithmetic.

A different equivalence between s- and z-domains can be based on the finite-difference approximation of simple s-domain functions. For example, the integral of a continuous function $x(t)$ from $t = 0$ to some arbitrary time point $t = N \cdot T$ can be approximated by the sum of the discrete signal values:

$$\int_0^t x(\tau)\mathrm{d}\tau \approx \sum_{k=0}^{N} x_k \cdot T \tag{4.26}$$

The multiplication with the sampling interval in Eq. (4.26) corresponds to the multiplication with dt in the integral. Sine the integral value is likely stored in memory anyway, Eq. (4.26) can be rewritten as a summation where each new value depends on the current sample and the previous sum,

$$y_k = \sum_{k=0}^{N} x_k \cdot T = y_{k-1} + x_k \cdot T \tag{4.27}$$

allowing for an elegant recursive formulation. Eq. (4.27) belongs in the category of *digital filters*. The z-domain transfer function of the filter can be determined by considering that a one-step delay (i.e., y_{k-1}) corresponds to a multiplication with z^{-1}. The z-transform of the integrating filter therefore becomes

$$Y(z) = z^{-1}Y(z) + X(z) \cdot T \tag{4.28}$$

Since this filter has an input (namely, $X(z)$) and an output (namely, $Y(z)$), we can determine its transfer function by factoring out $Y(z)$ and applying Eq. (4.9),

$$H_{intg}(z) = \frac{Y(z)}{X(z)} = \frac{T}{1 - z^{-1}} = \frac{Tz}{z - 1} \tag{4.29}$$

Along the same lines, we could formulate the trapezoidal approximation of an integral as

$$y_k = y_{k-1} + \frac{(x_k + x_{k-1})}{2} \cdot T \quad \circ\!\!-\!\!\bullet \quad H_{intg}(z) = \frac{T}{2} \cdot \frac{z+1}{z-1} \tag{4.30}$$

The ideal integrator in the Laplace domain is $1/s$. Intuitively, we could claim that $1/s$ and $H_{intg}(z)$ describe the same functionality. Their equality provides an approximate relationship between s- and z-plane,

$$s = \frac{1}{H_{intg}(z)} = \frac{2}{T} \cdot \frac{z-1}{z+1} \tag{4.31}$$

Solving Eq. (4.31) for z leads to the first-order Padé approximation of a time delay (*cf.* Eq. (3.138) with $e^{-s\tau} = z^{-1}$). Converting transfer functions between the s- and the z-plane with Eq. (4.31) is known as the *bilinear transformation*. At the start of this chapter, we mentioned that a digital controller can be approximated by its time-continuous transfer function provided that the time lag T is very short compared to any dynamic response of the other system elements. As a rule of thumb, we can distinguish these three approaches:

- The sampling period of the digital system is two orders of magnitude shorter than the fastest time constant of the continuous system. In this case, the digital controller can be modeled as a continuous system, and Laplace-domain treatment is possible.
- When the sampling period of the digital system is faster than five to ten times of the fastest time constant of the continuous system, the bilinear transformation can be used to approximate Laplace-domain behavior. In this case, the z-domain transfer function of the controller is transformed into an s-domain approximation with Eq. (4.31), and the entire system is modeled in the Laplace domain. Care needs to be taken that all signals are band-limited, because the phase response of the dead-time delay $z = e^{sT}$ deviates strongly from its approximation at higher frequencies, and stability criteria such as the gain margin could become inaccurate in the approximation.
- When the sampling period of the digital system is less than five to ten times of the fastest time constant of the continuous system, the continuous components of the system are converted into their z-domain counterparts, and the entire system is modeled as a time-discrete system.

4.4 The *w*-Transform

Sometimes, it is desirable to know whether a pole in the z-plane is inside or outside of the unit circle, but the exact location of the pole is not important. s-plane methods can be applied when we define the w-transform as a mapping function that maps the unit circle of the z-plane onto the imaginary axis of a w-plane, and that maps any location of the z-plane inside the unit circle onto the left half-plane of the w-plane. Such a mapping function can be defined as

$$z = \frac{1+w}{1-w}; \quad w = \frac{z-1}{z+1} \tag{4.32}$$

Like s, w is a complex variable. However, the w-transform (Eq. (4.32)) is not an inversion of $z = e^{sT}$, and the s- and w-planes are not identical. However, any location z inside the unit circle (that is, $|z| < 1$) is mapped to a location in the w-plane with $\Re(w) < 0$. In some cases, the w-transform can provide a quick stability assessment of a z-domain transfer function. To examine a z-domain transfer function, for example,

$$H(z) = \frac{z-1}{z^2+2z+0.5} \tag{4.33}$$

we take the denominator polynomial $q(z) = z^2 + 2z + 0.5$ and substitute z to obtain

$$\begin{aligned} q(w) &= \left(\frac{1+w}{1-w}\right)^2 + 2\frac{1+w}{1-w} + 0.5 \\ &= \frac{1}{2}\frac{-w^2+w+7}{w^2-2w+1} \end{aligned} \tag{4.34}$$

If the roots of $q(z)$ lie inside the unit circle, the zeros of $q(s)$ lie in the left half-plane. In Eq. (4.34), we find zeros at -2.193 and $+3.193$, and we know that one pole of $H(z)$ lies outside of the unit circle and $H(z)$ is unstable. We make use of the w-transform in stability analysis (Chapter 10) and Bode analysis (Chapter 11).

4.5 Building Blocks for Digital Controllers

In analogy to Section 3.6, several low-order digital filters are introduced here. Without covering digital filter theory in detail, any digital filter can be seen as a weighted discrete sum of present and past input and output values as described by Eq. (4.35):

$$y_k = \sum_{n=1}^{N} a_n y_{k-n} + \sum_{m=0}^{M} b_m x_{k-m} \tag{4.35}$$

Two examples are the integrator formulas given in Eqs. (4.27) and (4.30). The first sum is referred to as the *recursive* part of the filter, and nonzero a_n give rise to infinite impulse response filters. The second sum represents the nonrecursive part. Filters with

$a_n = 0$ for all n are finite impulse response filters, because the impulse response returns to zero after no more than M discrete time steps T. The z-domain transfer function can be found by using the delay property of the z-transform, that is, a delay by one sampling interval T is represented by a multiplication with z^{-1} in the z-domain: when the z-transform of the sequence y_k is $Y(z)$, the delayed sequence y_{k-1} corresponds to $z^{-1}Y(z)$. The z-transform of Eq. (4.35) can therefore be written as

$$Y(z) - \sum_{n=1}^{N} a_n z^{-n} Y(z) = \sum_{m=0}^{M} b_m z^{-m} X(z) \tag{4.36}$$

By factoring out $X(z)$ and $Y(z)$, the transfer function of the digital filter becomes

$$H(z) = \frac{Y(z)}{X(z)} = \frac{\sum_{m=0}^{M} b_m z^{-m}}{-\sum_{n=0}^{N} a_n z^{-n}} \tag{4.37}$$

The negative sign in Eq. (4.37) is a result of moving the first summation term in Eq. (4.35) to the left-hand side. As a consequence we also get $a_0 = -1$.

Any recursive filter can lead to instability. As a general rule, **all poles of the z-domain transfer function must lie inside the unit circle in the z-plane**. We can demonstrate the rule with a simple first-order system,

$$y_k = a \cdot y_{k-1} + (1-a) \cdot x_{k-1} \tag{4.38}$$

Clearly, when $a > 1$, each subsequent value of y_k is larger than the previous one. The corresponding z-domain transfer function is

$$H(z) = \frac{1-a}{z-a} \tag{4.39}$$

and has one pole at $z = a$. Only for $a < 1$ is the stability requirement met.

4.5.1 Gain Block

The time-discrete analog of the gain block with gain g is realized by multiplying the current input value with g and applying it to the output. The only nonzero coefficient in Eq. (4.35) is $b_0 = g$ (and implicitly $a_0 = -1$).

- Time-domain function:

$$y_k = g \cdot x_k \tag{4.40}$$

- Z-domain transfer function:

$$H(z) = \frac{Y(z)}{X(z)} = g \tag{4.41}$$

- Impulse response:

$$y_{imp,k} = g \cdot \delta_k \tag{4.42}$$

In digital systems, the bit-discrete representation of numbers can play an important role, particularly when fast integer arithmetic is used. To provide an extreme example, assume an 8-bit value subjected to a gain of 1/64. The result only uses 2 bits. Rescaling this result back to 8 bits amplifies the digitization noise, and the output can only carry the discrete values 192, 128, 64, and 0. For integer implementations of digital filters, a careful implementation of scaling operations is needed that considers and avoids possible rounding errors.

4.5.2 Differentiator

The output signal of a differentiator approximates the first derivative of the input signal by applying a finite-difference formula. The finite difference can take three forms, backward, forward, and central difference. For real-time processing, only the backward difference can be realized as the other formulations depend on future input signals. Off-line processing allows using forward and central differences, and the central difference is attractive, because it does not introduce a phase shift.

- Time-domain function:

$$\begin{aligned} y_k &= \frac{x_k - x_{k-1}}{T} && \text{Backward difference} \\ y_k &= \frac{x_{k+1} - x_k}{T} && \text{Forward difference} \\ y_k &= \frac{x_{k+1} - x_{k-1}}{2T} && \text{Central difference} \end{aligned} \tag{4.43}$$

- Z-domain transfer function (backward difference):

$$H(z) = \frac{Y(z)}{X(z)} = \frac{z-1}{Tz} \tag{4.44}$$

- Impulse response:

$$y_{imp,k} = \{1/T, -1/T, 0, 0, \ldots\} \tag{4.45}$$

4.5.3 Integrator

The integrator was briefly introduced in the previous section. Time-discrete integration involves summation of the input values. A recursive formulation, where the previous output value represents the previous sum of input values, yields the filter function. Similar to the differentiator, a forward, backward, and central (i.e., trapezoidal) approximation can be used. However, all three integrators are causal and do not depend on future values.

- Time-domain function:

$$\begin{aligned} y_k &= y_{k-1} + T \cdot x_{k-1} && \text{Forward rectangular approximation} \\ y_k &= y_{k-1} + T \cdot x_k && \text{Backward rectangular approximation} \\ y_k &= y_{k-1} + \frac{T}{2}(x_k + x_{k-1}) && \text{Trapezoidal approximation} \end{aligned} \tag{4.46}$$

- Z-domain transfer function (trapezoidal approximation):

$$H(z) = \frac{T}{2} \cdot \frac{z+1}{z-1} \tag{4.47}$$

- Impulse response (trapezoidal approximation):

$$y_{imp,k} = \{T/2, T, T, T, \ldots\} \tag{4.48}$$

4.5.4 PID Controller

It is straightforward to combine gain (k_p) with an integrator and a differentiator to obtain a simple *PID* formulation. By setting $k_D = 0$, the *PID* controller is reduced to a *PI* controller. In its simplest form, the combination of backward difference and backward rectangular integration yields.

- Time-domain function:

$$y_k = k_I y_{k-1} + \left(k_p + k_I T + \frac{k_D}{T}\right) x_k - \frac{k_D}{T} x_{k-1} \tag{4.49}$$

- Z-domain transfer function:

$$H(z) = \frac{z(k_p + k_I T + k_D/T) - k_D/T}{z - k_I} \tag{4.50}$$

To accelerate the computation, some coefficients would typically be calculated beforehand, such as $k_1 = k_p + k_I T + k_D/T$ and $k_2 = k_D/T$, which leads to a simpler form of Eq. (4.50):

$$H(z) = \frac{k_1 z - k_2}{z - k_I} \tag{4.51}$$

- Impulse response:

$$y_{imp,k} = \begin{cases} k_1 & \text{for} \quad k = 0 \\ k_1 \cdot k_I^k - k_2 \cdot k_I^{k-1} & \text{for} \quad k > 0 \end{cases} \tag{4.52}$$

An alternative formulation uses first-order finite differences in which the *change* of the output, $y_k - y_{k-1}$ depends on the change of the input $x_k - x_{k-1}$. By using the more complex (but more accurate) central difference and the trapezoidal rule, the *PID* controller can be described as follows:

- Time-domain function:

$$y_k - y_{k-1} = \left(k_p + \frac{k_I T}{2} + \frac{k_D}{T}\right) x_k - \left(k_p + \frac{2k_D}{T} - \frac{k_I T}{2}\right) x_{k-1} + \frac{k_D}{T} x_{k-2} \tag{4.53}$$

Once again it is convenient to precalculate the coefficients for x_k, x_{k-1}, and x_{k-2} as k_0, k_1, and k_2, respectively. With this definition, the time-domain function is simplified to

$$y_k - y_{k-1} = k_0 x_k - k_1 x_{k-1} + k_2 x_{k-2} \tag{4.54}$$

- Z-domain transfer function:

$$H(z) = \frac{k_0 z^2 - k_1 z + k_2}{z(z-1)} \tag{4.55}$$

- Impulse response:

$$y_{imp,k} = \{k_0, k_0 - k_1, k_0 - k_1 + k_2, k_0 - k_1 + k_2, \ldots\} \tag{4.56}$$

The superiority of the second formulation can also be seen in the fact that the impulse response converges to a finite value, which is the expected behavior of the *PID* system. The first formulation does not show convergent behavior.

4.5.5 Time-Lag System

Emulating a first-order time-lag system (first-order lowpass with a continuous-domain pole at $s = -|\sigma|$) in a time-discrete system is possible by directly using the correspondence tables for the step response function $f(t) = 1 - e^{-\sigma t}$. We find the correspondence

$$f(t) = 1 - e^{-\sigma t} \quad \circ\!\!-\!\!\bullet \quad \frac{z}{z-1} \cdot \frac{1 - e^{-\sigma T}}{z - e^{-\sigma T}} \tag{4.57}$$

for the sampling rate T. The first term, $z/(z-1)$ represents the step input, while the second term is the actual filter transfer function. With the definition $a = e^{-\sigma T}$, we can now describe the filter:

- Time-domain function:

$$y_k = a y_{k-1} + (1-a) x_{k-1} \tag{4.58}$$

- Z-domain transfer function:

$$H(z) = \frac{1-a}{z-a} \tag{4.59}$$

- Impulse response:

$$y_{imp,k} = \{0, (1-a), (1-a)a, \ldots, (1-a)a^{k-1}\} \tag{4.60}$$

Note that $-1 < a < 1$ is required for stability.

4.5.6 Time-Lead System

In off-line digital signal processing and digital filters, a true phase-lead is achievable by accessing future input values x_{k+n}. For real-time processing, the idea of the first-order phase-lead system can be applied where the change (i.e., the first derivative) is added to the input signal. The idea behind this approach is to extrapolate the recent change to the present. The continuous equation $y(t) = x(t) + \tau_D \dot{x}(t)$ therefore leads the following digital filter (a is the normalized delay $a = \tau_D/T$):

- Time-domain function:

$$y_k = x_k + a(x_k - x_{k-1}) \tag{4.61}$$

- Z-domain transfer function:

$$H(z) = \frac{z(1+a) - a}{z} \tag{4.62}$$

- Impulse response:

$$y_{imp,k} = \{1 + a, -a, 0, 0, \ldots\} \tag{4.63}$$

4.5.7 Lead-Lag Compensator

We can combine the time-lead and time-lag compensators into a flexible lead-lag compensator that approximates the differential equation

$$\dot{y}(t) + ay(t) = bx(t) + \dot{x}(t) \tag{4.64}$$

where a and b determine the location of the pole and the zero, respectively, in the Laplace domain. A finite-difference approach is straightforward, with $(y_k - y_{k-1})/T$ on the left-hand side and $(x_k - x_{k-1})/T$ on the right-hand side. More elegantly, the differential Eq. (4.64) can be rearranged to require a discrete integration:

$$y(t) - x(t) = \int_0^t \big(bx(\tau) - ay(\tau)\big)\,\mathrm{d}\tau \tag{4.65}$$

The advantage of the form in Eq. (4.65) is that the more accurate trapezoidal integration can be applied instead of the backward difference. The discrete approximation of Eq. (4.65) can be written as

$$y_k - x_k = [y_{k-1} - x_{k-1}] + B(x_k + x_{k-1}) - A(y_k + y_{k-1}) \tag{4.66}$$

where $A = a \cdot T/2$ and $B = b \cdot T/2$. The two terms in square parentheses on the right hand side represent the accumulator for the integration; the following two terms are the trapezoidal finite areas added to the integral. Rearranging the terms yields the time-domain function and its z-transform:

- Time-domain function:

$$y_k = \frac{1}{1+A}\left[y_{k-1}(1-A) + x_k(1+B) + x_{k-1}(B-1)\right] \tag{4.67}$$

- Z-domain transfer function:

$$H(z) = \frac{1}{1+A} \cdot \frac{z(1+B) + B - 1}{z(1+A) + A - 1} \tag{4.68}$$

Because of the wide variety of configurations, the impulse response is not provided for this filter.

The above examples provide some digital filter functions that can be used as controllers or compensators in time-discrete systems. For the software implementation of a filter function, the time-domain equation would typically be used. To provide one example, the software implementation of Eq. (4.67) requires setting aside three memory locations to store the digital values for y_k, y_{k-1}, and x_{k-1}. Upon reset, these locations are initialized to zero. Furthermore, the coefficients A and B are computed from the given a, b, and T. In fact, it is even more convenient to compute $(1-A)/(1+A)$, $(B-1)/(1+A)$, and $(1+B)/(1+A)$ beforehand as can be seen in the listing below. Finally, a timer is started that causes the computation of Eq. (4.67) at each period T.

For this example, we assume that the conversion time of the ADC and the computation time are very short compared to T, meaning, the time between the start of the computation and the moment y_k is applied to the output is negligible. This assumption is usually not valid, and an additional time lag for conversion and processing time needs to be taken into account. We will examine this effect in more detail in Chapter 14.

The following steps (Algorithm 4.1) represent the pseudo-code to compute the filter output of the lead-lag compensator. This function is called in intervals of T. Since only immediate past values are stored, we use the shorthand `X` for x_k, `Y` for y_k, `Y1` for y_{k-1}, and `X1` for x_{k-1}.

Note that in line 2, `Y` is initialized and overwritten, whereas lines 3 and 6 add values to the existing `Y`. In lines 5 and 7, the old values for `X1` and `Y1` are updated and made available for the next sampling period. The ADC step in line 4 is not necessary if `X` is made available from outside this function.

Algorithm 4.1. Pseudocode algorithm to compute the output Y of a lead-lag compensator. This function needs to be called in regular intervals T, likely driven by a timer interrupt.

1 *Start of computation*;
2 $Y \longleftarrow Y1 \cdot (1-A)/(1+A)$;
3 $Y \longleftarrow Y + X1 \cdot (B-1)/(1+A)$;
4 Perform ADC and read X;
5 $X1 \longleftarrow X$;
6 $Y \longleftarrow Y + X \cdot (1+B)/(1+A)$;
7 $Y1 \longleftarrow Y$;
8 Apply Y to output;
9 *End of computation*;

5 First Comprehensive Example: The Temperature-Controlled Waterbath

Abstract

To make the mathematical methods that describe linear systems and linear feedback controls more accessible, a very simple example is introduced and explained exhaustively in this chapter: a temperature-controlled waterbath. First, a mathematical model of the waterbath itself is sought, with the temperature as the output signal and the heating power as the input signal. A first-order differential equation is found that bears strong similarity to the one governing the RC-lowpass. Methods are introduced how the process constants (and with those, the coefficients of the differential equation) can be determined by measurement. Lastly, a feedback control system is introduced and the behavior of the waterbath with and without feedback control compared.

In this chapter, we introduce a simple example. The example process is a waterbath where electrical power is introduced to heat a certain amount of water above room temperature. We will recognize this system as a first-order system that shows some similarities to the RC-lowpass in Chapters 2.1 and 3.3.

Initially, the water is equilibrated at room temperature. When the experiment starts (at $t = 0$), heating power is introduced. As the water heats up, some of that power is lost to the environment. Energy loss to the environment is proportional to the difference between water temperature and environmental temperature, and losses to the environment therefore increase as the temperature rises. We can intuitively see that a temperature exists where the heater power is just enough to compensate for the loss to the environment. This is the equilibrium point. In this chapter, we will describe the process mathematically (that is, we will develop a *model*) and examine the behavior of the process—both without and with feedback control—with the help of this model.

This example is split in two parts. In this chapter, we will examine the system strictly in the time domain, and one goal of this chapter is to explain how the process constants can be determined from laboratory measurements. In Chapter 6, we revisit the example, but apply Laplace-domain methods.

Linear Feedback Controls. http://dx.doi.org/10.1016/B978-0-12-405875-0.00005-X
© 2013 Elsevier Inc. All rights reserved.

5.1 Mathematical Model of the Process

In an electrically-heated waterbath, the water integrates the electrical power with an integration constant k_w:

$$k_w = \frac{1}{c\rho V} \tag{5.1}$$

where c is the specific thermal capacity of the water, ρ is the specific mass, and V is the volume. We assume that the heat capacity of the water is much larger than that of the bath container itself. As the water heats up, some of the heating power is lost to the environment. Energy loss to the environment is proportional to the difference between water temperature T and environmental temperature T_{env} with a proportionality constant k_e. Figure 5.1 shows the block diagram of the waterbath model.

The water temperature is governed by Eq. (5.2),

$$T(t) = k_w \int P_{in}(t) - P_{env}(t)\mathrm{d}t \tag{5.2}$$

where P_{in} is the power introduced by the heating element and P_{env} is the power lost to the environment. The latter is described by Eq. (5.3):

$$P_{env}(t) = k_e(T(t) - T_{env}(t)) \tag{5.3}$$

Note the difference between *variables*, such as the temperature and the heating power, which are functions of time, and *process constants*, such as k_w and k_e, which do not vary with time. By substituting Eq. (5.3) into (5.2), differentiating toward time, and collecting all terms that contain T, we arrive at the differential equation for the entire waterbath, Eq. (5.4):

$$\frac{1}{k_w k_e}\dot{T}(t) + T(t) = \frac{1}{k_e}P_{in}(t) + T_{env}(t) \tag{5.4}$$

At this point, we have obtained a mathematical *model* for the waterbath. Equation (5.4) allows us to predict—within certain limits—how the waterbath temperature reacts

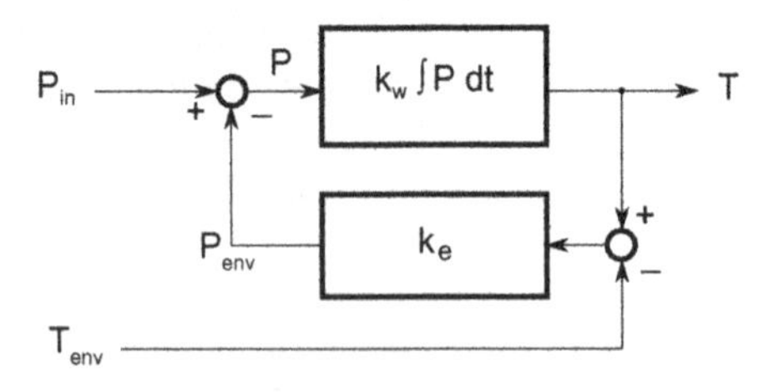

Figure 5.1 Block diagram model of the waterbath process. The water integrates the electrical heating power P, and the temperature T rises. With higher temperature, however, losses to the environment P_{env} increase linearly. The lost energy is no longer available to raise the water temperature. The system reaches an equilibrium when the introduced energy and the energy losses are equal.

under specific environmental conditions and when a specific heating power is applied. Like all mathematical models, this model has limits. For example, the model fails when the water starts to boil, or if it is frozen. Whenever we create a mathematical model, we must ensure that it describes the behavior of the system with sufficient accuracy. Here, *sufficient* is usually defined within the design problem. For example, if the waterbath normally operates around 37 °C, the model is sufficient. However, if the waterbath is part of a boiler system for steam generation, the model needs to be extended when the water starts to boil.

5.2 Determination of the System Coefficients

Equation (5.4) contains several coefficients, specifically k_e and k_w. Sometimes, process coefficients are known (for example, when electronic components are used), and often these coefficients can be computed. There are cases, however, where they need to be determined experimentally. To determine the unknown coefficients of the waterbath, we are interested in the special case where T_{env} is constant, and where a constant input power P_{in} is applied at $t = 0$ under the initial condition that the water is equilibrated at $T = T_{env}$ for $t < 0$. The function $P_{in}(t)$ that jumps from 0 to a constant nonzero value at $t = 0$ is called a step function (*cf.* Chapter 2.3):

$$P_{in}(t) = \begin{cases} 0 & \text{for} \quad t < 0 \\ P_{in} = const & \text{for} \quad t \geq 0 \end{cases} \tag{5.5}$$

The general solution for this special form of first-order ODE with constant coefficients (Eq. (5.4)) and with a step function on the right-hand side is

$$T(t) = A + B(1 - e^{-t/\tau}) \tag{5.6}$$

and its first derivative

$$\dot{T}(t) = \frac{B}{\tau} e^{-t/\tau} \tag{5.7}$$

We now need to determine the unknown constants A, B, and τ from the measured step response that is shown in Figure 5.2. In this special case, three specific points on the plot of temperature over time are important:

1. The equilibrium temperature: When $t \to \infty$, the exponential term in Eq. (5.6) vanishes. Furthermore, equilibrium implies that $\dot{T}(t) = 0$.
2. The initial temperature: When $t \to 0$, the exponential terms in Eqs. (5.6) and (5.7) approach unity, and some constants are isolated.
3. The time where the temperature rise reaches about 63% of its final value: This is the time where $t \approx \tau$ and serves to obtain the time constant of the system.

First, let us look at the equilibrium for $t \to \infty$ when the exponential term vanishes and $\dot{T}(t) = 0$ holds. From Eqs. (5.4) and (5.6), we immediately obtain Eq. (5.8),

$$A + B = \frac{P_{in}}{k_e} + T_{env} \tag{5.8}$$

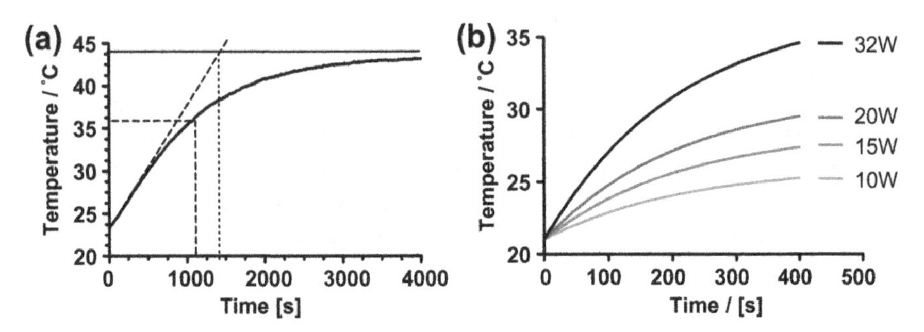

Figure 5.2 Temperature response of the waterbath process. (a) Shows laboratory data where 100 mL water were heated with a 15 W heating element. Auxiliary lines help identify the equilibrium temperature (horizontal line at 44 °C), the initial slope (dashed line that intersects 44 °C at $t = 1450$ s), and the point where 63% of the temperature rise has occurred (at $t = 1100$ s). (b) Shows the temperature as a function of time computed from Eq. (5.4) for different heater power settings. As stipulated by Eq. (5.4), the equilibrium temperature increases with increasing heater power, but the time constant, i.e., the point in time where 63% of the temperature rise has occurred, remains the same.

At the other extreme, when $t \rightarrow 0$, the exponential terms approach unity. By substituting Eq. (5.7) into (5.4), we obtain Eq. (5.9). Equation (5.9) follows from Eq. (5.8). Therefore,

$$\frac{B}{k_w k_e}\frac{1}{\tau} + T(t \rightarrow 0) = \frac{P_{in}}{k_e} + T_{env} \tag{5.9}$$

When we consider Eq. (5.8), Eq. (5.9) can only be valid when

$$\tau = \frac{1}{k_w k_e} \tag{5.10}$$

To obtain B, we need to make use of the observation that there are almost no losses to the environment when $t \rightarrow 0$. This approximation leads to $P_{env}(t \rightarrow 0) = 0$. From this condition, we know that the initial rate of temperature increase (i.e., the initial slope) is $\dot{T}(t \rightarrow 0) = k_w P_{in}$. By taking the first derivative of Eq. (5.6) (see Eq. (5.7)) and setting it equal to $k_w P_{in}$ (note that the exponential term becomes unity again), we obtain

$$\frac{B}{\tau} = Bk_w k_e = k_w P_{in} \tag{5.11}$$

which provides the solution for B,

$$B = \frac{P_{in}}{k_e} \tag{5.12}$$

and from which immediately follows that $A = T_{env}$ (see Eq. (5.8)). We would arrive at the same point by using $T(t \rightarrow 0) \approx T_{env}$ in Eq. (5.4). In this case, $T(t \rightarrow 0)$ and T_{env} cancel out. By substituting $\dot{T}(t = 0) = B/\tau$, we readily obtain Eq. (5.11).

By replacing the coefficients A, B, and τ in Eq. (5.6), we now obtain the specific response to a constant input power P_{in}, applied at $t = 0$:

$$T(t) = T_{env} + \frac{P_{in}}{k_e}(1 - e^{-k_w k_e t}) \tag{5.13}$$

Let us perform some sanity checks. First, at $t = 0$, we obtain $T(t \to 0) = T_{env}$. This agrees with the initial condition. Second, we require that the argument of the exponential term has no units. The units of k_w are Kelvin per Joule and the units of k_e are Watts per Kelvin. Therefore, the product $k_w k_e$ has units of inverse seconds which cancel out against the unit of the time variable. Lastly, we obtain the equilibrium temperature,

$$T(t \to \infty) = T_{env} + \frac{P_{in}}{k_e} \tag{5.14}$$

which makes intuitive sense as the equilibrium temperature increases with the environmental temperature and with the input power, but decreases with the constant that describes the losses to the environment. Note that we can use K and °C interchangeably, because we are only considering temperature differences. We can now plot the function described by Eq. (5.13) for different heater powers (Figure 5.2B). Let us assume that we try to heat 100 g of water (specific heat capacity $c = 4200$ J/kg K) with a heater that has different settings. The losses to the environment are 2 W/K, and the environmental temperature is 21 °C. From these values, we can compute $k_w = 0.0024$ K/J and $\tau = 210$ s. With a 10 W heater, the equilibrium temperature is 26 °C. With a 20 W heater, we get 31 °C at equilibrium. To achieve an equilibrium of 37 °C, we need 32 W heating power. The temperature curves for various heater setting are shown in Figure 5.2B. Note that the dynamic response, that is, the time to reach a certain percentage of the final temperature, does not depend on the input power. This is a characteristic constant of the system.

We can use these considerations to obtain numerical values for the constants from laboratory data, that is, a measured temperature timecourse. Consider the data in Figure 5.2A. At $t = 0$, heater power of 15 W was applied to the unknown waterbath, and the temperature monitored from its initial equilibrium $T(t < 0) = T_{env} = 23$ °C until it reached an approximate new equilibrium after approximately one hour. From the equilibrium temperature, we obtain the first key value, that is, the temperature rise above room temperature. From the diagram, we estimate the equilibrium to be at 44 °C (horizontal line), and the temperature rise is therefore 21 °C. Equation (5.14) can be solved for k_e,

$$k_e = \frac{P_{in}}{T(t \to \infty) - T_{env}} \tag{5.15}$$

and we obtain $k_e = 0.72$ W/K. Next, we determine the initial slope (dashed line in Figure 5.2). If we have curve fitting software,[1] the time constant can be obtained

[1] Many statistical packages offer nonlinear regression with user-defined equations. Examples include the R Project, which is a major statistical package, and Qtiplot, which is a data analysis and visualization package.

directly from the nonlinear regression ($\tau = 1061$ s). In the plot, we can use a ruler to draw the tangent to get an estimate. In Figure 5.2, the temperature tangent rises 21 °C in 1450 s or 0.0145 °C per second. We recall Eq. (5.2) with $P_{env} \approx 0$ for $t \rightarrow 0$. Since P_{in} is constant, the slope $\dot{T}$ provides k_w through

$$k_w = \left. \frac{\dot{T}}{P_{in}} \right|_{t \rightarrow 0} \tag{5.16}$$

We determine $k_w = 0.001$ K/Ws. Lastly, we have seen (Eq. (5.10)) that the time constant τ equals $1/k_e k_w$. Therefore, $\tau = 1440$ s. The dotted lines in Figure 5.2 show how the time constant can be directly determined. We estimate an equilibrium temperature rise of 21 °C. At time $t = \tau$, $1 - e^{-1} \approx 63\%$ of this rise (13 °C) has taken place. The curve crosses 36 °C (23 °C + 13 °C) at approximately 1100 s.

Why do the two values of τ deviate by approximately 30 percent between the direct estimate of the time constant from the plot and the computation from $1/(k_w k_e)$? This example gives us a sense of measurement inaccuracies, especially when "graphical differentiation" with a ruler is involved. The initial slope is particularly difficult to determine, and the estimate of the 63% rise is likely to be more accurate. This observation is confirmed by the result from nonlinear regression. When we use $\tau = 1100$ s, we obtain $k_w = 1/(\tau k_e) = 0.0013$ K/Ws.

Moreover, from 100 mL water, we expect $k_w = 1/c\rho V = 0.0024$ K/Ws. The lower value of 0.0013 K/Ws would indicate a volume of 184 mL. The discrepancy is attributable to the waterbath container itself, and this comparison shows that the initial assumption of a negligible contribution from the waterbath container and the heating element is not correct. One method to accurately determine the contribution of the empty waterbath is to measure the apparent k_w for different fill volumes and extrapolate to a hypothetical volume of 0 mL. Depending on the application, considerable effort may be required to obtain an accurate model of the process.

5.3 Determining the Transfer Function—General Remarks

It is theoretically possible to determine the transfer function of a system with all its constants with suitable measurements. Often, a step input reveals the system dynamics, although the analysis of the step response becomes more uncertain with higher-order systems. Measurements are associated with some degree of error. The example above demonstrates how the process constants can be determined in a simple first-order system. The general ideas, however, are applicable for all systems: obtaining information from the magnitude and the derivative of the step response, the equilibrium value of the response when $t \rightarrow \infty$, and, for resonant higher-order systems, the frequency and decay time of oscillations.

In practice, the output of the system will be connected to some measurement device. Often, a sensor is necessary to convert the output (such as temperature, pressure, or displacement) into a voltage. It is also useful to have electronic control over the input, for example, with a relay or an actuator.

If the step response is fast, that is, in the range of micro to milliseconds, the output can be monitored with an oscilloscope, whereby the input is connected to a square-wave generator. The input period would be chosen to be significantly longer than the step response to allow the system to reach a quasi-equilibrium. Conversely, slow systems, such as the waterbath, would be connected to a data logger, and the output variable recorded in regular intervals after a single application of a step input. In both cases, a reasonable approximation of $t \rightarrow \infty$ needs to be made. Often, waiting for an equilibrium is impractical. Since all measurements are subject to some error, the measurement can be aborted when the output variable reaches a certain percentage of the suspected final value, or if the relative change drops below a predefined threshold. A first-order system with the time constant τ, for example, reaches 90% of its equilibrium value after approximately 2.3τ and 95% of its equilibrium value after 3τ.

If a complete feedback control system is present, it is often necessary to open the feedback loop. **Performance problems of closed-loop systems are hard to identify and diagnose**. Poor performance or a nonfunctioning control system can have a large number of causes, and the presence of a feedback loop that performs some inexplicable action complicates the diagnosis of the problem. When the loop is opened and feedback control disabled, it becomes possible to measure the responses of individual elements that comprise the loop. To open the loop, the connection between the controller and the process can be interrupted, and the controller signal replaced by some adjustable signal (e.g., from a potentiometer). This configuration allows (a) to apply a defined signal, such as a step change, a square wave, or a sinusoidal oscillation, to the process, (b) measure the error signal, that is, the difference between process output and setpoint, and (c) measure the control action. In many cases, the analysis of the open-loop behavior can provide crucial clues why a feedback control system is not operating properly. Figure 5.3 shows example configurations where a closed-loop system has been converted into an open-loop configuration. Such a configuration allows to observe the system behavior when full feedback control has *not* been established.

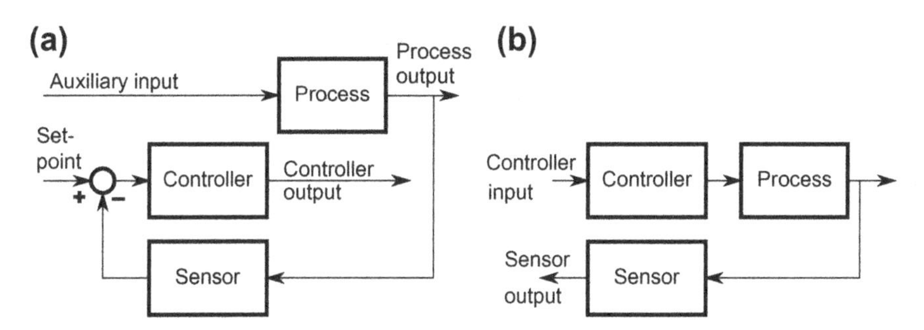

Figure 5.3 Two possible configurations with an opened feedback loop (*cf.* Figure 1.3). (a) The feedback loop has been opened between controller and process. The process can be driven by an auxiliary signal (e.g., a constant signal or a step change), and the error signal and the controller's response can be observed. (b) The feedback loop has been opened at the controller input, and the controller is driven with a simulated error signal, allowing to observe the combined response of controller and process.

5.4 Introducing Feedback Control

Now, we will introduce feedback control to the waterbath process with the relatively simple goal to achieve an equilibrium temperature of 37 °C. For this purpose, we will attach a temperature sensor that outputs $k_s = 0.1$ V/°C. At 37 °C, the sensor therefore outputs 3.7 V. This voltage is subtracted from the setpoint voltage, and then amplified by a factor k_p to yield the input power P_{in}. The block diagram for the complete closed-loop control system is shown in Figure 5.4.

At this point, P_{in} is no longer the input variable, but rather a dependent variable. P_{in} is now the *control action*, determined by the control deviation $\epsilon = V_{set} - V_{temp}$. It is reasonable to use Eq. (5.4) as a starting point and substitute $P_{in} = k_p(V_{set} - k_s T)$. Rearranging the equation so that all terms that contain T are on the left-hand side yields Eq. (5.17).

$$\frac{1}{k_w k_e}\left(\frac{k_e}{k_e + k_s k_p}\right)\dot{T}(t) + T(t) = \frac{k_p}{k_e}\left(\frac{k_e}{k_e + k_s k_p}\right)V_{set}(t) + \frac{k_e}{k_e + k_s k_p}T_{env}(t) \tag{5.17}$$

This equation can be simplified by giving the recurring unitless gain factor a new symbol, g,

$$g = \frac{k_e}{k_e + k_s k_p} \tag{5.18}$$

which leads to the differential equation of the closed-loop feedback system, Eq. (5.19):

$$\frac{g}{k_w k_e}\dot{T}(t) + T(t) = \frac{k_p}{k_e} g V_{set}(t) + g T_{env}(t) \tag{5.19}$$

Interestingly, this differential equation has the same general form as the open-loop differential equation of the process (the waterbath) in Eq. (5.4). One input is

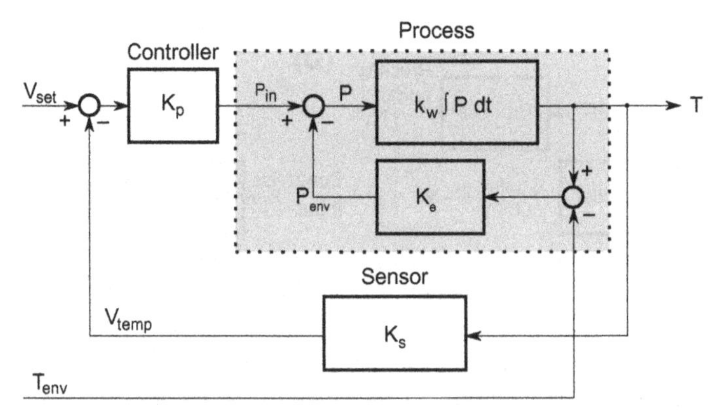

Figure 5.4 Waterbath (shaded section) with a feedback control loop. The feedback controller is a P controller type that multiplies the control deviation (i.e., $V_{set} - V_{temp}$) with a constant factor k_p and uses the resulting signal as control action. The gray-shaded area is the process that was introduced in Figure 5.1.

different—it is now the setpoint V_{set} rather than the heating power P_{in}. We can now determine the dynamics of a step change in the setpoint analogously to the waterbath without feedback control. When V_{set} is changed at $t = 0$, the general solution, Eq. (5.6) still holds, but with different constants A, B, and τ.

With the same reasoning that we applied in Eq. (5.9), we obtain

$$\tau = \frac{g}{k_w k_e} = \frac{1}{k_w(k_e + k_p k_s)} \tag{5.20}$$

To find B, we use the initial condition of $T(t \to 0) \approx T_{env}$. Unlike in Eq. (5.4), however, T_{env} does not cancel out. Rather, we obtain Eq. (5.21) for $t \to 0$:

$$\frac{g}{k_w k_e} \dot{T}(0) = \frac{k_p}{k_e} g V_{set} + (g-1) T_{env} \tag{5.21}$$

Substituting $\dot{T}(0) = B/\tau$ into Eq. (5.21) provides the constant B:

$$B = \frac{k_p}{k_e + k_p k_s} V_{set} - \frac{k_p k_s}{k_e + k_p k_s} T_{env} \tag{5.22}$$

Finally, we use the equilibrium ($t \to \infty$) once more to obtain A, because the exponential terms and $\dot{T}$ vanish when $t \to \infty$. The equilibrium equation is

$$A + B = \frac{k_p}{k_e} g V_{set} + g T_{env} \tag{5.23}$$

and by using B from Eq. (5.22), we obtain $A = T_{env}$ as expected. This leads us to the step response of the closed-loop system, Eq. (5.24)

$$T(t) = T_{env} + \left(\frac{k_p}{k_e + k_p k_s} V_{set} - \frac{k_p k_s}{k_e + k_p k_s} T_{env} \right) \left(1 - e^{-\frac{t}{\tau}} \right) \tag{5.24}$$

with τ defined in Eq. (5.20).

5.5 Comparison of the Open-Loop and Closed-Loop Systems

Let us compare the behavior of the open-loop and closed-loop systems. B is the temperature increase over the initial condition, T_{env}. From Eq. (5.22), we can see that B increases with the setpoint voltage, but decreases when the environment temperature is higher. Note that we assume a step function for V_{set}, but T_{env} is constant for $0 \le t < \infty$. Whereas each degree increase of the environment temperature raises the final temperature by one degree in the open-loop system (Eq. (5.13)), the influence of the environment temperature is reduced by a factor of g when feedback control is present. Feedback control can therefore reduce the influence of a disturbance. Moreover, the dynamic response of the closed-loop system is faster. Whereas the time constant τ is only influenced by process constants in the open-loop case (Eq. (5.10)), the closed-loop system has a shorter time constant because of the additional term with k_p in the denominator (Eq. (5.20)). Closed-loop feedback control leads to a faster response time.

Usually, a large k_p is chosen when pure P control is used (as in this example). When $k_s k_p \gg k_e$, we can see that $g \rightarrow 0$. A large k_p therefore influences the system response in two important ways. First, the dynamic response is improved, because τ becomes shorter. Second, the influence of T_{env} on the equilibrium temperature is reduced. Conversely, when $k_p \rightarrow 0$, $g \rightarrow 1$ follows. In the limiting case, the open-loop equation, Eq. (5.4), emerges from Eq. (5.19).

Figure 5.5 shows the system response of the closed-loop feedback system. We use the same values as before, $k_e = 2$ W/K, $T_{env} = 21$ °C, and $k_w = 0.0024$ K/J. In addition, we use $k_s = 0.1$ V/°C and the step function $V_{set} = 3.7$ V at $t \geq 0$. The dynamic response depends strongly on k_p. For k_p values of 10 W/V, 50 W/V, 200 W/V, and 1000 W/V, the equilibrium temperature T_∞, time constant τ, the gain factor g, and the final control deviation ϵ_∞ are listed in Table 5.1.

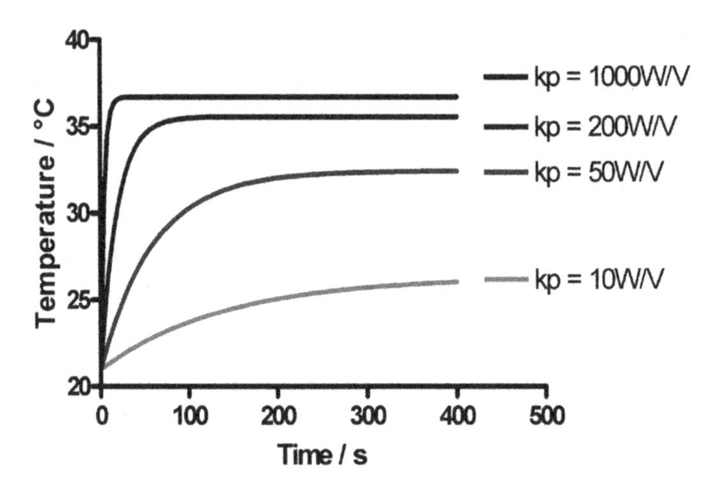

Figure 5.5 Temperature response of the closed-loop system at $V_{set} = 3.7$ V and different settings for k_p. Compared to Figure 5.2B, we can see that a higher k_p not only improves the steady-state response (i.e., how accurately the setpoint is reached), but also improves the time constant. With a large k_p, the system reacts faster.

Table 5.1 Closed-loop system response for different settings of k_p.

k_p (W/V)	g	T_∞ (°C)	τ (s)	ϵ_∞ (V)
10	0.67	26.3	140	1.07
50	0.29	32.4	60	0.46
200	0.09	35.6	19	0.15
1000	0.02	36.7	4.1	0.031

At this point, we need to note several shortcomings of our control strategy. First, it becomes apparent from Figure 5.5 that we never reach the setpoint temperature. In fact, this behavior can be derived from Eq. (5.24) when we let $t \to \infty$ and obtain for the equilibrium temperature:

$$T(t \to \infty) = \frac{1}{k_e + k_p k_s} \left(k_p V_{set} + k_e T_{env}\right) \tag{5.25}$$

The equilibrium temperature depends on a weighted sum of the setpoint and the environmental temperature. Increasing k_p increases the weight of the setpoint, and decreasing k_e (i.e., better insulation) decreases the weight of the environmental temperature. Unfortunately, physical constraints limit both coefficients. Notably, a heater with 1000 W/V or even 200 W/V is completely out of proportion for 100 mL water. A different control strategy is needed: including an integral component in the controller will eventually drive the control deviation to zero as we will show in the next chapter.

6 Laplace- and z-Domain Description of the Waterbath Example

Abstract

The example provided in Chapter 5 is continued. This chapter follows the same steps that were taken in Chapter 5, but systems and signals are treated in the Laplace domain. By following the same steps as in the previous chapter (description of the process, description of the feedback control system, comparison of the system with and without control), the more abstract Laplace-domain models are made more accessible, and the simplifications brought by the Laplace transform can be better appreciated. A new controller type with an integral component is introduced. Moreover, the example concludes with the introduction of a time-discrete control and its description in the z-domain, highlighting the parallels between Laplace- and z-domain models.

In this chapter, we continue the analysis of the waterbath example, but we perform the analysis in the Laplace domain.

6.1 Laplace-Domain Description of the Process

We begin the analysis of the waterbath example with the open-loop differential equation of the process (*cf.* Eq. (5.4)),

$$\frac{1}{k_w k_e}\dot{T}(t) + T(t) = \frac{1}{k_e} P_{in}(t) + T_{env}(t) \tag{6.1}$$

where k_w reflects the water's heat storage capacity (given in °C/Ws) and k_e (given in W/°C) describes the losses to the environment. Furthermore, we assume that the water has the temperature T_0 at the beginning of our experiment, i.e., at $t = 0$. We first examine the Laplace transform of Eq. (6.1). Recall that the Laplace transform of the first derivative includes the initial condition,

$$\frac{\mathrm{d}}{\mathrm{d}t} f(t) \quad \circ\!\!-\!\!\bullet \quad sF(s) - f(0^+) \tag{6.2}$$

where $f(0^+)$ is the function value at $t = 0$, approached from the positive time axis. By multiplying Eq. (6.1) by $k_w k_e$ and applying the corresponding Laplace-domain variables for each summation term, we arrive at Eq. (6.3):

$$s \cdot T(s) - T_0 + k_w \cdot k_e \cdot T(s) = k_w \cdot P_{in}(s) + k_w \cdot k_e \cdot T_{env}(s) \tag{6.3}$$

Linear Feedback Controls. http://dx.doi.org/10.1016/B978-0-12-405875-0.00006-1

© 2013 Elsevier Inc. All rights reserved.

The constant T_0 can be interpreted as an input signal, therefore, we move T_0 to the right-hand side. Also, we factor out $T(s)$ on the left-hand side:

$$T(s)\left(s + k_w \cdot k_e\right) = k_w \cdot P_{in}(s) + k_w \cdot k_e \cdot T_{env}(s) + T_0 \tag{6.4}$$

Rather than a differential equation, we are now looking at a polynomial of s. The first-order polynomial $s + k_w k_e$ is called the *characteristic polynomial* of the process, and because the highest power of s is one, the process is a first-order process. In the Laplace domain, we can now isolate the output variable $T(s)$ by dividing both sides of Eq. (6.4) by the characteristic polynomial:

$$T(s) = \frac{k_w}{s + k_w k_e} P_{in}(s) + \frac{k_w k_e}{s + k_w k_e} \cdot T_{env}(s) + \frac{1}{s + k_w k_e} T_0 \tag{6.5}$$

To examine the equilibrium values, we can use the initial and final value theorems. $P_{in}(s)$ and $T_{env}(s)$ are functions of s and must be treated as such. The initial value T_0 is a constant and not a function of s.[1] As a simplified rule of thumb, if we want to keep a variable input signal at the constant value c, it must be represented by a step function c/s in the Laplace domain. Thus, if we apply a constant heater power P_H and have a constant room temperature T_R, the above equation becomes

$$T(s) = \frac{k_w}{s + k_w k_e} \frac{P_H}{s} + \frac{k_w k_e}{s + k_w k_e} \cdot \frac{T_R}{s} + \frac{1}{s + k_w k_e} T_0 \tag{6.6}$$

Because of the linearity of the Laplace transform, we may look at the initial values separately and superimpose them. The initial and final value theorems state:

$$\begin{aligned} \lim_{t \to 0} f(t) &= \lim_{s \to \infty} s \cdot F(s) \\ \lim_{t \to \infty} f(t) &= \lim_{s \to 0} s \cdot F(s) \end{aligned} \tag{6.7}$$

For the initial value, we obtain from Eq. (6.6),

$$T(t \to 0) = \lim_{s \to \infty} \left(s \frac{k_w}{s + k_w k_e} \frac{P_H}{s} + s \frac{k_w k_e}{s + k_w k_e} \cdot \frac{T_R}{s} + s \frac{1}{s + k_w k_e} T_0 \right) = T_0 \tag{6.8}$$

which makes intuitive sense as we *know* that the initial temperature is T_0. Conversely, when $s \to 0$ for the final value theorem, we obtain

$$T(t \to \infty) = \frac{1}{k_e} P_H + T_R \tag{6.9}$$

which matches our time-domain result (Eq. (5.14)). For the dynamic response, and thus for the solution of the differential Eq. (6.1), we rearrange Eq. (6.6) to easily recognize the s-terms in our Laplace correspondence tables:

$$T(s) = P_H k_w \frac{1}{s(s + k_w k_e)} + T_R k_w k_e \frac{1}{s(s + k_w k_e)} + T_0 \frac{1}{s + k_w k_e} \tag{6.10}$$

[1] Another interpretation is that T_0 is valid only at the moment our experiment starts, i.e., at $t = 0$. In the time domain, this function is represented as $T_0 \delta(t)$, and its Laplace transform is T_0.

Now, we apply the correspondences

$$\frac{1}{s+a} \quad \bullet\!\!-\!\!\circ \quad e^{-at} \tag{6.11}$$

$$\frac{1}{s(s+a)} \quad \bullet\!\!-\!\!\circ \quad \frac{1}{a}\left(1-e^{-at}\right) \tag{6.12}$$

for the inverse transform of each summation term in Eq. (6.10)[2]:

$$\begin{aligned} T(t) = {} & \frac{P_H}{k_e}\left(1-e^{-k_w k_e t}\right) \\ & + T_R\left(1-e^{-k_w k_e t}\right) \\ & + T_0 e^{-k_w k_e t} \end{aligned} \tag{6.13}$$

The same time constant $1/k_w k_e$ appears in all exponential terms. This time constant is, in fact, directly related to the characteristic polynomial $s + k_w k_e$. Notice how this term appears in the denominators in Eq. (6.5). It implies that there is a location in the s-plane where the transfer functions go toward infinity, namely, at $s = -k_w k_e$. This location is called a *pole*. Any pole of the transfer function on the real axis of the s-plane has an exponential response. If a system has a pole at $-\sigma$, then the solution of the differential equation will have an exponential term that contains $e^{-\sigma t}$. Let us look at a few implications of this:

- The exponential term $e^{-\sigma t}$ vanishes for $t \to \infty$, and all solutions of the differential equation reach a steady state after a long time.
- A larger negative value of σ means that the exponential term vanishes faster. *If we could move a pole to the left—further away from the origin—we'd obtain a faster system response.*
- Poles on the right half-plane are problematic. For $\sigma > 0$, the exponential term is $e^{\sigma t}$. This term grows exponentially over time. Any system with poles in the right half-plane is unstable.

6.2 The Closed-Loop System

We will now examine the waterbath with a controller (Figure 6.1). For now, we will use a controller that amplifies the error signal ϵ by a factor k_p, with the resulting value being the heater power. k_p therefore has units of W/V. This controller type generates a corrective action that is proportional to the control error and is therefore referred to as *P* controller.

To simplify the math and to reduce the number of constants, we will assume a sensor with unit gain (i.e., $k_s = 1$), which implies that our setpoint is also interpreted as

[2]Important: You may perform the Laplace transform in either direction on each summation term and then superimpose the results (superposition principle). The same is *not* valid for multiplicative terms, because the transform of a multiplication is a convolution!

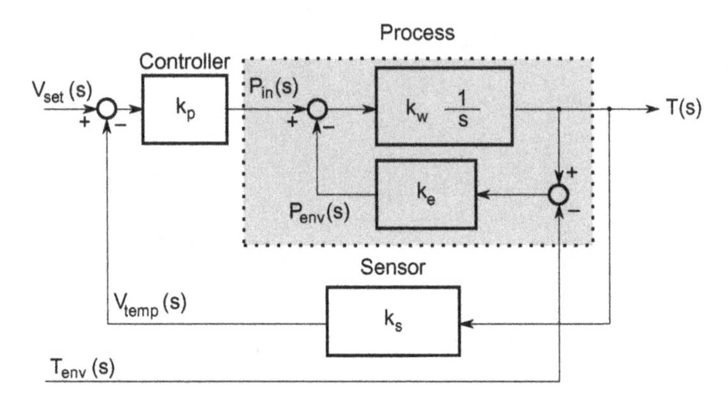

Figure 6.1 The waterbath with a *P* controller. The output temperature is monitored by a sensor with the sensor function k_s. An error signal is obtained as the difference between the setpoint and the sensor output. This error signal ϵ serves as the input for the controller.

temperature, T_{set}. Furthermore, we will ignore any initial condition (i.e., we arbitrarily set $T_0 = 0$. Strictly, this is valid only when the system is at steady-state near the setpoint), and we lose the ability to observe the behavior far from equilibrium.

Through loop elimination (explained in Chapter 7), we can rearrange the blocks in Figure 6.1 to yield the equivalent system in Figure 6.2. Through block diagram manipulation or straightforward algebraic solution of the loop relations, we then obtain the transfer function of the closed-loop system:

$$T(s) = \frac{k_w k_p}{s + k_w k_e + k_w k_p} T_{set}(s) + \frac{k_w k_e}{s + k_w k_e + k_w k_p} T_{env}(s) \tag{6.14}$$

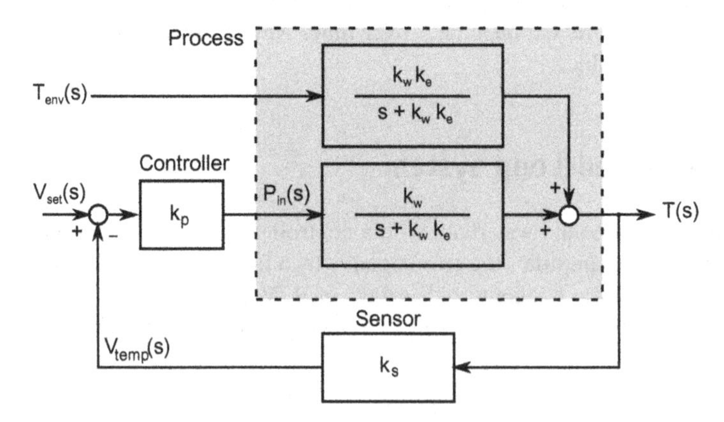

Figure 6.2 Simplified block diagram of the waterbath with a *P* controller. The process is highlighted by the gray rectangle. Note that we further simplify this arrangement with $k_s = 1$, which implies that the error signal is a temperature error rather than a voltage error, $V_{temp} \rightarrow T$, and $V_{set} \rightarrow T_{set}$.

The characteristic polynomial is now $s + k_w k_e + k_w k_p$. Since all system parameters are positive numbers, our pole at $s = -k_w k_e - k_w k_p$ has moved to the left compared to the open-loop system in Eq. (6.5). The system responds faster to changes in the environment and to the setpoint when k_p is large. Conversely, when $k_p \to 0$, we obtain the open-loop system with the same dynamics as in Eq. (6.5). Furthermore, we can approximate the system for very large k_p through Eq. (6.15):

$$T(s) \approx \frac{k_w k_p}{s + k_w k_p} T_{set}(s) + \frac{k_w k_e}{s + k_w k_p} T_{env}(s) \approx T_{set}(s) \tag{6.15}$$

Application of the final value theorem provides us with the equilibrium value T_∞ by using $T_{set}(s) = T_S/s$ and $T_{env}(s) = T_R/s$:

$$T_\infty = \frac{k_p}{k_e + k_p} T_S + \frac{k_e}{k_e + k_p} T_R \tag{6.16}$$

We can see from Eq. (6.16) that the environment temperature still influences the waterbath temperature, but the influence is attenuated by k_p. Let us define the steady-state tracking error $E_\infty = T_S - T_\infty$. Through substitution of T_∞ with Eq. (6.16), we obtain

$$E_\infty = \frac{k_e}{k_e + k_p} \left(T_S - T_R \right) \tag{6.17}$$

which clearly vanishes as $k_p \to \infty$. Very large k_p are realized in operational amplifiers. Unfortunately, most systems (heaters, motors, etc.) have a limited energy and the error amplification k_p is limited by practical considerations. In such cases, pure P control is associated with a significant steady-state tracking error, which may be unacceptable and calls for a different controller design.

6.3 Sensitivity and Tracking Error

Let us expand on the idea of the steady-state tracking error and gather some general observations. The closed-loop system in Figure 6.2 is shown in a more generalized form in Figure 6.3. A unit-gain feedback path is assumed, and the controller and process transfer functions are generalized as $H(s)$ and $G(s)$, respectively. We define the tracking error as the difference between setpoint and output signal:

$$E(s) = R(s) - Y(s) = \frac{1}{1 + G(s)H(s)} \cdot R(s) - \frac{G(s)}{1 + G(s)H(s)} \cdot D(s) \tag{6.18}$$

In this case, the tracking error is identical to the error deviation $\epsilon(s)$. A recurring term is the *loop gain* $L(s) = G(s)H(s)$, which is the product of all transfer functions along the (opened) loop. The loop gain determines the *sensitivity* of the system $S(s)$,

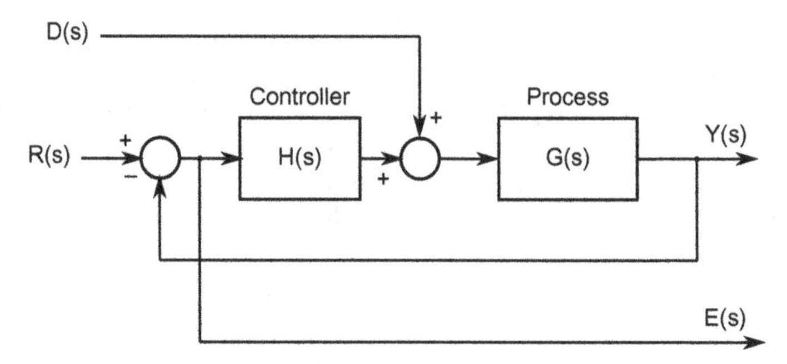

Figure 6.3 Generalization of the feedback system in Figure 6.2 where the environmental temperature acts as a disturbance $D(s)$ and the setpoint is designated $R(s)$. A unit-gain feedback path is assumed. The control deviation, that is, the difference between setpoint signal and output signal, is observed separately as tracking error $E(s)$.

defined as

$$S(s) = \frac{1}{1 + L(s)} \tag{6.19}$$

We can express the tracking error in terms of the sensitivity:

$$E(s) = S(s)R(s) - S(s)G(s)D(s) \tag{6.20}$$

In the example of P control, the sensitivity becomes $S(s) = 1/(1 + k_p G(s))$, and we can see that the sensitivity diminishes with increasing controller gain k_p. As evident from Eq. (6.20), the tracking error $E(s)$ also diminishes. In the above example (Eq. (6.14)), loop gain and sensitivity are

$$L(s) = k_p \frac{k_w}{s + k_w k_e}; \quad S(s) = \frac{s + k_w k_e}{s + k_w(k_p + k_e)} \tag{6.21}$$

Here, the tracking error becomes

$$E(s) = S(s)T_{set}(s) - S(s)G(s)k_e T_{env}(s) \tag{6.22}$$

and we can see that application of the final value theorem to Eq. (6.22) leads to the steady-state tracking error in Eq. (6.17): as k_p increases, so does the loop gain; the sensitivity diminishes, and with it the tracking error.

Interestingly, a large loop gain and the associated low sensitivity can also improve the robustness of a system against variations in the process itself. Equation (6.16) indicates that the controlled steady-state temperature depends on both the disturbance $T_{env}(s)$ and the quality of the insulation k_e. It is evident from Eq. (6.16) that a large k_p reduces the effect of k_e. More generally, we can examine the tracking error as the process experiences a change in one or more of its coefficients. Let us model this case as an additive function, that is, the process is now $G(s) + \Delta G(s)$. The tracking error experiences a similar change and becomes $E(s) + \Delta E(s)$. We can rewrite the setpoint component of Eq. (6.18) as

$$E(s) + \Delta E(s) = \frac{1}{1 + (G(s) + \Delta G(s))H(s)} \cdot R(s) \tag{6.23}$$

By substituting $E(s) = S(s)R(s)$ and isolating $\Delta E(s)$ under the assumption that $\Delta G(s) \ll G(s)$, we arrive at

$$\Delta E(s) = -S^2(s)H(s)\Delta G(s)R(s) \tag{6.24}$$

The influence of a changing process parameter on the tracking error is proportional to the square of the sensitivity and approximately inversely proportional to the square of the loop gain, and we can conclude that feedback control can make a closed-loop system more robust against changes within the process.

6.4 Using a *PI* Controller

We can improve (i.e., decrease) the steady-state tracking error by introducing an integrator in the controller. Combining a *P* controller with an integrator results in the *PI* (proportional-integral) controller. Reducing the tracking error becomes particularly important when the proportionality constant k_p of a *P* controller is limited by practical considerations. The output of an ideal integrator can only reach steady-state *if the input is zero*. Let us replace the k_p-block in Figure 6.2 by a *PI* controller, that is, a controller with the response function $PI(s)$ to an error input $\epsilon(s)$:

$$PI(s) = \left(k_p + k_I \cdot \frac{1}{s}\right) \cdot \epsilon(s) \tag{6.25}$$

The now-familiar solution approach provides us with the Laplace-domain temperature response $T(s)$:

$$\begin{aligned} T(s) = &\frac{k_w k_p s + k_w k_I}{s^2 + (k_w k_e + k_w k_p)s + k_w k_I} T_{set}(s) \\ &+ \frac{k_w k_e s}{s^2 + (k_w k_e + k_w k_p)s + k_w k_I} T_{env}(s) \end{aligned} \tag{6.26}$$

The system has two integrators (energy storage). The first integrator is the water itself, and the second integrator can be realized with some suitable circuitry that integrates the error signal. Consequently, the closed-loop system is a second-order system. This is evidenced by the occurrence of s^2 in the characteristic polynomial.

Does the *PI* controller keep its promise? For $s \to 0$, the transfer function for T_{env} vanishes, and the transfer function for T_{set} becomes unity. More formally, we obtain by applying the final value theorem to Eq. (6.26) and with functions for the step input signals:

$$
\begin{aligned}
T(t \to \infty) &= \lim_{s \to 0} s \cdot \left(\frac{k_w k_p s + k_w k_I}{s^2 + (k_w k_e + k_w k_p)s + k_w k_I} \frac{T_{set}}{s} \right. \\
&\quad \left. + \frac{k_w k_e s}{s^2 + (k_w k_e + k_w k_p)s + k_w k_I} \frac{T_{env}}{s} \right) \\
&= T_{set}
\end{aligned}
\tag{6.27}
$$

A *PI* controller *completely* suppresses the disturbance in this configuration and lets the steady-state tracking error E_∞ vanish.

Is there a price to pay for this behavior? Yes, the system can overshoot. The location of the poles can be determined by solving

$$
s^2 + (k_w k_e + k_w k_p)s + k_w k_I = 0 \tag{6.28}
$$

which provides us with the location of the poles $p_{1,2}$

$$
\begin{aligned}
p_{1,2} &= -\frac{k_w k_e + k_w k_p}{2} \pm \sqrt{\frac{(k_w k_e + k_w k_p)^2}{4} - k_w k_I} \\
&= -k_w \left(\frac{k_e + k_p}{2} \pm \sqrt{\frac{(k_e + k_p)^2}{4} - \frac{k_I}{k_w}} \right)
\end{aligned}
\tag{6.29}
$$

Overshoot occurs when the expression under the square root turns negative, and the two poles thus becomes a complex conjugate pair. We can see that a large k_p reduces the overshoot tendency, whereas a large k_I increases the tendency. The dynamics of the system can be optimized by carefully balancing k_p and k_I with the process, namely k_w and k_e.

How can a *PI* controller be realized? To control a process with a relatively fast response, specifically, systems with time constants in the millisecond range, a simple op-amp circuit can be used. This circuit is shown in Figure 6.4. By using the complex

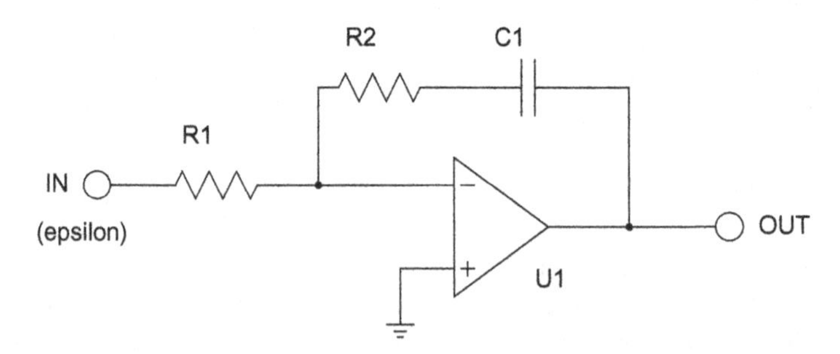

Figure 6.4 Simple *PI* controller realized with an op-amp. The external components determine $k_p = -R_2/R_1$ and $k_I = -1/(R_1 C_1)$. Component limitations restrict this circuit to processes with response times in the millisecond range. The error signal ϵ is fed into the input of the circuit, and the output signal V_{out} is proportional to the corrective action. Note the sign change!

impedance of the capacitor C_1, namely, $Z(C_1) = (sC)^{-1}$, we can obtain the transfer function

$$H_{PI}(s) = \frac{V_{out}(s)}{\epsilon(s)} = -\frac{R_2 + 1/(sC_1)}{R_1} = -\frac{R_2}{R_1} - \frac{1}{R_1 C_1 s} \tag{6.30}$$

from which we immediately obtain $k_p = -R_2/R_1$ and $k_I = -1/(R_1 C_1)$. The sign change is inherent in this circuit. Appropriate design of the error amplifier (the subtraction stage) can correct the sign.

For large time constants and for additional "tricks," such as preventing *integral windup*, *PI* controllers are often realized in software. Furthermore, integrated *PID* chips exist, such as the Maxim MAX1978 for *PID* control of a Peltier module. The *PID* controller is covered in detail in Chapter 13.

The example is expanded upon in Figure 6.5, which shows a possible realization of the entire control circuit. A resistive heating coil (RL) is used to provide heating power, and RL is switched by Q_1. To avoid losses in Q_1, it is used as a digital element (on or off), whose duty cycle determines the heating power. A simple pulse-width modulator is realized with an integrated timer chip (U_1, LM555) that converts an input voltage to a pulse-width modulated output voltage. The input to the pulse-width modulator is the analog signal for the corrective action.

The waterbath temperature is sensed by a temperature-dependent Zener diode (D_2, LM335). Together with R_7, the sensor ensures that the signal labeled Ⓐ is calibrated to 10 mV/K, that is, at room temperature, it carries a voltage of 2.93 V. A setpoint voltage with the same calibration needs to be provided, for example, through a voltage divider with potentiometer, and used as the input to the control system. Op-amp A_1 acts as difference amplifier and outputs the control deviation $\epsilon(t)$ at its output (labeled Ⓒ). Since the gain of the sensor is relatively low, A_1 can provide additional gain. When $R_1 = R_3$ and $R_2 = R_4$, the gain is $\alpha = R_1/R_2$.

Op-amp A_2 is the central element of the *PI* controller. It has the same transfer function as the circuit in Figure 6.4 (Eq. (6.30)), and the analog corrective action $V_c(t)$ (labeled Ⓒ) obeys the Laplace-domain relationship

$$V_c(s) = \left(\frac{R_6}{R_5} - \frac{1}{R_5 C_1 s}\right) \cdot \frac{R_1}{R_2} \cdot \left(V_{set}(s) - V_{sensor}(s)\right) \tag{6.31}$$

The pulse-width modulator is more complex, and slightly nonlinear. Diode D_1 prevents negative voltages at the output of A_2 that could destroy the timer circuit. Consequently, no heating takes place when $V_{set} < V_{sensor}$, which is desirable. The heating power in turn depends on the heater voltage V_{cc}, the on-time τ of transistor Q_1, and the period T of the timer chip,

$$P_H(s) = \frac{V_{cc}^2}{R_L} \cdot \frac{\tau}{T} \tag{6.32}$$

where τ increases monotonically with V_c, provided that $V_c > 0$. The LM555 chip is very popular, however, it has some nonlinear behavior and it saturates easily. As an alternative, specialized integrated PWM chips are available. Often, they already

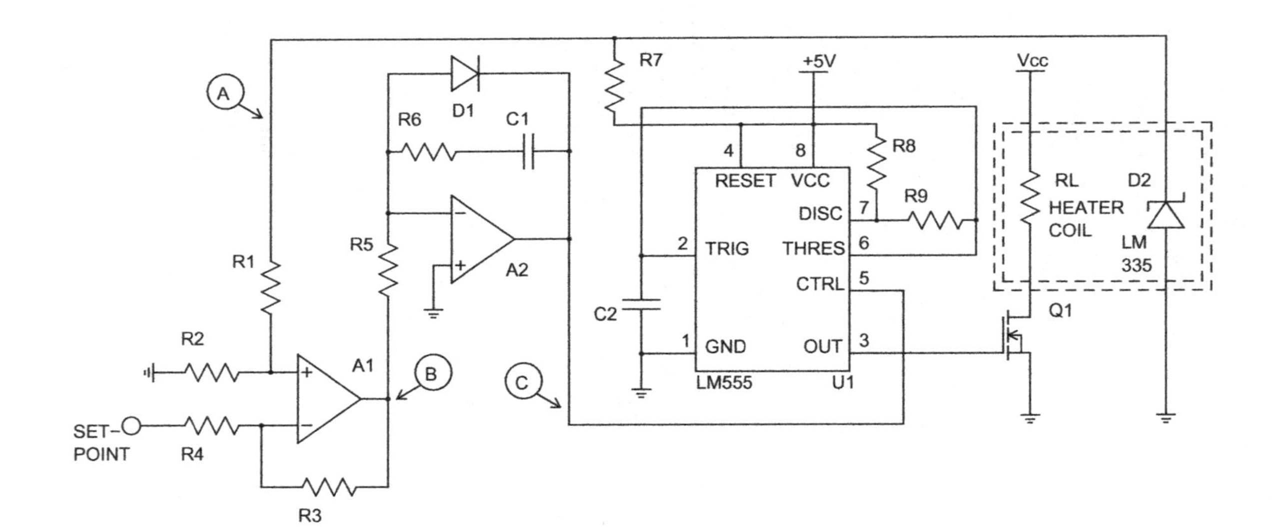

Figure 6.5 Circuit schematic of a complete temperature control unit for a waterbath with *PI* control. The process (i.e., the waterbath itself) is indicated by a double dashed box. Heating is provided by the current through RL. D_2 is a temperature-dependent Zener diode and acts as the sensor. The signal labeled Ⓐ is the temperature, scaled by 10 mV/K, and the setpoint input needs to carry a similarly scaled voltage. The difference amplifier around A_1 provides the error signal (labeled Ⓑ), but with a negative sign. The *PI* controller is built around A_2 as shown in Figure 6.4. The linear control action (labeled Ⓒ) is converted into a digital signal with a proportional pulse width, and Q_1 is used as switch to control the heater power.

include a P-type error amplifier, and some tweaks are required to achieve PI control. One example is the popular UC3842 PWM controller, which features oscillator, pulse generator, and error amplifier on one chip. The error amplifier is flexible enough to configure it as PI controller, thus allowing to replace A_2 in Figure 6.5. Alternatively, a microcontroller can read an analog input voltage and convert it into a proportional PWM signal. However, when a microcontroller is used anyway, it is tempting to integrate the PI control and the computation of the control deviation in software and dispense with the operational amplifiers entirely. This thought leads us to time-discrete control of the waterbath.

6.5 Time-Discrete Control

The waterbath is a good example where the process dynamics are very slow compared to the sampling time of most digital control systems. In the example, we found the time constant of the first-order process in the neighborhood of 20 min. Its corresponding pole in the s-plane is therefore very close to the origin, perhaps at $s = -1 \times 10^{-3}\ \text{s}^{-1}$. Any reasonable digital control system can be expected to process at least 10–100 samples per second, and we can build this example on a very conservative sampling rate of $T_s = 0.1$ s. To avoid confusion with the temperature, the sampling rate is denoted T_s in this section.

Figure 6.6 shows the closed-loop block diagram of the waterbath with digital control. To simplify the example, the sensor transfer function has been set to unity, and no initial conditions are being considered.

First, we examine the z-domain transfer function, beginning with the simple case of pure P control. The z-domain correspondence of the waterbath transfer function with zero-order hold is found in Table B.5. We obtain the open-loop z-domain transfer function

$$\begin{aligned} L(z) &= k_p k_w \left(1 - z^{-1}\right) \mathscr{Z}\left\{\frac{1}{s(s + k_w k_e)}\right\} \\ &= k_p k_w \frac{1 - e^{-k_T T_s}}{z - e^{-k_T T_s}} \end{aligned} \tag{6.33}$$

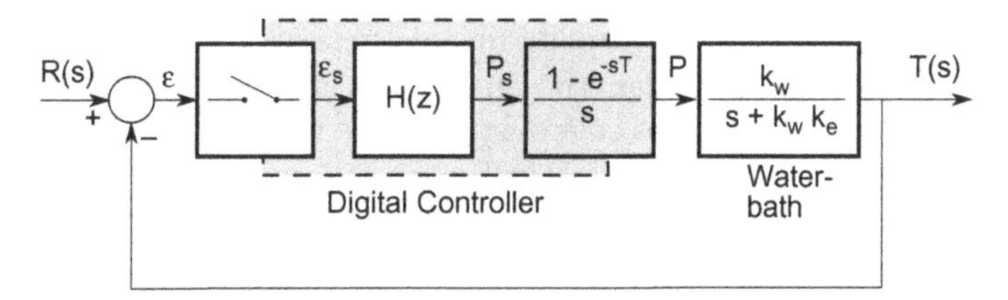

Figure 6.6 Simplified closed-loop block diagram of the waterbath with time-discrete control. In analogy to Figure 4.6, the block diagram is a mixed Laplace- and z-domain system, and suitable transformations are necessary to compute the transfer function in either of the domains.

where $k_T = k_w k_e$ is the location of the process pole in the left half-plane of the Laplace domain. The closed control loop causes the z-transformed waterbath temperature $T(z)$ to follow the input signal $R(z)$ as

$$T(z) = \frac{k_p k_w (1 - e^{-k_T T_s})}{z - e^{-k_T T_s} + (1 - e^{-k_T T_s}) k_w k_p} \cdot R(z) \tag{6.34}$$

Note that $e^{-k_T T_s} \approx 1$ due to the orders-of-magnitude difference between the sampling interval and the waterbath dynamics. With $T_s = 0.1$ s, $k_T = 10^{-3}$ s^{-1}, and $k_w = 0.001$ K/Ws, the numerator is approximately $10^{-7} k_p$. Intuitively, we can conclude that the change of the process output is very small between each sampling interval. This observation can further be confirmed by interpreting Eq. (6.34) as a digital filter and by computing the finite-difference response to an input sequence r_k:

$$\begin{aligned} T_k &= T_{k-1}\left(e^{-k_T T_s} - k_p k_w (1 - e^{-k_T T_s})\right) + r_{k-1} k_p k_w \left(1 - e^{-k_T T_s}\right) \\ &= E \cdot T_{k-1} + (r_{k-1} - T_{k-1}) \cdot k_w k_p \cdot (1 - E) \end{aligned} \tag{6.35}$$

In the second line of Eq. (6.35), $E = e^{-k_T T_s}$ was used. E is just smaller than 1, and $(1 - E)$ a small positive number. The second line of Eq. (6.35) therefore illustrates how the discretely sampled temperature T_k depends to more than 99.99% on the previously sampled temperature T_{k-1} and to less than 0.01% on the control deviation $r_{k-1} - T_{k-1}$.

By applying the same method, a digital *PI* controller can be employed. The starting point is Eq. (6.33), but the *P* controller gain k_p is replaced by the z-domain transfer function of the *PID* controller (Eq. (4.50)) with $k_D = 0$. The resulting open-loop transfer function is:

$$L(z) = \frac{z(k_p + k_I T_s)}{z - k_I} \cdot \frac{k_w(1 - e^{-k_T T_s})}{z - e^{-k_T T_s}} \tag{6.36}$$

The closed-loop transfer function for the setpoint (input signal) $R(z)$ and the process output $T(z)$ is therefore

$$T(z) = \frac{k_w(1 - E)(k_p + k_i T_s) z}{z^2 + z[k_w k_p (1 - E) - k_w k_I T_s (1 - E) - k_I - E] + k_I E} \cdot R(z) \tag{6.37}$$

where the same shorthand notation $E = e^{-k_T T_s}$ has been used. It is interesting to note at this point that the stability behavior of the time-discrete system differs fundamentally from that of the continuous system. The denominator in Eq. (6.37) has two roots in the z-plane. Unlike the continuous control system (Eq. (6.29)), where no roots of the transfer function can be found in the right half-plane, the time-discrete system can have roots outside the unit circle, notably for large k_p or k_I.

6.5.1 Time-Discrete Control with the Bilinear Transform

The orders-of-magnitude difference between process time constant and sampling period can be used to justify the conversion of the time-discrete controller function to the

Laplace-domain with the help of the bilinear transform. We recall that the bilinear transform provided us with a relationship between s- and z-domains through

$$s = \frac{2}{T} \cdot \frac{z-1}{z+1}; \quad z = \frac{2+sT}{2-sT} \tag{6.38}$$

When we use the z-domain transfer function of the *PID* controller (Eq. (4.55)) and set $k_D = 0$, the z-domain transfer function of the *PI* controller becomes

$$H_{PI}(z) = \frac{(2k_p + k_I T)z - (2k_p - k_I T)}{2(z-1)} \tag{6.39}$$

We can substitute z with Eq. (6.38) to obtain the s-domain approximation of the controller:

$$H_{PI}(s) \approx \frac{(2k_p + k_I T)(2+sT) - (2k_p - k_I T)(2-sT)}{2(2+sT) - 2(2-sT)} = \frac{k_p s + k_I}{s} \tag{6.40}$$

This result should come as no surprise since we used the trapezoidal discrete integration rule as our basis for developing the bilinear transform. Equation (4.55) is also based on the trapezoidal rule. The simpler Eq. (4.50) leads to an s-domain denominator polynomial of $sT(1+k_I) + 2 - 2k_I$, and the bilinear transform leads to an s-domain approximation that no longer acts as an integrator unless $k_I = 1$. This example is particularly well suited to highlight the shortcomings of the bilinear transform.

Often, it is advisable to approach the problem from the other direction and begin with a desired s-domain controller function. With Eq. (6.38), the z-domain approximation is computed, which in turn leads to the digital filter that is used as the time-discrete controller.

7 Block Diagrams: Formal Graphical Description of Linear Systems

Abstract

The first part of the book focuses on the mathematical foundations of linear feedback control systems. This chapter, together with most of the subsequent chapters, add to the design engineer's toolbox for control systems. Many analysis and design problems can be treated by following specific recipes, much like a cookbook. In this chapter, the formal description of linear systems with block diagrams is covered. Block diagrams represent a linear system as a block with a transfer function, an input and an output signal. In the Laplace- and z-domains, the output signal is obtained by multiplying the input signal with the transfer function. Linear systems, including feedback control systems, can be represented by block diagrams, much like an electronic circuit diagram. As an alternative to the algebraic solution, rules for block diagram manipulation and simplification are introduced in this chapter. Signal flow graphs, which are a parallel representation, are also briefly introduced.

We have been using block diagrams to graphically represent functional blocks of a system. It is now necessary to formalize these block diagrams and introduce rules to manipulate and simplify them. Block diagram simplification provides a convenient alternative to algebraically resolving a loop equation. In the context of this chapter, we assume all block diagrams to be treated in the Laplace-domain. The same rules apply to block diagrams in the z-domain. Therefore, all rules described in this chapter can be used for the z-transforms of time-discrete systems.

Note that there is a parallel graphical representation, the *signal flow graph*. Signal flow graphs and block diagrams convey the same information, just in a different manner. For both, the same rules apply. For reasons of uniformity, we will focus on block diagrams and only briefly introduce signal flow graphs.

7.1 Symbols of a Block Diagram

Block diagrams are composed of very few functional elements. These are shown in Figure 7.1. Linear, time-invariant systems only need the summation operation and the multiplication, either with a constant or with a transfer function, i.e., a function of s. For convenience, a take-off point is also defined where the same signal is routed to two different destinations. All signals (lines) have arrows to indicate the direction of signal flow. These three elements are sufficient to graphically describe any system that follows the general differential equation (Eq. (2.1) in Chapter 2) and, more specifically,

Linear Feedback Controls. http://dx.doi.org/10.1016/B978-0-12-405875-0.00007-3
© 2013 Elsevier Inc. All rights reserved.

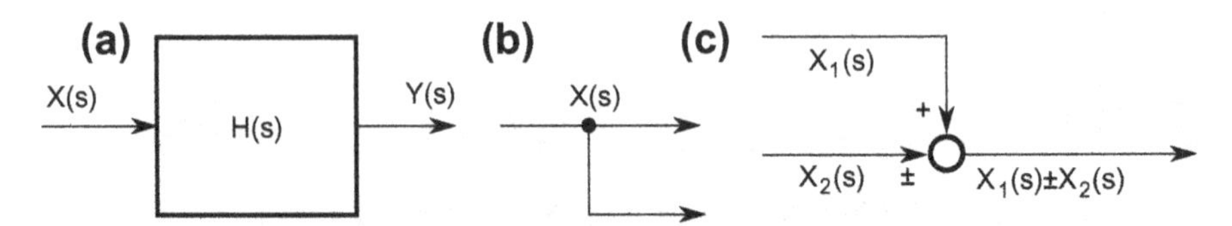

Figure 7.1 Basic building blocks of block diagrams. (a) A box indicates a linear, time-invariant system with the transfer function $H(s)$. In this example, the input and output signals are labeled $X(s)$ and $Y(s)$, respectively, and $Y(s) = H(s) \cdot X(s)$. (b) A branchoff point splits a signal and routes it to two different destinations. Every link of the branchoff point carries the signal $X(s)$. (c) Summation point. Two or more signals are added at this point. Signs near the arrows indicate whether the signal is being added or subtracted. Summation points may have more than two inputs.

its Laplace transform where the derivatives with respect to time can be described as a polynomial of s.

7.2 Block Diagram Manipulation

We have seen that we can describe linear systems graphically with only three elements: the addition, the multiplication with a complex transfer function, and the take-off point, where a signal is routed to multiple nodes. A block diagram can provide much detail with blocks representing simple functions, or it can be less detailed, with more complex functions inside the blocks. Ultimately, any system with one input and one output can be represented by one single block that contains its transfer function. A number of rules exist to transform block diagrams. The goal of each of these rules is to apply a small change to the graph and still obtain the same output signal. Some rules are rather intuitive. For example, two consecutive addition points can be flipped (commutativity). Two chained blocks with the transfer functions $G(s)$ and $H(s)$ can be combined into one single block with the transfer function $G(s) \cdot H(s)$, provided that no take-off point or summation point exists between the two blocks (Figure 7.2b), and two parallel blocks can be combined by addition (Figure 7.2a). The three most important rules are moving a summation point to the other side of a block (Figure 7.3), moving a take-off point to the other side of a block (Figure 7.4), and loop elimination (Figure 7.5).

Proof for the loop elimination rule can readily be provided. The feedback signal (i.e, the output of the feedback block) is $H(s)Y(s)$. Let us denote the signal after the summation point $\epsilon(s)$. We find two equations, one for $\epsilon(s)$ and one for $Y(s)$:

$$\begin{aligned} Y(s) &= G(s) \cdot \epsilon(s) \\ \epsilon(s) &= X(s) \pm H(s) \cdot Y(s) \end{aligned} \tag{7.1}$$

The symbol $\pm$ indicates alternative positive or negative feedback. We eliminate the dependent variable $\epsilon(s)$ by substitution and obtain

$$Y(s) = G(s) \cdot \big(X(s) \pm H(s) \cdot Y(s)\big). \tag{7.2}$$

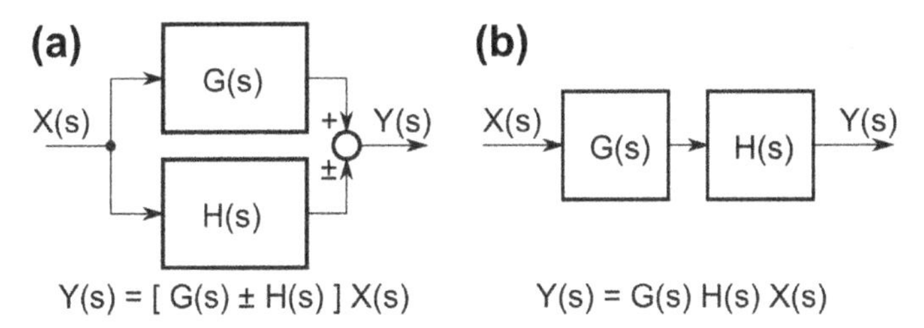

Figure 7.2 Rules to join parallel blocks (a) and serial blocks (b). Two parallel blocks, whose outputs are added, can be represented by a single transfer function that is the sum of the individual transfer functions. In this example, the two blocks in (a) have the overall transfer function of $G(s) \pm H(s)$. Two serial blocks have a transfer function that is the product of the individual transfer functions (in this case, $G(s) \cdot H(s)$), provided that no summation or take-off point exists between the two chained blocks.

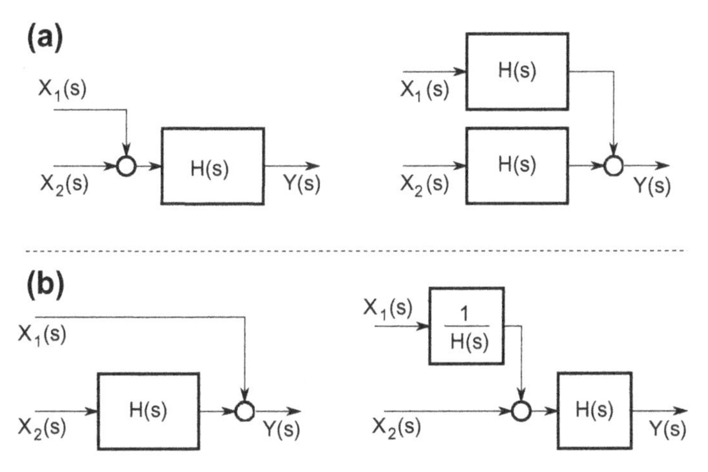

Figure 7.3 Rules to move a summation point (a) to the right of a block, and (b) to the left of a block. The inputs to the summation may carry any sign, and the sign does not change.

The equation needs to be rearranged to collect the output variable on the left-hand side and the independent variables on the right-hand side:

$$Y(s) \mp G(s) \cdot H(s) \cdot Y(s) = G(s) \cdot X(s) \tag{7.3}$$

We now factor out $Y(s)$ on the left-hand side and rearrange the terms to obtain the transfer function of the combined block:

$$\frac{Y(s)}{X(s)} = \frac{G(s)}{1 \mp G(s) \cdot H(s)} \tag{7.4}$$

which is the transfer function indicated in Figure 7.5. A negative feedback loop creates a positive sign in the denominator of the transfer function and *vice versa*.

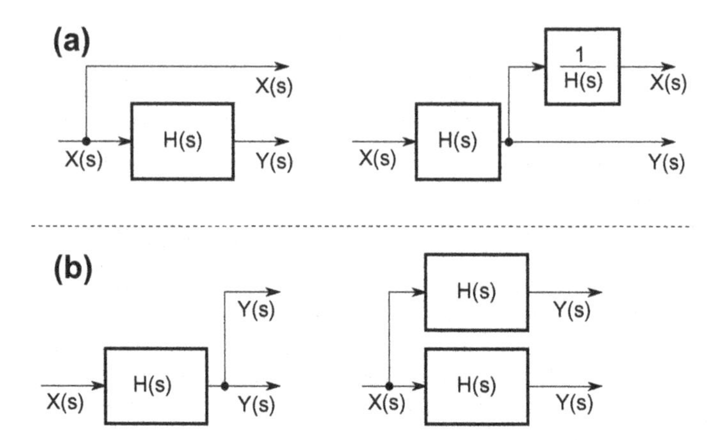

Figure 7.4 Rules to move a take-off point (a) to the right of a block, and (b) to the left of a block.

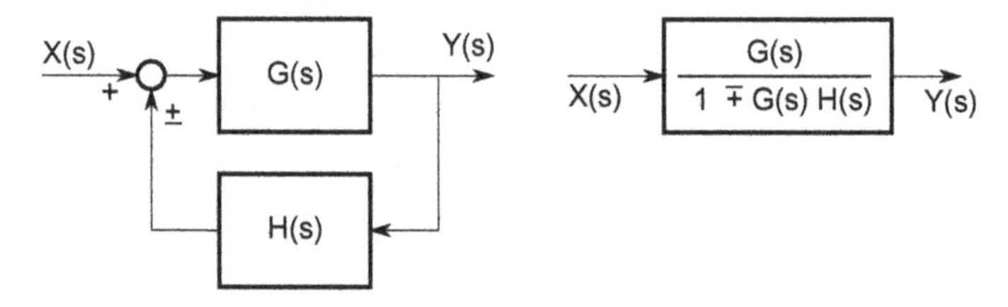

Figure 7.5 Loop elimination rule. This rule allows to reduce a feedback loop into a single block. Note the sign change from the feedback path to the final transfer function.

The overall strategy of block diagram simplification (often with the goal to reduce a complex block diagram to a single block) is to move summation and take-off points in such a way that it creates feedback loops. With the third rule, those feedback loops are then reduced to single blocks. Any two chained blocks can now be merged. Then the process is repeated.

7.3 Block Diagram Simplification Examples

A simple example for block diagram manipulation is given in Figure 7.6. The example shows a frequently occurring feedback loop to control a process $G(s)$ with a controller $H(s)$ in the presence of a disturbance $D(s)$. The setpoint is $X(s)$, and the sensor in the feedback path has been normalized to have a unit transfer function.

The example can be solved with arithmetic. A good start is to use $\epsilon(s)$ as the input to the controller with $\epsilon(s) = X(s) - Y(s)$. The upper path yields the expression for $Y(s)$,

$$\begin{aligned} Y(s) &= G(s) \cdot \big(D(s) + H(s) \cdot \epsilon(s)\big) \\ &= G(s) \cdot D(s) + G(s) \cdot H(s) \cdot X(s) - G(s) \cdot H(s) \cdot Y(s) \end{aligned} \tag{7.5}$$

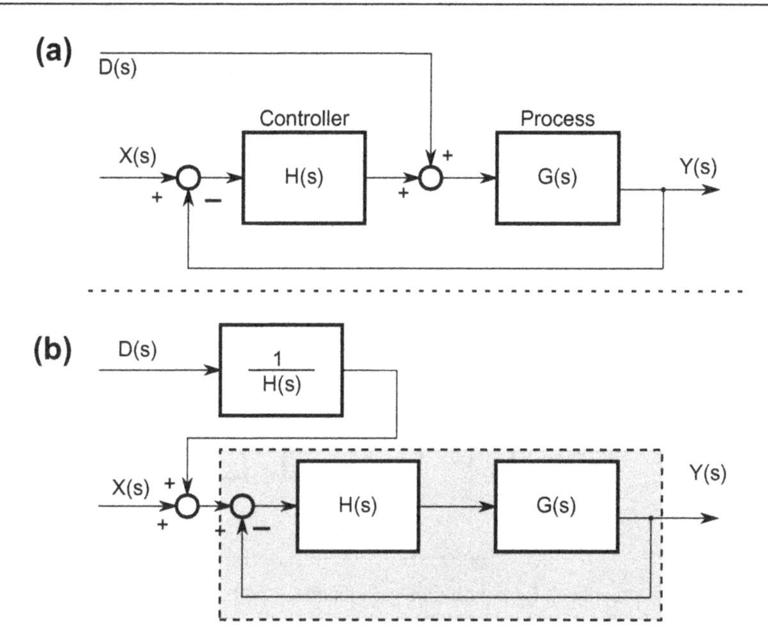

Figure 7.6 A frequently occurring feedback loop (a) with process $G(s)$, controller $H(s)$, and unit-gain feedback path can be manipulated to quickly provide the transfer functions for $D(s)$ and $X(s)$. The disturbance summation point is moved to the left of the controller. Since the addition is commutative, the neighboring summation points can be exchanged, revealing a pure feedback loop (shaded area). The feedback loop can now be eliminated (Figure 7.5).

By collecting all terms with the controlled variable $Y(s)$ on the left-hand side of the equation and factoring out $Y(s)$, the equation can be divided by the term $1+G(s)\cdot H(s)$, and $Y(s)$ emerges as the sum of two signals, multiplied with their respective transfer functions:

$$Y(s) = \frac{G(s)\cdot H(s)}{1+G(s)\cdot H(s)}\cdot X(s) + \frac{G(s)}{1+G(s)\cdot H(s)}\cdot D(s). \tag{7.6}$$

The same result can be achieved by moving the disturbance summation point out of the feedback loop (Figure 7.6). The summation point is moved across the controller to the left. Because the order of a summation does not matter, the summation point can be moved even further to the left of the feedback summation point. This leaves a pure feedback loop that can be eliminated with the loop elimination rule (gray shaded area in Figure 7.6), and the feedback loop can be seen to have the transfer function $G(s)\cdot H(s)/(1+G(s)\cdot H(s))$. Both the setpoint path and the disturbance path pass through the new simplified block and are superimposed to yield $Y(s)$:

$$Y(s) = \frac{G(s)\cdot H(s)}{1+G(s)\cdot H(s)}\cdot X(s) + \frac{1}{H(s)}\cdot\frac{G(s)\cdot H(s)}{1+G(s)\cdot H(s)}\cdot D(s). \tag{7.7}$$

In the disturbance term, $H(s)$ cancels out, and Eq. (7.6) emerges.

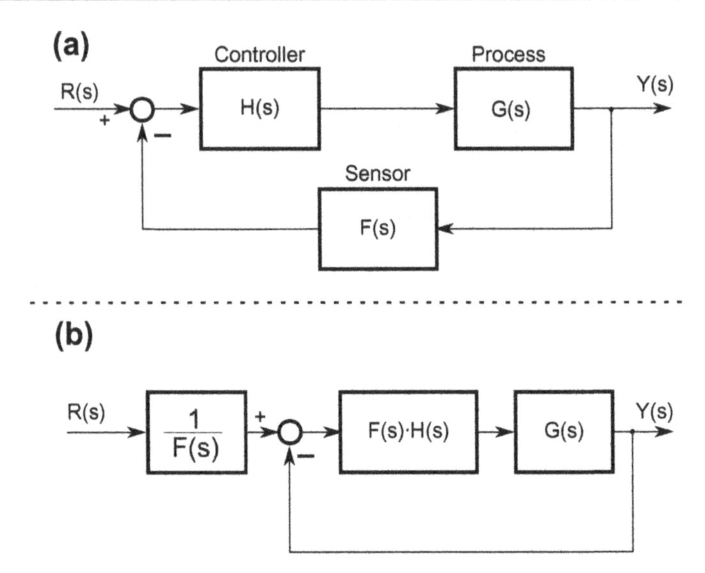

Figure 7.7 Creating a feedback loop with unit-gain feedback path. (a) A general feedback loop with a sensor in the feedback path that has the transfer function $F(s)$. (b) The block with $F(s)$ can be moved across the summation point into the feedforward path, where the transfer function $F(s)$ can be combined with the controller transfer function $H(s)$. The original control deviation $\epsilon(s)$ is no longer directly accessible, and the signal after the summation point is $\epsilon(s)/F(s)$. Consequently, the setpoint signal $R(s)$ needs to be subjected to $1/F(s)$ as well.

The second example for block diagram manipulation follows a similar idea. In many examples, the transfer function of the feedback path is somewhat arbitrarily set to unity (such as in the previous example). In feedback control loops where the feedback path has a nonunity transfer function, for example, $F(s)$ in Figure 7.7a, the loop element $F(s)$ can be moved to the feedforward path as shown in Figure 7.7b. The original control deviation $\epsilon(s)$, defined as the difference between setpoint $R(s)$ and the sensor output, is no longer accessible, and the signal after the summation point in Figure 7.7b is $\epsilon(s)/F(s)$. The original setpoint needs to be multiplied with $1/F(s)$.

An example is the waterbath sensor in Chapter 6, which provides a voltage of 0.1 V/°C. In the block diagram model, the summation point now carries the temperature signal instead of a voltage. Since the setpoint is a voltage, it needs to be scaled by the reciprocal (10°C/V) for the units and the magnitude to match. In a second simplification step, the setpoint can be directly interpreted as a temperature, for example, by labeling a potentiometer with a temperature scale.

We will now illustrate more complex block diagram manipulation and simplification steps with the example given in Figure 7.8. The goal is to (1) obtain a canonical feedback configuration (Figure 1.3) with one feedforward and one feedback path, and (2) obtain the overall transfer function $Y(s)/X(s)$. The individual steps are explained in Figures 7.8–7.12. To improve readability, the transfer functions are shown as F_k in place of the more complete $F_k(s)$.

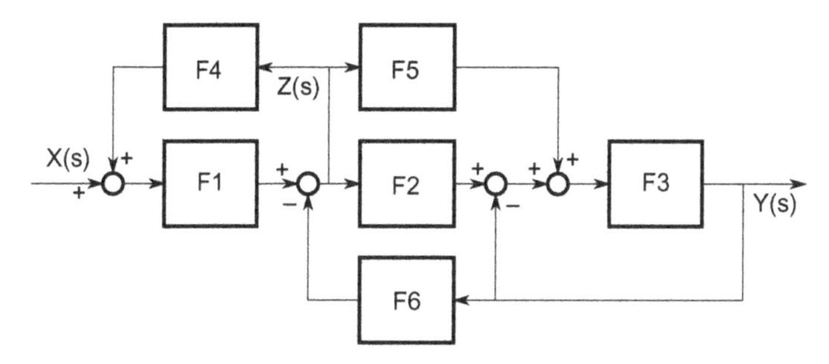

Figure 7.8 A block diagram with multiple nested feedback loops needs to be analyzed. There are no straightforward simplification steps possible, therefore we "untangle" the rightmost loop by exchanging the two consecutive summation points.

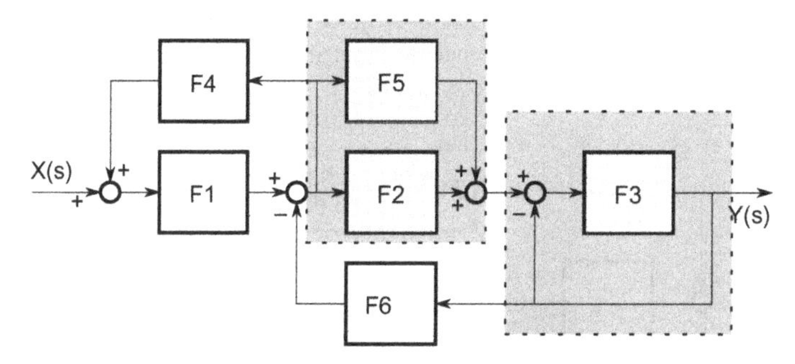

Figure 7.9 Exchanging two summation points has created two parallel blocks ($F_2(s)$ and $F_5(s)$) and one feedback loop with $F_3(s)$, which are highlighted with gray rectangles. Both can now be eliminated.

We can see from Figure 7.12 that the system consists of the controller $F_1(s)$ and the process $F_C(s)$ in the feedforward path and $F_B(s)$ in the feedback path. The feedback loop is positive. In terms of the initial transfer functions, $F_B(s)$ and $F_C(s)$ are

$$\begin{aligned} F_B(s) &= \frac{F_4(s)\cdot(1+F_3(s))}{F_3(s)\cdot(F_2(s)+F_5(s))} \\ F_C(s) &= \frac{F_3(s)\cdot(F_2(s)+F_5(s))}{1+F_3(s)+F_6(s)F_3(s)(F_2(s)+F_5(s))}. \end{aligned} \tag{7.8}$$

We can finally compute the overall transfer function by loop elimination from Figure 7.12,

$$\begin{aligned} H(s) = \frac{Y(s)}{X(s)} &= \frac{F_1(s)\cdot F_C(s)}{1-F_1(s)\cdot F_B(s)\cdot F_C(s)} \\ &= \frac{F_1(s)F_3(s)(F_2(s)+F_5(s))}{1+F_3(s)+F_6(s)F_3(s)(F_2(s)+F_5(s))-F_1(s)F_4(s)(1+F_3(s))}. \end{aligned} \tag{7.9}$$

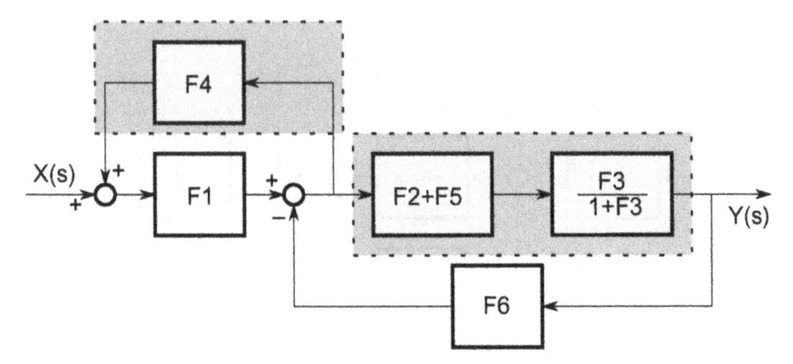

Figure 7.10 The parallel block created one new block with the transfer function $F_2(s) + F_5(s)$, and the feedback loop is now represented by a single block with the transfer function $F_3(s)/(1 + F_3(s))$. These two blocks are now chained and can be joined by multiplying the transfer functions (gray rectangle). By moving the take-off point to the input of F4, we can isolate F6 in a pure feedback loop. Note that it is not possible to simply move the take-off point that feeds F4 in front of the summation, because this would change the input signal to F4.

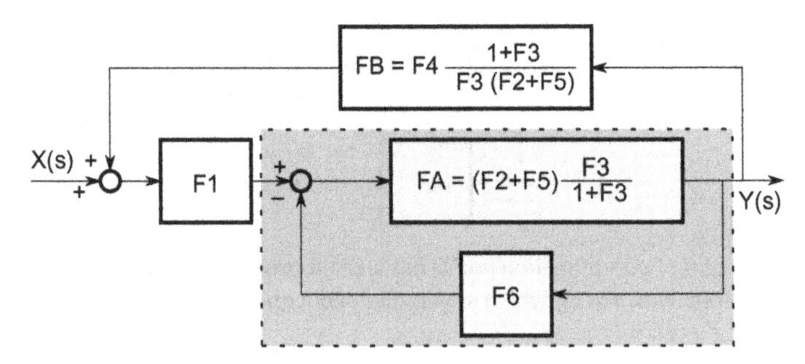

Figure 7.11 To improve readability, the chained block from Figure 7.10 is now abbreviated $F_A(s)$. Multiplying its reciprocal with $F_4(s)$ creates a new block, $F_B(s)$, with $Y(s)$ as its input. Therefore, the take-off point must be moved to the right of the block $F_A(s)$. Moving the take-off point creates a feedback loop with $F_A(s)$ and $F_6(s)$ (highlighted with gray rectangle).

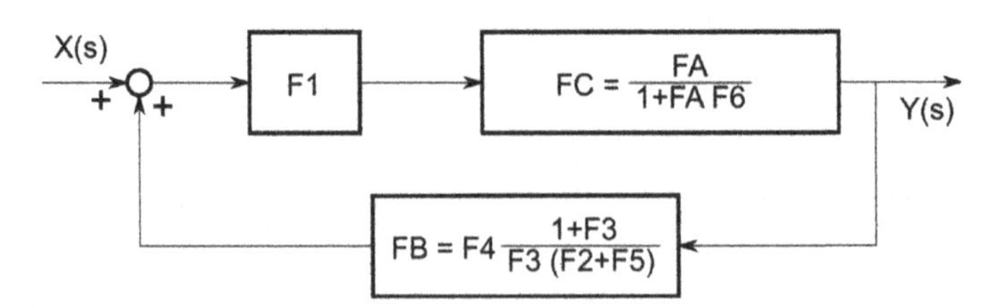

Figure 7.12 Eliminating the inner feedback loop in Figure 7.11 creates a system with one single loop in the canonical feedback configuration. One more application of the loop elimination rule yields the overall transfer function.

Note that we can arrive at the same end result by applying straightforward arithmetic. To illustrate this path, let us refer back to Figure 7.8. We denote the input to the block $F_2(s)$ as $Z(s)$. The input to $F_1(s)$ is therefore $X(s) + F_4(s) \cdot Z(s)$, and we can describe $Z(s)$ with

$$Z(s) = F_1(s) \cdot \left(X(s) + F_4(s) \cdot Z(s)\right) - F_6(s) \cdot Y(s), \tag{7.10}$$

which requires us collect all terms with $Z(s)$ on the left-hand side of Eq. (7.10) and all other terms on the right-hand side. With known $Z(s)$, we can describe $Y(s)$ through

$$Y(s) = F_3(s) \cdot \left(F_2(s) \cdot Z(s) + F_5(s) \cdot Z(s) - Y(s)\right). \tag{7.11}$$

By substituting $Z(s)$ from Eq. (7.10) and rearranging all terms with the dependent variable ($Y(s)$) on the left-hand side and all terms with $X(s)$ on the right-hand side (note that we no longer have any other dependent variables), we arrive at Eq. (7.9).

7.4 Signal Flow Graphs

Block diagrams are related to electrical circuit diagrams, where each line carries a time-variable signal, and each block serves as a component or building block. Conversely, signal flow graphs represent the building blocks (i.e., the transfer functions) as branches, whereby the nodes represent the input and output signals, respectively. Multiple signals leading into the same node are implicitly added, and multiple signals emerging from the same node carry the same signal. Figure 7.13 shows the basic elements in analogy to Figure 7.1. To create identical functions in Figures 7.13b and 7.1b, the branch transfer functions $G(s)$ and $H(s)$ would be unity transfer functions (likewise for Figures 7.13c and 7.1c).

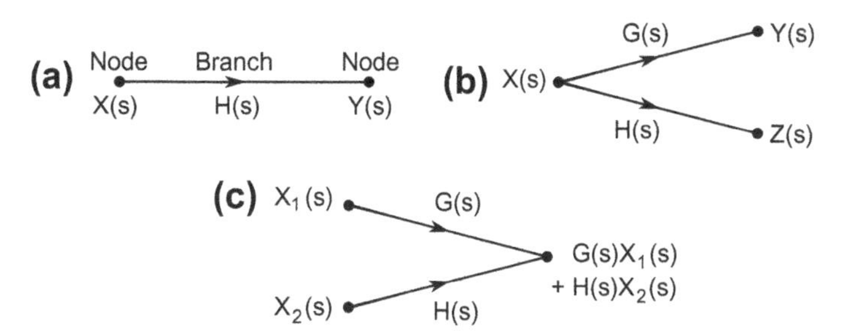

Figure 7.13 Basic building blocks of signal flow graphs. (a) A branch indicates a linear, time-invariant system with the transfer function $H(s)$. In this example, the input and output nodes are labeled $X(s)$ and $Y(s)$, respectively, and $Y(s) = H(s) \cdot X(s)$. Note that the arrow in the middle of the branch indicates the signal flow from the input node to the output node. (b) Signals emerging from one node are the same. The branches may have different transfer functions, though, and in this example $Y(s) = G(s) \cdot X(s)$ and $Z(s) = H(s) \cdot X(s)$. (c) Signals leading into the same node are implicitly added, after being multiplied by the transfer functions of the feeding branches.

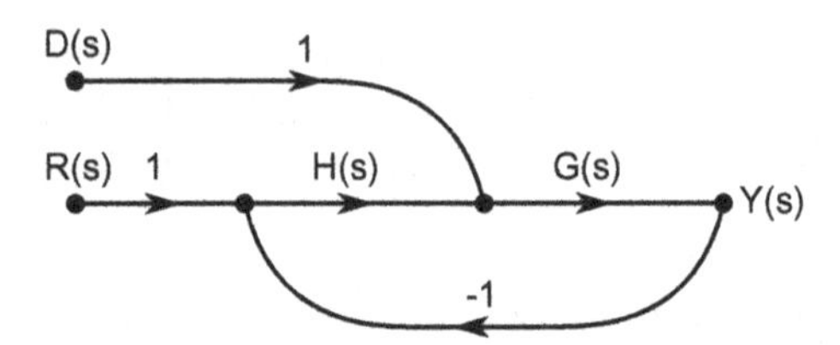

Figure 7.14 Example for a signal flow graph. The system shown here corresponds to the block diagram in Figure 7.6a. Note that the summation point with one negative input has been replaced by a branch with the transfer function -1, since difference operations are not possible in signal flow graphs.

The same simplification rules apply for block diagrams and for signal flow graphs. However, it is common practice to number the nodes (e.g., x_1, x_2, ...) and to label the transfer functions according to their nodes. For example the branch A_{ij} takes the signal x_i as input and generates the output signal x_j. It can be seen that signal flow graphs are more amenable to matrix representations and thus to state variable representations of linear systems.

To provide one example for a complete signal flow graph, consider Figure 7.14. The system depicted in Figure 7.14 is the same as the one in Figure 7.6a, which is its block diagram equivalent. Since signal flow graphs do not have a defined subtraction operation, a sign change in the feedback path is necessary.

8 Linearization of Nonlinear Components

Abstract

The assumption of a linear system allows to apply the elegant solution methods in the Laplace- and z-domain. However, in many instances, system components do not exhibit linear behavior. When a feedback control system is in operation and keeps the system near a defined operating point, a linear approximation for nonlinear components in the operating point can be derived. In this chapter, three approaches to linearize a component are presented: first, when the nonlinear characteristic equation of that component is known, second, when the characteristic of the component is provided as tabular data, and third, when the characteristic of the component is present only in graphical form, for example, a figure in a datasheet. Furthermore, this chapter covers saturation as a very common occurrence of nonlinear behavior and explains why a feedback system where a component saturates behaves like an open-loop system without control. Linearization of nonlinear components is another tool in the design engineer's toolbox.

Sometimes, components of a process or a controller have a nonlinear relationship between the input and output signal. Some examples are:

- Valves for flow control where flow increases rapidly when the valve is just opened, and only small changes in flow rate are seen when the valve is almost fully open.
- Nonlinear friction elements, particularly when the transition between static friction and dynamic friction is considered.
- A pendulum where the force that pulls the pendulum to the center position is proportional to $\cos\theta$.
- Semiconductors with exponential voltage-current characteristics.

Linearization of a nonlinear component is based on two assumptions:

1. Feedback control exists and keeps your system operating near the operating point with only small deviations from the operating point.
2. The curve of the nonlinear element is continuous near the operating point.

We need to distinguish between three cases. In the first case, the curve is known in an analytical fashion, that is, the equation that relates output and input signals is known. In the second case, some measured data pairs of the output signal at some discrete input signal are available in tabular form. In the third case, the curve is available as a diagram, for example, provided in a datasheet.

In all cases, the goal is to determine the slope of the nonlinear element near the operating point. The operating point needs to be determined or estimated in advance. Parameters obtained through linearization are only valid for small deviations from the

Linear Feedback Controls. http://dx.doi.org/10.1016/B978-0-12-405875-0.00008-5

© 2013 Elsevier Inc. All rights reserved.

operating point, because the slope—and with it the loop gain—may vary significantly over the range of the nonlinear component. We need to keep in mind that major changes of the setpoint may cause major changes in the loop gain and therefore in the closed-loop system itself. If we operate a system near the limits of stability, such a change in the operating point may cause undesirable responses and even instability.

8.1 Linearization of Components with Analytical Description

As an example, let us consider the power dissipation of a resistor (or a resistive load, such as heating coils). When a voltage V is applied, the resistor R dissipates a power of

$$P = \frac{V^2}{R} \tag{8.1}$$

Let us further assume that we know the operating point from analyzing the system, and that the feedback control system requires power dissipation of approximately $P_0 = 3$ W at equilibrium near the operating point. This requires approximately $V_0 = 2.6$ V across a given resistor of 2.2 Ω. A graphical representation of the nonlinear curve (Eq. (8.1)) is shown in Figure 8.1.

To linearize the behavior of the resistive heater, we try to approximate Eq. (8.1) with a linear equation,

$$P = m \cdot V + P_I|_{V \approx V_0} \tag{8.2}$$

where m is the slope of Eq. (8.1) in the operating point (dashed line in Figure 8.1). The offset P_I is chosen such that Eq. (8.2) meets the nonlinear curve in the operating

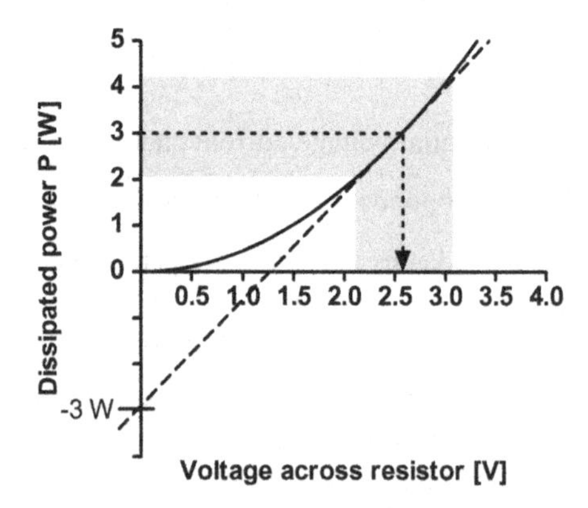

Figure 8.1 A resistive heating coil described by its nonlinear power dissipation curve. The dotted line with the arrow indicates the operating point, and the gray shaded region indicates the allowable deviation from the operating point. The dashed line is the tangent of the curve in the operating point, and its slope is the linearized gain of the nonlinear element. It can be seen that the tangential line does not deviate much from the original curve in this range.

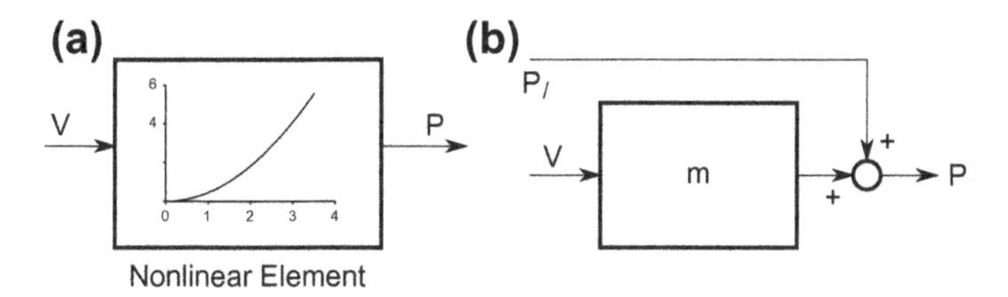

Figure 8.2 Block representation of a nonlinear element (a) (the nonlinearity is indicated by the sketched curve) and the linearized equivalent (b). In the operating point, the linearized element behaves like a gain block with gain m and an additive component P_I from the intercept with the vertical axis (Figure 8.1 and Eq. (8.2)).

point, that is, $P_0 = m \cdot V_0 + P_I$. Equation (8.2) is therefore the linear approximation of Eq. (8.1) in the operating point. We can analytically find the slope m as

$$m = \left.\frac{\mathrm{d}P}{\mathrm{d}V}\right|_{V=2.6\,\mathrm{V}} = \left.\frac{2V}{R}\right|_{V=2.6\,\mathrm{V}} \approx 2.34\,\frac{\mathrm{W}}{\mathrm{V}} \tag{8.3}$$

and the intercept of the dashed line with the vertical axis is $P_I = -P_0 = -3\,\mathrm{W}$. We can now describe small deviations from the operating point as

$$P = 2.34\,\frac{\mathrm{W}}{\mathrm{V}} \cdot V - 3\,\mathrm{W} \tag{8.4}$$

The offset (in this case $P_I = -3\,\mathrm{W}$) can be treated as an additive constant signal that gets added to the output of the linearized element (Figure 8.2). When we only consider the *deviations* from the operating point ΔU and ΔP, the element becomes a gain block,

$$\Delta P = 2.34\,\frac{\mathrm{W}}{\mathrm{V}} \cdot \Delta V \tag{8.5}$$

Note that the values in Eqs. (8.4) and (8.5) are valid only near the previously-defined operating point. If we need to operate the system with a different setpoint, the gain factor m is likely to change. In this example, if our operating point changes to $P = 1$ W, an input voltage of 1.5 V is required, and the slope becomes $m = 1.36$ W/V. Since m is usually part of a feedback loop and thus contributes to the overall loop gain $L(s)$, the loop gain in this example would have been reduced to half the value of the previous setpoint; sensitivity would have doubled. Such a major change of the loop gain has a strong influence on both static and dynamic behavior of a feedback control system.

Linearization introduces a small error. The error grows with the magnitude of the deviation from the setpoint. The maximum error is simply the deviation of the linearized component from the actual component when the largest deviation from the operating point is assumed. In the example above, let us assume that the input voltage can deviate by $\pm 0.5\,\mathrm{V}$ from the setpoint $V_0 = 2.6\,\mathrm{V}$ (gray shaded region in Figure 8.1). The linearized system yields $\Delta P = 1.2$ W for $\Delta V = 0.5\,\mathrm{V}$. A comparison of the power dissipation computed from the nonlinear equation (Eq. 8.1) to the one computed from

Table 8.1 Linearization error of the resistive heater in the range 2.6 ± 0.5 V. The second column was calculated with Eq. (8.1), and the third column with Eq. (8.4). The error in the operating point itself (second row) should be zero due to the definition of Eq. (8.2).

Input Voltage (V)	Power (exact) (W)	Power (linearized) (W)	Error(%)
2.1	2.0	1.9	–5
2.6	3.1	3.1	0
3.1	4.4	4.3	–2.3

the linearized system (Eq. (8.5) with $P_0 = 3$ W in the operating point) is shown in Table 8.1. It can be seen that the error is reasonably low.

8.2 Linearization of Tabular Data

In some cases, a datasheet or similar product information provides the characteristic curve in tabular form. Table 8.2 lists the resistance values of a NTC (resistor with negative temperature coefficient) for several discrete temperature values. In the third column of Table 8.2, the output voltage of a voltage divider is given where the NTC is the upper resistor, the lower resistor has 10 kΩ, and the supply voltage is 5 V. Once again, we are looking for a linear relationship between temperature T and voltage divider output V_{out},

$$V_{out} = m \cdot T + V_0|_{T \approx T_0} \tag{8.6}$$

where T_0 is the operating-point temperature, m the slope of the voltage divider in the operating point, and V_0 the voltage offset that is required for the nonlinear and linear curves to meet in the operating point.

Table 8.2 Resistance values of a NTC and of a voltage divider with the NTC as lower resistor at several discrete temperatures.

Temperature (°C)	NTC Resistance (kΩ)	Divider Output Voltage (V)
0	33.6	1.15
10	20.2	1.65
20	12.6	2.22
25	10	2.50
30	8.0	2.78
40	5.2	3.28
50	3.5	3.70
60	2.4	4.03

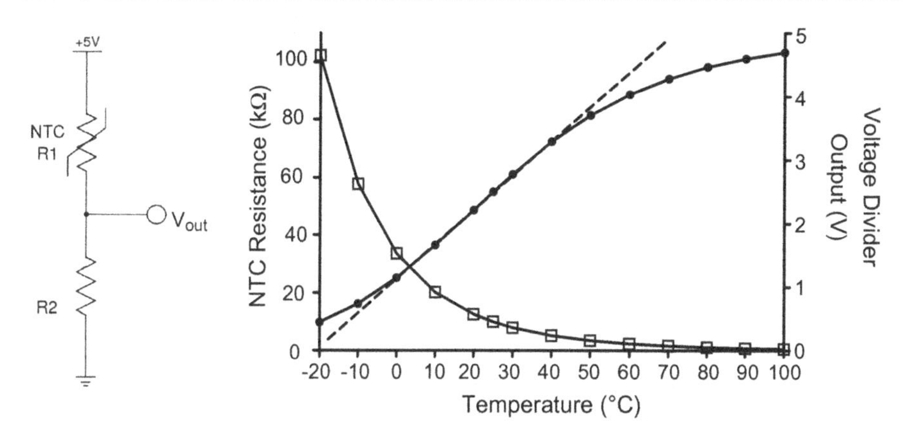

Figure 8.3 Temperature sensing with a temperature-dependent resistor. The resistance of the NTC decreases approximately exponentially with temperature (left axis, □). If the NTC is used as the upper resistor in a voltage divider, the output voltage (right axis, •) shows a much less pronounced nonlinearity than the NTC resistance as indicated by the dashed line.

To illustrate these relationships, the two curves (NTC resistance and voltage divider output) are plotted against temperature in Figure 8.3. The NTC resistance decreases by about 40%/10 °C temperature increase, although a single-exponential model would not describe the curve well enough for high-accuracy applications. The voltage divider output is also nonlinear, but its nonlinearity is much less pronounced, especially in the range from 10 °C to 50 °C, than the nonlinearity of the NTC itself. The example demonstrates that the nonlinear behavior can sometimes be reduced with simple measures, which greatly extend the range of the linear approximation and its accuracy.

Returning to Table 8.2, we now assume an operating point $T_0 = 37\,°\mathrm{C}$. This temperature is not given in the table, and we find (a) the output voltage by linear interpolation of the neighboring values as $V_0 = 2.93\,\mathrm{V}$, and (b) the slope m from rise over run of the two neighboring values. The slope between 30 and 40 °C is $m = 0.05\,\mathrm{V}/°\mathrm{C}$. At the operating point, $m \cdot T_0 = 1.85\,\mathrm{V}$, and an offset of 1.08 V needs to be added to obtain V_0. We can now approximate the output voltage of the NTC-based voltage divider in the operating point with Eq. (8.7) or—if only changes near the operating point are relevant—with Eq. (8.8).

$$V = 0.05\frac{V}{°\mathrm{C}} \cdot T + 1.08\ \mathrm{V} \tag{8.7}$$

$$\Delta V = 0.05\frac{V}{°\mathrm{C}} \cdot \Delta T \tag{8.8}$$

8.3 Linearization of Components with Graphical Data

Now let us examine a valve that controls fluid flow. With a quarter turn of the handle, the valve can be brought from the fully closed to the fully open position with a maximum

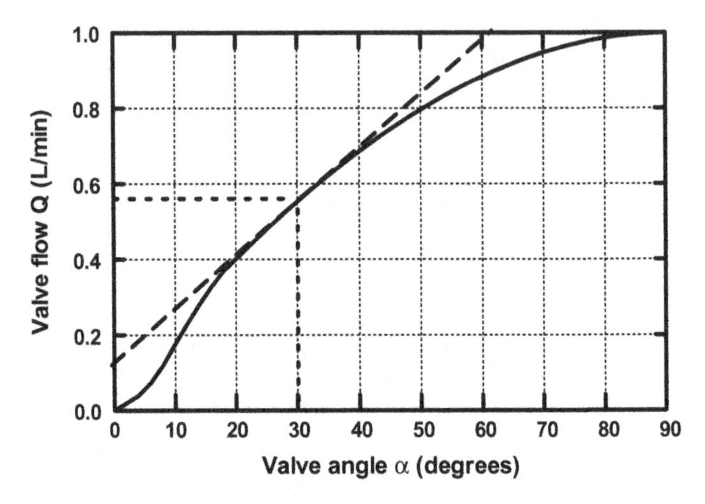

Figure 8.4 Flow rate as a function of the turn angle of the handle in a flow valve. The operating point is indicated by the dotted lines. By drawing the tangent (dashed line) to the curve with a ruler, the slope and the intercept can be estimated.

flow of 1 L/min. Figure 8.4 was taken from the datasheet of the valve, and it shows the flow rate when the handle is turned by a specific angle α from the fully closed position.

In an example system, the equilibrium point is approximately at $\alpha = 30°$ with a corresponding flow rate of 0.55 L/min, determined from the graph. To obtain the tangent, we will have to use a ruler and simply draw the tangent through the operating point (dashed line in Figure 8.4), and then determine its slope as rise divided by run. With this method, we can determine the slope to be $m = 0.014$ L/min/° and the intercept $Q_I = 0.13$ L/min. Clearly, the graphical method is the least accurate of the three. This inaccuracy needs to be taken into account when a system that contains such a nonlinear component is operated near the stability boundary, or if it needs to be run within tight tolerances.

8.4 Saturation Effects

Saturation effects occur when any part of a feedback control system reaches a physical limit. These limits can have many forms: a spring that is compressed to the limit, a positioner or potentiometer that reaches its stops, a motor driver that reaches the maximum allowable current, or an operational amplifier where the output voltage is limited to the supply voltage. A saturation function can be symmetrical (one example is the output voltage of an operational amplifier) or asymmetric. The waterbath is a good example for an asymmetrical saturation function: the heater power has an upper limit dictated by the heating element and the driver power, but the element can only heat. If the waterbath temperature is above the setpoint, the linear system theory would demand a

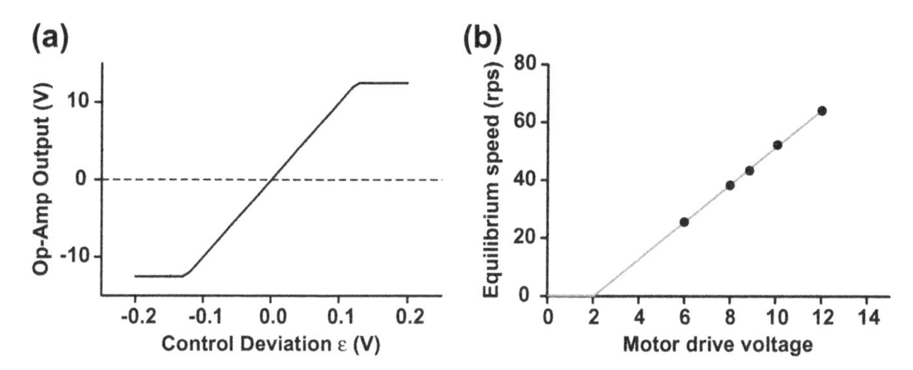

Figure 8.5 Two typical saturation functions. (a) shows the static response of a P controller, set to $k_p = 100$ and realized with an op-amp. The supply voltage of the operational amplifier is ± 15 V, and the output usually does not fully reach the voltage level of the supply rails. The output saturates at approximately ± 12.5 V, corresponding to an error deviation at the input of more than ± 0.125 V. (b) shows some data points of the equilibrium speed of a DC motor (in revolutions per second), measured with different drive voltages. Linear regression (gray line) indicates that the motor would not move below 2 V due to stick friction.

negative power (i.e., cooling) as control action, which is impossible for a resistive heater element.

Another equally disadvantageous nonlinear function is the threshold function: the nonlinear block requires a minimum input for the output to change. Examples are diodes, including the base-emitter diodes of transistors: A minimum bias voltage of 0.7 V is required before a nonzero collector current becomes possible. Stick friction is another example for this type of nonlinearity. Two examples are shown in Figure 8.5.

When the operating point of a feedback control system moves into the saturation range of any of its components, the slope is horizontal, and therefore the *gain is zero.* A change of the input does not cause any change of the output of a saturated component. This also means that the loop gain is zero: The system behaves like an open-loop configuration, and *no control is available.*

For this reason, elements that reach saturation cannot be linearized. The operating points of any feedback control system need to be placed in regions where no element saturates. A robust device would move toward its operating point even when one or more elements are saturated. One example is the waterbath, which heats up (albeit at a slower rate) even when the P controller or heater power driver are at their maximum. An example of a system that is not robust is the motor with stick friction (Figure 8.5B). If this motor were controlled with a PI controller at low speeds, the motor would not move, and the integral component would add up the control deviation until the motor starts to overcome the stick friction and suddenly starts to move. The result would be an irregular oscillatory movement.

Sometimes, it is required to open the loop and measure the open-loop behavior to identify possible saturation effects (*cf.* Figure 5.3). With the help of an auxiliary signal, the entire operating range can be covered and any element that reaches saturation within the operating range of the control system needs to be redesigned. Often, adding or subtracting offsets can shift the operating point into the linear region. For the stick friction, a double control loop could be devised where the inner loop with pure *P* control overcomes the nonlinearity, and the outer loop with *PI* or *PID* control provides the actual control toward the desired static and dynamic response. In software solutions, the motor drive voltage can be subject to the inverse of the nonlinear curve to ameliorate the stick friction effects to some extent.

9 A Tale of Two Poles: The Positioner Example and the Significance of the Poles in the *s*-Plane

Abstract

In this chapter, a new example system is introduced that has very different behavior than the waterbath example in Chapters 5 and 6. The focus of this chapter lies on the dynamic response. The relationship between polynomial roots in the Laplace domain and the transient, dynamic behavior is established. In the toolbox analogy, the design engineer gains the ability to quickly assess the dynamic behavior by merely looking at the locations of poles in the s-plane. Exponential and oscillatory responses are linked to the Laplace-domain transfer function. To the steady-state performance criteria (Chapters 5 and 6) are added new transient-response performance criteria based on the time-integrated control deviation. The example is then continued by introducing a digital controller and obtaining the dynamic response from the z-domain transfer function. Dynamic behavior of continuous and time-discrete controls are compared.

In this chapter, we examine the significance of the poles in the s-plane and their relationship to a system's dynamic response. Specifically, we study pole pairs that are associated with second-order systems. Poles are those locations in the s-plane where the characteristic polynomial becomes zero. To explain the information contained in the location of the poles, we now introduce a second-order example process, the positioner. We will then examine the positioner under feedback control and study the relationship between the pole location and the dynamic response.

Positioning tasks occur frequently. Some examples include print head positioning in ink-jet printers, arm positioning for robots (see Figure 1.5), scribe systems, servo systems, cranes, and other forms of mechanical load management, CNC machines, or even avionic control elements, such as flap and rudder positioners. Positioners have in common a rapidly variable setpoint, and the design task is to let the positioning system follow the setpoint both accurately and rapidly. In many cases, disturbances play a secondary role, and we will develop the example in this chapter without considering disturbances.

Linear Feedback Controls. http://dx.doi.org/10.1016/B978-0-12-405875-0.00009-7
© 2013 Elsevier Inc. All rights reserved.

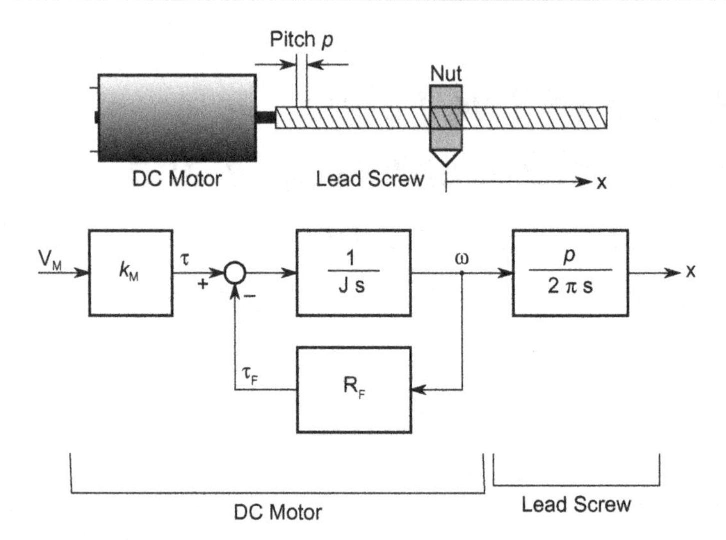

Figure 9.1 Sketch and block diagram of the positioner. A DC motor drives a lead screw, which in turn actuates a nut. We define the position of the nut as x with $x = 0$ at the center of the travel range. The lead screw has the pitch p, and each turn ($\phi = 2\pi$) of the screw advances the nut by p. For the DC motor, we use the simplified armature motor model where a voltage-dependent torque τ accelerates the rotor (rotational inertia J) and overcomes the velocity-dependent friction R_F.

9.1 A Head-Positioning System

The example process in this chapter is a lead screw connected to a DC motor. The motor turns the lead screw, which in turn moves a nut, which could then actuate the scribe or print head. The positioner is sketched in Figure 9.1.

The DC motor model in this example is a simplified model of an armature motor where only the rotational inertia J and the friction of the rotor R_F are considered as dominant factors.[1] Magnetic forces produce a torque τ that depends on the current through the motor coils. If the coils are predominantly resistive, the current is proportional to the motor input voltage. Therefore, we can approximate the magnetic torque τ as the product of the input voltage V_M and the motor's efficiency k_M. In turn, k_M combines many factors, such as the number of windings, coil resistance, strength of the permanent magnets, number of magnetic poles, and the air gap between stator and rotor. At low angular velocities ω, the torque predominantly accelerates the rotor. As the rotor builds up speed, the friction also increases and balances the torque. For a constant V_M, the motor reaches an equilibrium velocity ω_∞ where the available torque is equal to the frictional torque τ_F. We obtain the differential equation of a first-order system,

$$\frac{J}{R_F}\dot{\omega}(t) + \omega(t) = \frac{k_M}{R_F} V_M(t) \tag{9.1}$$

[1] The major component that acts opposite to the magnetic force is a reverse electromagnetic force, but in this context, it can be modeled as a velocity-dependent friction.

The lead screw advances the nut by p when the lead screw rotates by $\phi = 2\pi$ (i.e., one revolution). Since the angular position is the integral of the angular velocity (note that we assume $\phi_0 = 0$ to keep the example simple), the equation of the lead screw is

$$\dot{x}(t) = \frac{p}{2\pi}\omega(t) \tag{9.2}$$

In the Laplace domain (and again under assumption that all initial conditions are zero), we obtain the overall transfer function of the process in normalized form:

$$\frac{X(s)}{V_M(s)} = \frac{k_M}{R_F} \cdot \frac{p}{2\pi} \cdot \frac{1}{s\left(\frac{J}{R_F}s + 1\right)} \tag{9.3}$$

The individual components in Eq. (9.3) can be recognized. The term k_M/R_F determines the motor's equilibrium angular velocity. The term $p/2\pi$ represents the linear translation by the lead screw. The fraction J/R_F is the time constant of the motor. The transfer function has one pole at $-R_F/J$ and another pole in the origin.

Is the process stable? After all, we have a pole in the origin (the integrator pole), which is on the imaginary axis. According to the definition of stability, a system is stable if every *bounded* input produces a bounded output. This definition indicates a stable process. A pole (or pole pair) on the imaginary axis, however, indicates a *marginally stable* system. Stability is discussed in more detail in Chapter 10. In this example, marginally stable means that, if the motor keeps running under the influence of a step input voltage, the nut either reaches a stop or runs off the lead screw.

9.2 Introducing Feedback Control

Now we will examine the system with feedback control (Figure 9.2). We use a suitable position sensor, such as a linear potentiometer, that is conveniently calibrated to have a unity gain function. We can, for example, apply ± 1 V to the ends of the potentiometer to obtain a slider range from -1 V (leftmost position) to 0 V (center) to $+1$ V (rightmost position). This sensor voltage gets subtracted from the setpoint (X_{set}), which is also a voltage calibrated to the position. The resulting control deviation $\epsilon(s)$ is fed into a controller with a transfer function of $H(s)$, which in turn generates the corrective action $V_M(s)$.

For now, let us use a simple P controller with $H(s) = k_p$. We can now determine the closed-loop transfer function. The substitution of

$$V_M(s) = k_p \cdot \epsilon(s) = k_p \cdot \left(X_{set}(s) - X(s)\right) \tag{9.4}$$

into Eq. (9.3) followed by collecting all terms that contain the dependent variable ($X(s)$) on the left-hand side yields

$$X(s) = \frac{\alpha\, k_p}{s^2 + \frac{R_F}{J}s + \alpha\, k_p} X_{set}(s) \tag{9.5}$$

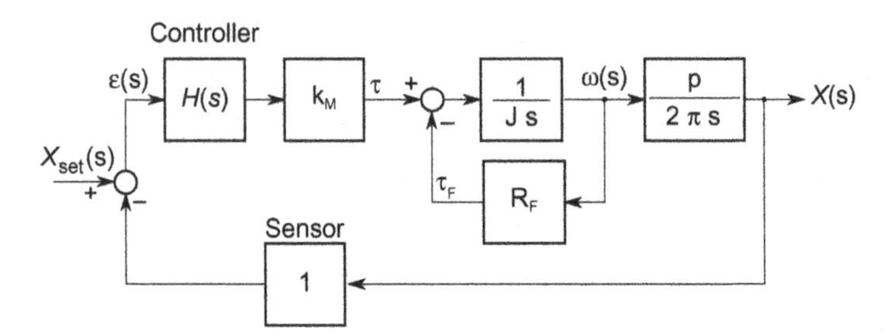

Figure 9.2 Block diagram of the positioner with a feedback control system. The position $X(s)$ is the controlled variable. A sensor, conveniently calibrated to have a unity transfer function, provides the position information. The control deviation $\epsilon(s)$ is the difference between setpoint $X_{set}(s)$ and actual position $X(s)$. A controller with the transfer function $H(s)$ generates the corrective action $V_M(s)$ and thus drives the motor.

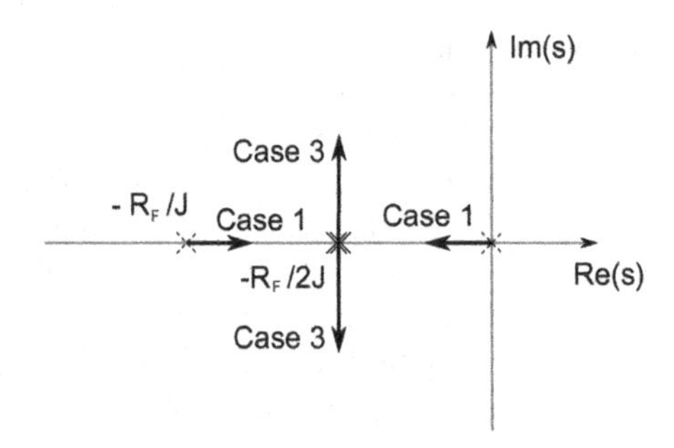

Figure 9.3 Root locus of the poles of the closed-loop system. With $k_p \to 0$, the poles start at the locations of the open-loop poles (dotted poles at $-R_F/J$ and in the origin). With increasing k_p, the poles move along the real axis toward each other (case 1 in Eq. (9.8)) until they form a real-valued double pole at $-R_F/2J$ (case 2 in Eq. (9.8)). If k_p is further increased, the poles form a complex conjugate pair with constant real-valued component and increasing imaginary component (case 3 in Eq. (9.8)).

where α combines several process-dependent constants:

$$\alpha = \frac{k_M\, p}{2\pi\, J} \tag{9.6}$$

The closed-loop system is still a second-order system, but the feedback control has moved the integrator pole away from the origin. In fact, we can now look at the location of the poles as we increase k_p (Figure 9.3). Starting with $k_p = 0$, we obtain the characteristic polynomial of the open-loop system, $s(s + R_F/J)$. With $k_p > 0$, we set the characteristic polynomial to zero to find the poles $p_{1,2}$:

$$p_{1,2} = -\frac{R_F}{2J} \pm \sqrt{\left(\frac{R_F}{2J}\right)^2 - \alpha\, k_p} \tag{9.7}$$

There are three important cases to distinguish. First, for small k_p, we obtain two real-valued poles. Second, when the term under the square root vanishes, the poles form a real-valued double pole at $-R_F/2J$. Third, with even larger k_p, the poles become a complex conjugate pair and spread out into the complex plane:

$$\begin{cases} k_p < \dfrac{\pi R_F^2}{2Jk_M p} & \text{Case 1: Real-valued poles} \\ k_p = \dfrac{\pi R_F^2}{2Jk_M p} & \text{Case 2: Real-valued double-pole} \\ k_p > \dfrac{\pi R_F^2}{2Jk_M p} & \text{Case 3: Complex conjugate pole pair} \end{cases} \tag{9.8}$$

9.3 Dynamic Response of the Closed-Loop System

In Chapters 5 and 6, we discussed the steady-state response to a step input of a process under feedback control. The steady-state response is the response after a system reaches its equilibrium. From the final value theorem, applied to Eq. (9.5) with a unit step at the input, we can see that the positioner system precisely reaches the setpoint in the absence of any disturbance and after a sufficiently long equilibration time. Often, it is important to know how fast the setpoint is reached, and how much overshoot we have to expect. An example application is illustrated in Figure 9.4. In the scribe system, it is important that the positioning speed of the pen (y-axis) is much faster than the paper feed (t-axis), otherwise, a step change of the input signal cannot be accurately represented as indicated in Figure 9.4. The design engineer can optimize the y-axis positioning system (e.g., low inert mass, large leadscrew pitch), but the general pole configuration—notably the integrator pole in the origin—remains unchanged.

At this point, the feedback control system can be optimized by adjusting the controller parameters to (1) achieve a certain rise time or (2) stay below an acceptable overshoot. The location of the poles tells us how the system will respond to a step change at the setpoint. To examine the relationship between the location of the poles and the transient response, let us change Eq. (9.5) into a normalized second-order format,

$$X(s) = \frac{\omega_n^2}{s^2 + 2\zeta\omega_n s + \omega_n^2} X_{set}(s) \tag{9.9}$$

where ω_n is the natural (undamped) frequency of the system and ζ is the damping efficiency. Equation (9.9) describes a typical spring-mass-damper model. We can compare the characteristic equation of the spring-mass-damper model to that of the closed-loop positioner system (Eq. (9.5)),

$$\begin{aligned} q(s) &= s^2 + 2\zeta\omega_n s + \omega_n^2 \\ q(s) &= s^2 + \frac{R_F}{J} s + \alpha k_p \end{aligned} \tag{9.10}$$

and find the corresponding coefficients $2\zeta\omega_n = R_F/J$ and $\omega_n^2 = \alpha \cdot k_p$. The closed-loop system described in Eq. (9.5) has the same dynamic behavior as a

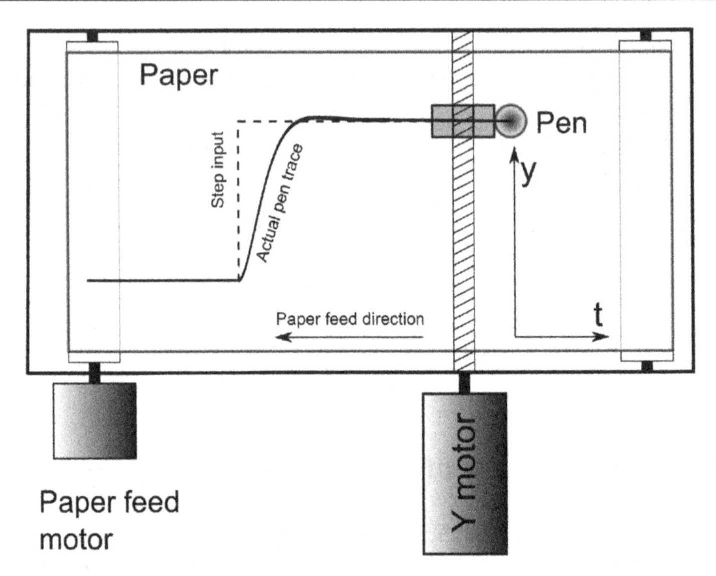

Figure 9.4 Positioner system in an example application as a $y - t$ scribe system. A paper feed motor moves the paper underneath the pen at a constant speed. The pen is positioned proportional to an input signal. Thus, the pen draws the input signal on the paper as a function of time. It is desirable that the pen follows a change of the input signal as quickly as possible, but mechanical limitations will cause a certain delay. A step input of y (dashed line) will therefore be plotted with a limited slope (thick line).

spring-mass-damper system, however, we can see that k_p in the P controlled scribe example influences both the frequency of oscillations ($\omega_n = \sqrt{\alpha k_p}$) and the damping factor with

$$\zeta = \frac{R_F}{2J\sqrt{\alpha k_p}} \tag{9.11}$$

With pure P control, we cannot adjust ζ and ω_n independently. In this case, we focus primarily on the damping factor ζ. More complex controller design, for example a *PD* controller, allows independent adjustment of ζ and ω_n. For ζ, several rules of thumb apply. First, overshoot only occurs when $\zeta < 1$. The overdamped case ($\zeta > 1$), the critically damped case ($\zeta = 1$) and the underdamped case with overshoot ($\zeta < 1$) must be treated separately, because the character of their response is fundamentally different. Let us begin with the overdamped case (low k_p, first introduced as Case 1 in Eq. (9.8)), where the two real-valued poles allow the denominator polynomial to be rewritten in product form as $(s + a)(s + b)$. The step function $1/s$ reflects the input signal $X_{set}(s)$. The step response is a double exponential equilibration:

$$\frac{1}{s} \cdot \frac{ab}{(s+a)(s+b)} \quad \bullet\!\!-\!\!\circ \quad 1 - \frac{b}{b-a} e^{-at} + \frac{a}{b-a} e^{-bt} \tag{9.12}$$

Here, a and b are the two real-valued solutions p_1 and p_2 in Eq. (9.7) and therefore depend on the process constants *and* on k_p. Equation (9.12) can be solved, for example, for the time until the output reaches a defined percentage of the final value.

The critically damped case (Eq. (9.8), Case 2) where a double-pole occurs, can be treated similarly. However, this case is special as k_p is chosen to cancel out the term under the square root, and consequently $p_1 = p_2$. Thus, the denominator polynomial becomes in product form $(s+a)^2$ where $a = p_1 = p_2$. With this special value of k_p, the parameter a depends only on the system constants, namely, $a = R_F/2J$. The step response still exhibits an exponential characteristic:

$$\frac{1}{s} \cdot \frac{a^2}{(s+a)^2} \quad \bullet\!\!-\!\!\circ \quad 1 - e^{-at} - a\,t\,e^{-at} \tag{9.13}$$

The third case, i.e., the underdamped case (large k_p, introduced as Case 3 in Eq. (9.8)) needs to be treated differently as the denominator polynomial can no longer be rewritten in product form with real-valued coefficients. Moreover, overshoot will occur. From the Laplace correspondence tables, we obtain the unit step response

$$\frac{1}{s} \cdot \frac{\omega_n^2}{(s^2 + 2\zeta\omega_n s + \omega_n^2)} \quad \bullet\!\!-\!\!\circ \quad 1 - \frac{1}{\xi} \cdot e^{-\zeta\omega_n t} \cdot \sin\left(\omega_n \xi t + \phi\right) \tag{9.14}$$

with $\xi = \sqrt{1-\zeta^2}$ and $\cos\phi = \zeta$. The underdamped case implies $\zeta < 1$. Equation (9.14) clearly describes an attenuated oscillation. The attenuation time constant (i.e., the time after which the oscillation amplitude decays to 37% of its initial amplitude) is $1/\zeta\omega_n$, and the oscillations are "stretched" from the natural frequency by the damper, that is, the oscillation frequency is $\omega_D = \omega_n\sqrt{1-\zeta^2}$. Note that we can often assume $\omega_D \approx \omega_n$ when we have very low damping coefficients ζ. Conversely, for values of ζ near one, the exponential decay occurs faster than the sinusoidal oscillation, and the step response may contain no more than a single overshoot.

9.4 Dynamic Response Performance Metrics

The dynamic response of a second-order system can be quantified by a number of different metrics that can guide the design engineer to find the optimum pole placement. The criteria presented here should be understood as approximations, and detailed analysis of the actual dynamic response may be necessary to fine-tune the control system. Key performance metrics for the dynamic response, explained in detail below, are:

- Rise time.
- Time to first overshoot.
- Percent overshoot.
- Settling time.

These criteria can be applied to higher-order systems, but the rule-of-thumb formulas given below cannot be used. Again, it will be necessary to compute the step response or obtain it through simulation.

Rise time: There are several possible definitions for the rise time, and the rise time can be obtained for underdamped as well as for overdamped systems. Possible definitions of the rise time are the time until the system's step response reaches 50%

of its final value, the time until the step response enters a specified tolerance band (for example, within 2% of the final value), or—more commonly used—the time it takes the output signal to swing from 10% of the final value to 90% of the final value. For a critically damped system with two poles at $s = -a$, the rise time T_{rise} from 10% to 90% can be estimated with

$$T_{rise} = \frac{3.36}{a} \tag{9.15}$$

For an underdamped system, a rough estimate for the rise time is given by

$$T_{rise} = \frac{2.16\zeta + 0.6}{\omega_n} \tag{9.16}$$

However, this estimate becomes inaccurate when $\zeta > 0.75$. For highly underdamped systems (for example, $\zeta < 0.3$), the rise time becomes somewhat irrelevant.

Peak overshoot: The time from the application of a step change ($t = 0$) to the peak of the first overshoot, T_P, can be obtained from Eq. (9.14) by setting its first derivative to zero. Fortunately, this can also be done in the Laplace domain: the first derivative of a system's step response is the system's impulse response. According to the transform correspondence tables, the *impulse* response of Eq. (9.9) is

$$x(t) = \frac{\omega_n}{\sqrt{1-\zeta^2}} \cdot e^{-\zeta\omega_n t} \cdot \sin \omega_n \sqrt{1-\zeta^2}\, t \tag{9.17}$$

which becomes zero if the argument of the sine function equals π. We obtain:

$$T_P = \frac{\pi}{\omega_n \sqrt{1-\zeta^2}} \tag{9.18}$$

Percent overshoot: By evaluating the step response at T_P, the magnitude M_P of the first (and highest) overshoot emerges as

$$M_P = 1 + \exp\left(-\frac{\pi\zeta}{\sqrt{1-\zeta^2}}\right) \tag{9.19}$$

which can be normalized to a more general *relative overshoot* M_R, given in percent:

$$M_R = 100\% \cdot \exp\left(-\frac{\pi\zeta}{\sqrt{1-\zeta^2}}\right) \tag{9.20}$$

Settling time: For highly underdamped systems ($\zeta < 1/\sqrt{2}$), multiple oscillations occur. The rise time becomes less relevant due to the multiple oscillations. Here, a more appropriate definition is that of the *settling time* T_s, defined as the time it takes for the step response to settle within a tolerance band around the final value. Often, a 2% tolerance band is used. A rough approximation for the time it takes until no oscillations exceed the 2% tolerance band can be derived by considering the exponential envelope only, $e^{-\zeta\omega_n T_s} < 0.02$:

$$T_s \approx \frac{4}{\zeta\,\omega_n} \tag{9.21}$$

The above relationships are important in the *design* of feedback control systems when feedback control allows us to influence where a complex conjugate pole pair lies. To use the scribe system as an example, we rewrite Eq. (9.3) as

$$\frac{X(s)}{V_M(s)} = \frac{k_M}{R_F} \cdot \frac{p}{2\pi} \cdot \frac{1}{s\left(\frac{J}{R_F}s+1\right)} = \alpha \cdot \frac{1}{s(s+\upsilon)} \tag{9.22}$$

where the process constants have been combined into two new constants $\alpha = k_M p/2\pi J$ and $\upsilon = R_F/J$. To provide a numerical example, we assume the motor time constant to be 20 ms ($\upsilon = 50\ \mathrm{s}^{-1}$) and the motor proportionality constant $\alpha = 0.25\ \mathrm{m/Vs}^2$. For the closed-loop system with *P* control (Figure 9.2), the transfer function is

$$\frac{X(s)}{X_{set}(s)} = \frac{\alpha k_p}{s^2 + \upsilon s + \alpha k_p} \tag{9.23}$$

and the roots of the characteristic equation are found at

$$p_{1,2} = -\frac{\upsilon}{2} \pm \sqrt{\left(\frac{\upsilon}{2}\right)^2 - \alpha\,k_p} \tag{9.24}$$

To obtain critical damping, we need $k_p = 2500\ \mathrm{V/m}$, and the rise time according to Eq. (9.15) is $T_{rise} = 134$ ms.

Next, let us assume that we want to achieve a faster rise time, but we are constrained to a maximum of 2% allowable overshoot. From Eq. (9.20), we obtain $\zeta = 0.78$. Our pole pair lies on two straight lines in the s-plane described by $p_{1,2} = -\omega_n\left(0.78 \pm 0.626j\right)$. To obtain this response, we need $k_p \approx 4000$ V/m, which places the pole pair at $-25 \pm 19.4j$. The rise time (obtained by explicit computation of the step response) has improved to approximately $T_{rise} = 75$ ms. Moreover, the settling time into the 2% tolerance band is now 120 ms instead of 230 ms for the critically damped system. For both time criteria, we see an almost two-fold improvement by allowing minor overshoot.

A further increase of the gain to $k_p = 5000\ \mathrm{V/m}$ yields $\zeta = 1/\sqrt{2}$. This damping coefficient represents a special, frequently used case where the maximum overshoot is approximately 4.6%. A pole pair for which $\zeta = 1/\sqrt{2}$ lies on the dotted diagonal lines where $\theta = 45°$ in Figure 9.5. Any pole pair lying inside the wedge outlined by the dashed lines (i.e., the gray shaded area in Figure 9.5) consequently has less than 4.6% overshoot. In the scribe example, the rise time has improved to approximately 60 ms, but the settling time is now 170 ms due to the overshoot that exceeds 2%. This special case, $\zeta = 1/\sqrt{2}$, represents the smallest damping coefficient where a step response does not "dip back" below the equilibrium value from the overshoot. Any further increase of the gain leads to multiple oscillations.

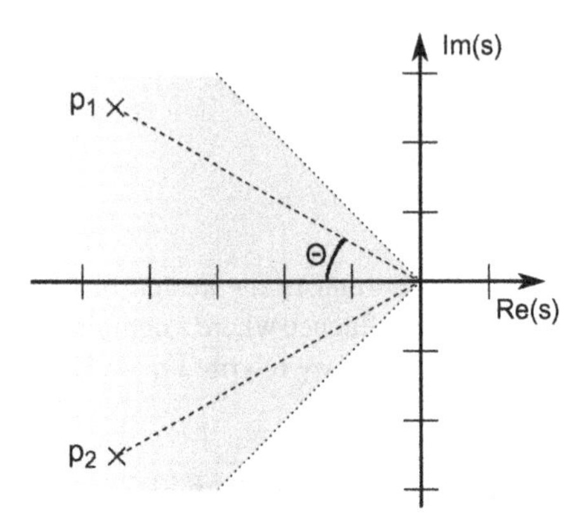

Figure 9.5 Location of a complex conjugate pole pair in the s-plane. The location of the poles $p_{1,2}$ is related to the coefficients ω_n and ζ in the second-order term. Note that for this configuration $\zeta < 1$ is required. The gray shaded area indicates possible locations for pole pairs with $\theta < -45°$, which corresponds to $1/\sqrt{2} < \zeta < 1$.

In the other direction, i.e., from a given pole location, we can determine the undamped resonance frequency and the damping coefficient. The transfer function in Eq. (9.9) describes a case that occurs frequently. We can determine the roots of the characteristic equation $s^2 + 2\zeta\omega_n s + \omega_n^2$ as

$$p_{1,2} = -\omega_n \left(\zeta \pm j\sqrt{1-\zeta^2}\right) \tag{9.25}$$

The pole pair is associated with a natural (undamped) frequency ω_n and a damping coefficient ζ as follows:

$$\omega_n = \sqrt{\Re(p)^2 + \Im(p)^2} = |p| \quad ; \qquad \zeta = \frac{\Re(p)}{\sqrt{\Re(p)^2 + \Im(p)^2}} = \frac{\Re(p)}{|p|} \tag{9.26}$$

Here, p is either of the two complex conjugate poles, because neither ω_n nor ζ depend on the sign of the imaginary part of p. Furthermore, we can relate ζ to the angle θ in Figure 9.5:

$$\tan\theta = \sqrt{\frac{1}{\zeta^2} - 1} \tag{9.27}$$

The effect of obtaining an accelerated rise time (and therefore an accelerated step response) by allowing for a minor overshoot is demonstrated in Figure 9.6. In the overdamped case $\zeta = 2$, the slow pole dominates the system response. The fastest response without overshoot occurs in the critically damped case when $\zeta = 1$. However, when a minor overshoot is allowed—in this case with $\zeta = 0.7$ where the system approaches the final value from above without dipping below the final value—the step response is even faster. These considerations apply only for second-order systems. Higher-order systems generally have more complex step responses. Moreover, other

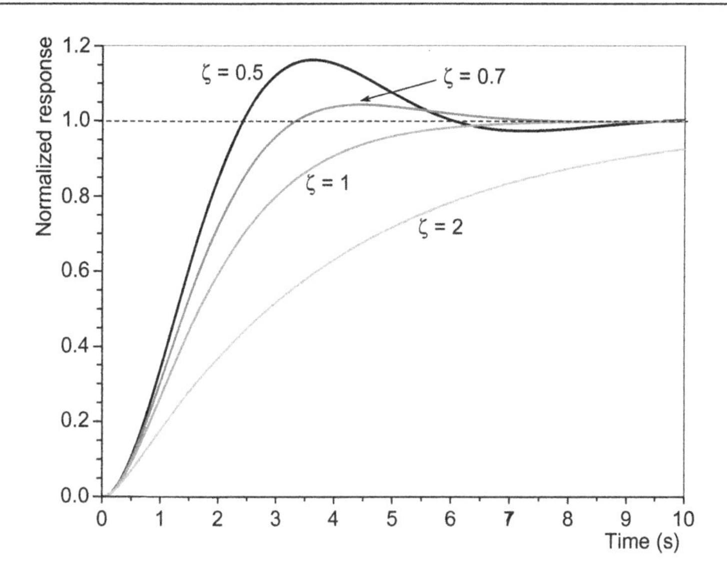

Figure 9.6 Effect of the location of the poles on the rise time of a second-order system. An overdamped case ($\zeta = 2$ with two real-valued poles) is dominated by the slower pole and has a correspondingly slow rise time. Critical damping ($\zeta = 1$ with a real-valued double-pole) elicits the fastest step response without overshoot. Further reducing the damping factor (in this case, $\zeta = 0.7$) improves both rise time and settling time, but at the cost of some overshoot. Further lowering ζ may actually increase the settling time because of repeated oscillations (note the undershoot for $\zeta = 0.5$).

input signals to the system (examples: ramp or oscillations) cause fundamentally different responses. Often, the computation or numerical determination of the system response is needed to optimize the controller parameters in an iterative design process: first, initial controller parameters are estimated, for example, by evaluating a pole-zero diagram. Next, the system response is computed, and modified controller parameters are determined. With the help of computer tools, this iterative process can be quite efficient.

9.5 Time-Integrated Performance Metrics

A frequently-used performance criterion is the time-averaged tracking error of a control system. In the case of the scribe, the tracking error $E(s)$ is the difference between setpoint $X_{set}(s)$ and the actual pen position $X(s)$. The tracking error is measured in the time-domain. In the absence of disturbances, the control deviation $\epsilon(s)$ can be used to represent the time-dependent error, and a number of quantitative descriptors exist that differ in the type of error emphasized, but not in principle. A general formulation for an integrated error index E is

$$E = \int_0^\infty \mathrm{f}\left(\epsilon(t)\right)\, \mathrm{d}t \tag{9.28}$$

Depending on the practical application, the input can be any function, but most frequently, $\epsilon(t)$ is understood as the tracking error following a step input. Functions $f(\epsilon(t))$ in Eq. (9.28) include $\epsilon^2(t)$ for the integrated squared error (ISE) and $|\epsilon(t)|$ for the integrated absolute error (IAE). By including the time as a weight factor, the error index can be made more sensitive toward long-term tracking errors and less sensitive toward errors occurring in the initial response. Such a value is known as the integrated time-weighted absolute error, or ITAE performance criterion,

$$E_{\mathrm{ITAE}} = \int_0^\infty t \cdot |\epsilon(t)| \; \mathrm{d}t \tag{9.29}$$

and a variation is the integrated, time-weighted squared error (ITSE), which more strongly emphasizes large error magnitudes,

$$E_{\mathrm{ITSE}} = \int_0^\infty t \cdot \epsilon(t)^2 \; \mathrm{d}t \tag{9.30}$$

In practical applications, the integration does not take place from 0 to ∞, but is rather cut off after a reasonable equilibration time that depends on the overall system dynamics.

To analytically design a system for optimum performance, the tracking error can be computed for a step input, and the integral solved for the performance index of choice. To optimize some controller parameter κ (in the example of the scribe, this could be the controller gain k_p), the first derivative $\partial E/\partial\kappa$ is set to zero, under the assumption that only one local minimum of E exists. The resulting equation can then be solved for κ. With this method, it can be shown that an underdamped second-order system with $\zeta = 0.7$ (Figure 9.6) has an optimal step response under the ITAE criterion.

For complex system responses, a computer simulation is more practical. The step response of a given system can be computed (e.g., with Scilab, see Appendix D) and subtracted from the setpoint. The discrete sum of the absolute value or square of those data points, multiplied with the simulated discrete time step, provides the discrete approximations of IAE and ISE, respectively. For the ITAE metric, the absolute error values need to be multiplied with time.

Like all performance indices, the integrated error indices may have limited use in some systems. The integrated errors IAE, ISE, and ITAE are shown in Figure 9.7a for a normalized second-order system (Eq. (9.9)) and a unit step input. The natural frequency was held constant at $\omega = 5\ \mathrm{s}^{-1}$, and the damping coefficient ζ was allowed to cover a wide range from weakly attenuated oscillations to an overdamped system. All performance indices show a clear minimum, with minimal values for IAE at $\zeta = 0.66$, for ISE at $\zeta = 0.5$, and for ITAE at $\zeta = 0.71$.

The situation is much less clear when we examine the influence of k_p on the P controlled scribe system (Figures 9.2 and 9.4). The closed-loop equation of the scribe system, first introduced as Eq. (9.5) above, is

$$X(s) = \frac{\alpha\, k_p}{s^2 + \frac{R_F}{J} s + \alpha\, k_p} X_{set}(s) \tag{9.31}$$

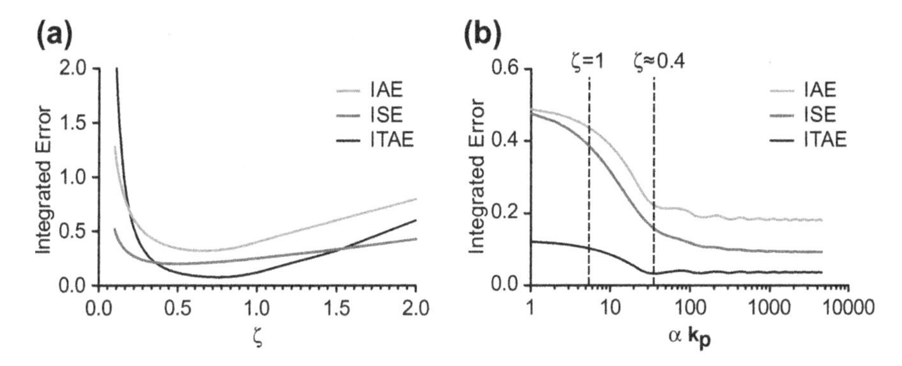

Figure 9.7 Integrated error indices for two different systems. (a) The first example shows the step response of a spring-mass-damper system (Eq. (9.9)), where ISE, IAE, and ITAE exhibit distinct minima in a somewhat underdamped system as the damping coefficient ζ is allowed to vary. (b) Conversely, the feedback-controlled scribe system (Figure 9.4 and Eq. (9.31)) exhibits a convergent behavior as k_p is increased. Note the differences in the abscissa: In (b), a logarithmic scale was used to show more detail at lower values of k_p. Furthermore, ζ increases from left to right in (a), whereas ζ decreases with increasing k_p from left to right in (b).

where the inverse time constant for the motor was chosen as $R_F/J = 5\ \mathrm{s}^{-1}$, and the combined coefficient $\alpha\, k_p$ varied over several orders of magnitude. As k_p increases in the lower range, the integrated error indices diminish, but then converge and remain constant as k_p is further increased. We know from the root locus (Figure 9.3) that increasing k_p beyond the critically damped case leads to an increase of the frequency of the oscillatory component, but the attenuation time constant (i.e., the real component of the poles) remains constant. This situation is different from the spring-mass-damper system with variable ζ (cf. Figure 3.3), where the poles for the underdamped system follow a quarter circle towards the imaginary axis, in other words, the frequency converges and the attenuation time constant becomes very low as $\zeta \to 0$.

In the scribe example, on the other hand, more and more oscillations are fit into the same exponentially attenuated envelope as $\zeta \to 0$, which changes the integrated error indices only marginally. It could be argued, however, that the qualitative performance of the scribe system deteriorates rapidly once multiple overshoots are allowed with $\zeta < 0.7$, yet IAE and ITAE diminish only down to $\zeta = 0.4$, and ISE for somewhat lower ζ. This example demonstrates that it is important to look at the system performance from different perspectives and not from one single criterion.

9.6 Feedback Control with a Time-Discrete Controller

A time-discrete, digital control for the positioner can be designed analogously to the continuous system. The starting point is the combined transfer function of the motor and the leadscrew (Eq. (9.3) in the simplified form of Eq. (9.22)),

$$\frac{X(s)}{V_M(s)} = \beta \cdot \frac{1}{s(s+v)} \tag{9.32}$$

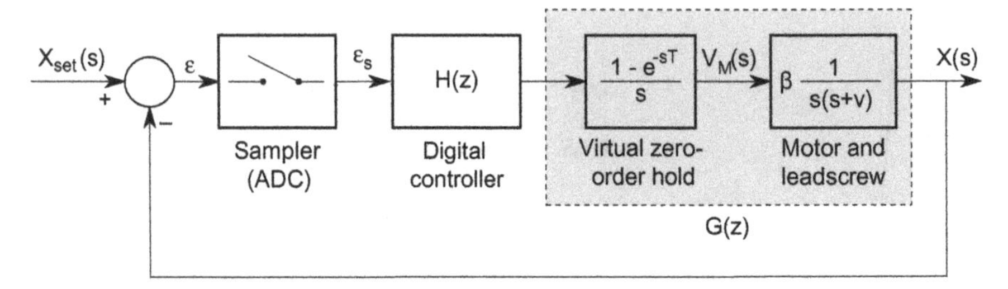

Figure 9.8 DC motor—positioner example with time-discrete (digital) control. As in the time-continuous example (Figure 9.2), we assume a unit-gain sensor. The process constants have been simplified and combined into the gain β and the inverse time constant v. The assumption of a virtual zero-order hold element is necessary to convert the sampled output of the controller into a continuous signal $V_M(s)$ for the time-continuous process.

The digitally controlled system is sketched in Figure 9.8 and follows the outline given in Section 4.1. Introduction of a virtual zero-order hold element is necessary to account for the interpretation of the sampled controller output as a continuous signal by the process. The input to the controller is the sampled control deviation ϵ_s.

In a practical control example, the sensor is not necessarily analog. Rather, a position encoder can provide either directly a digitally coded position signal or the quadrature-encoded speed. When a motor with a quadrature encoder is used, the controller needs to integrate the pulses to keep track of the position. Furthermore, the setpoint could also be generated digitally. In all cases, however, the discrete nature of the control remains, namely, that a control action is computed in regular time intervals T.

For the numerical treatment of the example, we assume the motor time constant to be 20 ms ($v = 50\ \mathrm{s}^{-1}$) and the sampling rate $T = 5$ ms. Furthermore, $\beta = 0.25\ \mathrm{m/V\ s}^2$. Realistically, a basic microcontroller could perform the analog-to-digital conversion and the computation of the corrective action in less than 50 μs, but the purpose of this example is to demonstrate the control behavior when the sampling rate is in the same order of magnitude as the process time constants. In this case, the approximation of the controller by its equivalent Laplace-domain function is not possible, and the entire transfer function needs to be determined in the z-domain.

The first step in computing the z-domain transfer function, and thus the system response, is to translate the process with its zero-order hold into the z-domain. To obtain low-order Laplace- and z-domain equivalences, a partial fraction expansion is performed (note that the same result for $G(s)$ can be found directly in Table B.5.):

$$\begin{aligned} G(z) &= \beta\left(1 - z^{-1}\right) \mathscr{Z}\left\{\frac{1}{s^2(s+v)}\right\} \\ &= \beta\left(1 - z^{-1}\right) \mathscr{Z}\left\{\frac{1}{v\, s^2} - \frac{1}{v^2\, s} + \frac{1}{v^2\,(s+v)}\right\} \\ &= \frac{\beta}{v^2}\left(\frac{z(e^{-vT} + vT - 1) - (vT+1)e^{-vT} + 1}{(z-1)(z - e^{-vT})}\right) \end{aligned} \tag{9.33}$$

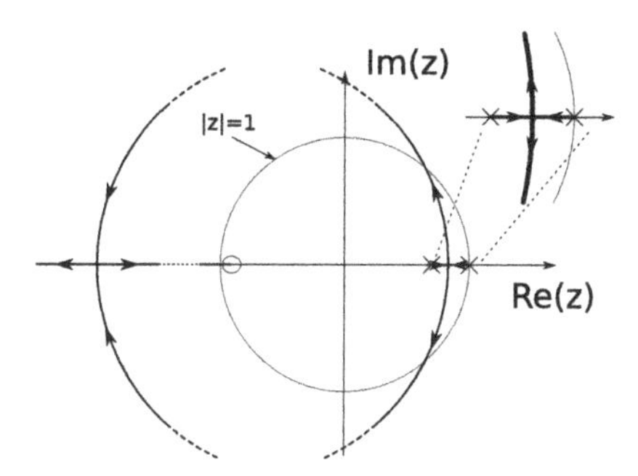

Figure 9.9 Root locus diagram of the digitally-controlled positioner. Indicated are the poles ($\times$) and zero ($\circ$) of the open-loop system (i.e., at $k_p = 0$). For low k_p, exponential responses are obtained similar to the ones obtained from the time-continuous system. At higher values of k_p, the response becomes complex with an oscillatory component. Unlike with time-continuous control, however, two poles leave the unit circle and cause an unstable system at even higher k_p.

At this point, it is convenient to introduce numerical values and reduce the somewhat unwieldy expression in Eq. (9.33) to

$$G(z) = 10^{-6}\frac{\mathrm{m}}{\mathrm{V}}\left(\frac{2.88z + 2.65}{z^2 - 1.7788z + 0.7788}\right) \tag{9.34}$$

Note that we omitted all units as they would make the partial fraction expansion more complex. Overall, $G(z)$ has units of m/V, as does the factor β/v^2.

In parallel to the analysis of the dynamic response of the continuous system (Section 9.3), we assume simple *P* control and compute the closed-loop transfer function (with the gain constant 10^{-6} factored into k_p):

$$\frac{X(z)}{X_{set}(z)} = \frac{k_p(2.88z + 2.65)}{z^2 + (2.88k_p - 1.7788)z + 2.65k_p + 0.7788} \tag{9.35}$$

There are several noteworthy similarities and differences to the transfer function of the continuous system:

- The open-loop integrator pole at $s = 0$ in the s-plane is mapped to $z = 1$ in the z-plane, and the motor pole is on the positive real axis of the z-plane.
- For low values of k_p, the system has a double exponential step response. A critically damped system can be achieved with both continuous and time-discrete control with very similar values for k_p.
- At larger values of k_p, the poles form a complex conjugate pair with an oscillatory response. The behavior of the continuous and discrete systems begin to diverge: In the continuous system, increasing k_p increases the frequency of the oscillations, but not their decay time. In the time-discrete system, oscillations take longer to decay.
- The pole pair lies on the unit circle for $k_p = 83000$. An undamped oscillation with a frequency of approximately 20 Hz occurs. At $k_p > 83000$, the complex pole pair leaves the unit circle. The time-discrete system becomes unstable for large k_p, which cannot happen with the continuous system.

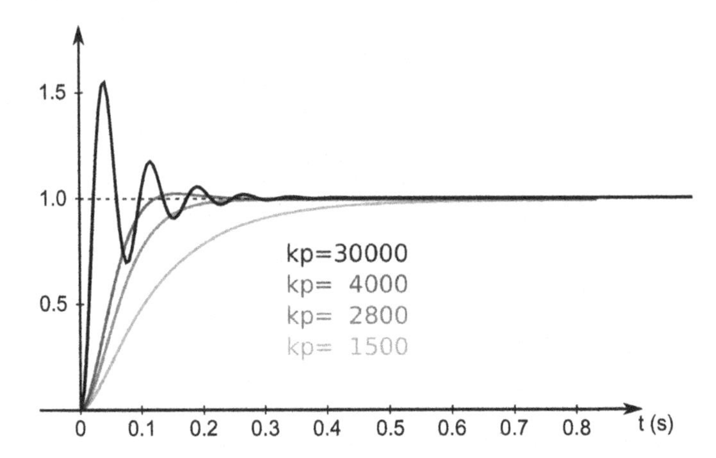

Figure 9.10 Step response of the digitally controlled positioner at increasing values of k_p. For values of k_p below 2,800, the controlled process responds similar to the overdamped continuous system, and at $k_p = 2800$, a critically damped system is achieved. For larger values of k_p, however, overshoot with oscillations develops rapidly, and with $k_p > 83000$, the system becomes unstable.

- For very high values of k_p, the poles become real-valued on the negative real axis of the z-plane (Figure 9.9), and one pole returns to the stable region inside the unit circle. This pole contributes an attenuated oscillation at a frequency of $1/2T$. The overall system, however, remains unstable.

The unit step response of the controlled system is shown in Figure 9.10. The high values for k_p reflect the overall low loop gain of the process, and they are similar to the continuous system. For the values for β and υ in this example, a continuous feedback control system requires $k_p = 2500$ for critical damping, and the digital system requires $k_p = 2800$. When k_p is increased, the behavior of the continuous and discrete systems diverge, and the time-discrete system develops an oscillatory response with stronger overshoot than the continuous system, and it can eventually turn unstable. This example highlights the different behavior of continuous and time-discrete control systems and emphasizes the need for careful design. In a practical realization, the motor is unlikely to oscillate with 20 Hz frequency, because the energy needed to overcome the inertia is enormous. Rather, the driver stage for the motor current is likely to saturate, and nonlinear effects begin to take hold. In a laboratory model, the motor indeed showed oscillations, but at a lower frequency than expected from the time-discrete model. Additional effects that the model does not consider are nonlinear friction and mechanical slack in the gearbox and wormgear. In the laboratory model, the controlled system did not show true instability in the sense of exponentially-increasing oscillations, but rather continuous oscillations of constant amplitude, limited by the capability of the motor drive stage. The drive currents, however, can become very large and can potentially damage the drive stage.

The behavior changes again when the microcontroller creates an additional delay T. Initially, we assumed that the analog-to-digital conversion and the computation of the corrective action are fast compared to T. Often, conversion and computation can run in parallel to further shorten T, and to provide fully synchronized input and output sampling points. The disadvantage of this approach is the fact that the computation relies on the conversion result from the previous sampling period. Therefore, the controller introduces an additional delay of z^{-1}. The consequence is an onset of oscillations at lower values of k_p and at a lower frequency.

Clearly, mathematical models and computer simulations have their limits and can capture a real-world system only to a certain extent. Simulation tools can be very helpful for the initial design stages. For the final design, however, the use of a physical prototype is unavoidable.

10 Stability Analysis for Linear Systems

Abstract

One of the most crucial design requirements for any feedback control system is *stability*. Stability means that a system has a bounded output signal for any bounded input signal. This chapter presents simple tools to quickly determine whether a given system is stable, and to determine the value range of coefficients (such as the controller gain) required to keep the feedback control system operating in a stable manner. For continuous systems, this algorithmic method is the *Routh-Hurwitz scheme*, and for time-discrete systems, the analogous method is the *Jury test*.

One of the most fundamental tasks of control design is to test whether a system is stable. This question translates directly into algorithmic tests that determine if any poles of a transfer function are in the right half of the s-plane or outside of the unit circle in the z-plane. Two common test are the *Routh-Hurwitz scheme* for continuous systems and the *Jury test* for time-discrete systems.

Several definitions of stability exist that lead to the same findings, but may differ in marginally stable cases. In general, **a system is stable if every *bounded* input produces a bounded output.** This stability criterion is referred to as *BIBO stable*. Common definitions are:

- A continuous system is stable if all poles of the transfer function have a negative real component (i.e., lie in the left half of the s-plane).
- A time-discrete system is stable if all poles of the transfer function lie inside the unit circle $|z| = 1$ of the z-plane.
- A time-continuous, linear, time-invariant system with the transfer function $H(s)$ is stable if its impulse response is absolutely integrable, that is,

$$\int_0^\infty |h(t)|\mathrm{d}t < \infty \tag{10.1}$$

  where $h(t) = \mathcal{L}^{-1}\{H(s)\}$.
- A time-discrete, linear, time-invariant system with the impulse response h_k is stable if the impulse response is absolutely summable, that is,

$$\sum_{k=0}^{\infty} |h_k| < \infty \tag{10.2}$$

The third and fourth definitions are problematic, because under those definitions, a marginally stable system (such as an integrator) counts as unstable. However, we can

Linear Feedback Controls. http://dx.doi.org/10.1016/B978-0-12-405875-0.00010-3
© 2013 Elsevier Inc. All rights reserved.

establish an intuitive stability criterion that helps us identify marginally stable systems as actually stable. Imagine a water tank (the integrator with a pole in the origin) and a pipe that allows water to flow into the tank. If we pour a bucket of water into the tank (this is our bounded signal, and we could perhaps describe it as $x(t) = u(t) - u(t - \tau)$), the water level rises a bit and then remains constant. However, if we turn on the faucet and allow water to continuously flow into the tank (this can be represented by a signal $x(t) = u(t)$), it will overfill.

The debate at this point is whether the unit step signal $u(t)$ is bounded. In fact, $u(t)$ itself is *not* absolutely integrable and therefore not bounded. The apparent instability of the integrator is therefore caused by an inadvertent unbounded input signal, rather than a true process instability. More closely examined, the integrator has a transfer function of $1/s$, and the step function has a Laplace transform of $1/s$. The output signal is therefore $1/s^2$ and is represented by a double pole in the origin. In a similar fashion, an undamped resonance system (poles at $\pm j\omega$) is stable, because its amplitude remains within bounds, *unless* it is excited at its resonance frequency. In that case, however, a complex conjugate *double* pole pair exists at $\pm j\omega$, and the amplitude increases indefinitely. Clearly, this problem only occurs with a marginally stable system where the input signal is represented by a pole or pole pair identical to the process pole or pole pair, resulting in double poles on the imaginary axis.

The shortcoming of the mathematical definition of *absolutely integrable* or *absolutely summable* is that it does not recognize the marginally stable cases correctly. We conclude that any linear system with poles on or to the left of the imaginary axis (continuous case) or with poles on or inside the unit circle (time-discrete case) is BIBO stable. However, systems with repeated poles on the imaginary axis (on the unit circle for discrete systems) are unstable.

10.1 The Routh-Hurwitz Scheme

A linear system with n poles that has the transfer function

$$H(s) = \frac{p(s)}{q(s)} = \frac{p(s)}{a_n s^n + a_{n-1} s^{n-1} \cdots a_1 s + a_0} \tag{10.3}$$

is stable if none of the roots of the characteristic polynomial $q(s)$ have positive real components, i.e., if none of the roots lie to the right of the imaginary axis in the s-plane. Often, but not always, numerical solutions for $q(s) = 0$ can be found. If the solution for $q(s) = 0$ is inaccessible, the Routh-Hurwitz scheme provides a fast and robust method to test whether any roots lie in the right half-plane. The Routh-Hurwitz scheme does not explicitly provide the roots, but it tells us the *number of roots* in the right half-plane. Since the Routh-Hurwitz scheme can be applied even if some of the coefficients a_k are unknown or undetermined, this method can be used to rapidly determine the stable range of unknown or variable coefficients.

We arrange the coefficients of $q(s)$ in Eq. (10.3) in a table as follows, whereby any coefficient a_{n-k} (and b_{n-k} and c_{n-k}, etc.) counts as zero for $k > n$:

$q(s) = a_n s^n + a_{n-1} s^{n-1} \cdots a_1 s + a_0$				
s^n	$\mathbf{a_n}$	a_{n-2}	a_{n-4}	$\cdots$
s^{n-1}	$\mathbf{a_{n-1}}$	a_{n-3}	a_{n-5}	$\cdots$
s^{n-2}	$\mathbf{b_{n-1}}$	b_{n-3}	b_{n-5}	$\cdots$
s^{n-3}	$\mathbf{c_{n-1}}$	c_{n-3}	c_{n-5}	$\cdots$
s^{n-4}	$\mathbf{d_{n-1}}$	d_{n-3}	d_{n-5}	$\cdots$
$\cdots$				
s^0	$\cdots$			

Starting at the line with s^{n-2}, each coefficient gets computed from the two lines above the coefficient, and from the leftmost column and the column to the right of the coefficient. The coefficients b_{n-1} and b_{n-3} serve as examples:

$$b_{n-1} = \frac{-1}{a_{n-1}} \begin{vmatrix} a_n & a_{n-2} \\ a_{n-1} & a_{n-3} \end{vmatrix}; \quad b_{n-3} = \frac{-1}{a_{n-1}} \begin{vmatrix} a_n & a_{n-4} \\ a_{n-1} & a_{n-5} \end{vmatrix} \tag{10.4}$$

For the coefficients c_i, the lines with s^{n-2} and s^{n-1} are used in a similar fashion:

$$c_{n-1} = \frac{-1}{b_{n-1}} \begin{vmatrix} a_{n-1} & a_{n-3} \\ b_{n-1} & b_{n-3} \end{vmatrix}; \quad c_{n-3} = \frac{-1}{b_{n-1}} \begin{vmatrix} a_{n-1} & a_{n-5} \\ b_{n-1} & b_{n-5} \end{vmatrix} \tag{10.5}$$

This schematic principle can be continued for the coefficients d_i, where the lines with s^{n-3} and s^{n-2} are used:

$$d_{n-1} = \frac{-1}{c_{n-1}} \begin{vmatrix} b_{n-1} & b_{n-3} \\ c_{n-1} & c_{n-3} \end{vmatrix}; \quad d_{n-3} = \frac{-1}{c_{n-1}} \begin{vmatrix} b_{n-1} & b_{n-5} \\ c_{n-1} & c_{n-5} \end{vmatrix} \tag{10.6}$$

The Routh array needs to be completely filled down to the row with s^0, and higher-order polynomials necessitate computation of coefficients $e_i, f_i, \ldots$

The polynomial q(s) has no roots in the right half-plane (and the system is stable) when all coefficients in the first column have the same sign. Conversely, each sign change in the first column indicates the existence of one pole in the right half-plane.

Sometimes, zero-valued elements in the array require special treatment:

1. One element (and only one element) in the first column is zero and causes a division by zero in one of the higher-order coefficients. This type of zero-valued element usually indicates a marginally stable system. For example, the case $b_{n-1} = 0$ makes it impossible to calculate the c_i-coefficients. In this case, we may set $b_{n-1} = \epsilon$ with ϵ being a very small positive value. Now, subsequent coefficients may be calculated, and the system is then analyzed for $\epsilon \to 0^+$.
2. More than one element in one row is zero. This generally leads to subsequent coefficients being undefined as a consequence of zero divided by zero. This case indicates symmetries of the type $(s - p)(s + p)$, where generally $p \in \mathbb{C}$. An auxiliary polynomial $U(s)$ can be formed from the coefficients of the line immediately preceding the row with the zeros. $U(s)$ has the order of the power of s

in that line. For example, $U(s)$ for the line with s^{n-2} is computed as $U(s) = b_{n-1}s^{n-2} + b_{n-3}s^{n-4} + b_{n-5}s^{n-6} \cdots$. The polynomial $U(s)$ is a factor of $q(s)$ and its roots are also roots of $q(s)$. We can examine the roots of $U(s)$ individually (often sufficient to detect instability), or divide $q(s)$ by $U(s)$ to reduce the order of the polynomial.

3. Repeated roots on the imaginary axis of the s-plane cause a system to be unstable. The Routh-Hurwitz scheme does not reveal this instability.

10.2 Routh Arrays for Low-Order Systems

Low-order systems occur frequently, and the special cases for second, third, and fourth order are listed here. For a polynomial $q(s) = a_2s^2 + a_1s + a_0$, the Routh array is:

$q(s) = a_2s^2 + a_1s + a_0$		
s^2	a_2	a_0
s^1	a_1	0
s^0	$b_1 = a_0$	

In the third row, b_1 is computed as:

$$b_1 = \frac{-1}{a_1} \begin{vmatrix} a_2 & a_0 \\ a_1 & 0 \end{vmatrix} = a_0 \tag{10.7}$$

When all primary coefficients $a_k > 0$, a second-order system *can never have unstable poles*. We can confirm this observation by using frequency-response methods (Chapter 11) and with the root locus method (Chapter 12).

A general third-order system has the characteristic polynomial $q(s) = a_3s^3 + a_2s^2 + a_1s + a_0$, and its Routh array is:

$q(s) = a_3s^3 + a_2s^2 + a_1s + a_0$		
s^3	a_3	a_1
s^2	a_2	a_0
s^1	b_2	0
s^0	a_0	

Here, b_2 needs to be explicitly computed:

$$b_2 = \frac{-1}{a_2} \begin{vmatrix} a_3 & a_1 \\ a_2 & a_0 \end{vmatrix} = \frac{a_2a_1 - a_0a_3}{a_2} \tag{10.8}$$

In the last row (s^0), we need to compute c_2; however, we can easily see that $c_2 = b_2a_0/b_2 = a_0$, which is similar to the last line of the second-order example. In third-order systems, b_2 can introduce a double sign change when $a_2a_1 < a_0a_3$. If b_2 turns negative, one sign change occurs from a_2 to b_2, and the second from b_2 to a_0, and this indicates a complex conjugate pole pair in the right half-plane. The order of three is the lowest order where instability can occur.

A general fourth-order system has the characteristic polynomial $q(s) = a_4 s^4 + a_3 s^3 + a_2 s^2 + a_1 s + a_0$. Compared to the previous two examples, we need one more column:

$q(s) = a_4 s^4 + a_3 s^3 + a_2 s^2 + a_1 s + a_0$			
s^4	a_4	a_2	a_0
s^3	a_3	a_1	0
s^2	b_3	a_0	
s^1	c_3	0	
s^0	a_0		

We need to explicitly compute multiple coefficients in the rows for s^2 and s^1:

$$\begin{aligned} b_3 &= \frac{-1}{a_3}\begin{vmatrix} a_4 & a_2 \\ a_3 & a_1 \end{vmatrix} = \frac{a_2 a_3 - a_1 a_4}{a_3} \\ b_1 &= \frac{-1}{a_3}\begin{vmatrix} a_4 & a_0 \\ a_3 & 0 \end{vmatrix} = a_0 \\ c_3 &= \frac{-1}{b_3}\begin{vmatrix} a_3 & a_1 \\ b_3 & a_0 \end{vmatrix} = \frac{a_1 b_3 - a_0 a_3}{b_3} \end{aligned} \tag{10.9}$$

As in the previous cases, the last column holds $d_3 = a_0$. In the fourth-order system, both b_3 and c_3 can introduce sign changes, and stability requires both $a_2 a_3 > a_1 a_4$ and $a_1 b_3 > a_0 a_3$ to be met. It is interesting to note that the fourth-order system can only have two sign changes (either b_3 is negative, or c_3 is negative, or both are negative) or none at all. We can see this observation confirmed with the root locus method, where the asymptote configuration for both third- and fourth-order systems allows only one complex conjugate pole pair to move into the right half-plane.

10.3 Stability of Time-Discrete Systems with the *w*-Transform

For time-discrete systems whose transfer function is given in the z-domain,

$$H(z) = \frac{p(z)}{q(z)} = \frac{p(z)}{a_n z^n + a_{n-1} z^{n-1} \cdots a_1 z + a_0} \tag{10.10}$$

stability requires that all poles of the transfer function $H(z)$, and thus all roots of the denominator polynomial $q(z)$ lie inside the unit circle in the z-plane. With $z = e^{sT}$, we have established a relationship between the Laplace- and z-domains, where the imaginary axis of the s-plane maps to the unit circle of the z-plane (*cf.* Section 4.3).

If we are only interested in the number of roots inside and outside the unit circle, we can use the w-transform (Section 4.4): With the w-transform (Eq. (4.32)), any location z inside the unit circle (that is, $|z| < 1$) is mapped to a location in the w-plane with $\Re(w) < 0$. With the w-transform, we therefore obtain a new polynomial $q(w)$ for which we can apply the Routh-Hurwitz scheme.

Example . A time-discrete system has a transfer function $H(z)$ with the denominator polynomial $q(z) = z^3+z^2+0.75z$. We use Eq. (10.11) to substitute z in the denominator polynomial

$$z = \frac{1+w}{1-w} \tag{10.11}$$

and obtain the new w-domain equation

$$\begin{aligned} q(w) &= \left(\frac{1+w}{1-w}\right)^3 + \left(\frac{1+w}{1-w}\right)^2 + 0.75 \cdot \left(\frac{1+w}{1-w}\right) \\ &= \frac{0.75w^3 + 1.25w^2 + 3.25w + 2.75}{-w^3 + 3w^2 - 3w + 1} \end{aligned} \tag{10.12}$$

We are only interested in the new *numerator* polynomial $0.75w^3 + 1.25w^2 + 3.25w + 2.75$, because we are looking for any w for which $q(w) = 0$. We can use the third-order Routh-Hurwitz scheme and compute $b_2 = (4.0625 - 2.0625)/1.25 > 0$ (Eq. (10.8)). There is no sign change from a_2 to b_2 to a_0, and $q(w)$ has no zeros in the right half-plane. Therefore, $q(z)$ has no roots outside the unit circle and $H(z)$ is stable.

10.4 The Jury Test

Our starting point is once again the transfer function given in Eq. (10.10), and we need to determine if $q(z)$ has any roots outside the unit circle. The Jury test is comprised of several steps. If the test fails any of the steps, the system is unstable and the subsequent steps do not need to be completed.

1. Compute $q(z)$ for $z = 1$. If $q(1) \leq 0$, the system is unstable.
2. Compute $q(z)$ for $z = -1$. If n (i.e., the order of $q(z)$) is even and $q(-1) \leq 0$, the system is unstable. If n is odd and $q(-1) \geq 0$, the system is unstable. We can alternatively formulate this stability requirement for any n as $(-1)^n \cdot q(-1) > 0$ and find the system to be unstable when $(-1)^n \cdot q(-1) \leq 0$.
3. If $|a_0| \geq a_n$, the system is unstable.
4. If the previous steps have not indicated an unstable system, we need to continue by creating the *Jury array*. Note that the coefficients in every even line are the same as in the line above, but in reverse order. For a nth order system, the Jury array has $2n - 3$ lines.

$q(z) = a_n z^n + a_{n-1} z^{n-1} \cdots a_1 z + a_0$

1	a_0	a_1	a_2	$\ldots$	$\ldots$	a_{n-1}	a_n
2	a_n	a_{n-1}	a_{n-2}	$\ldots$	$\ldots$	a_1	a_0
3	b_0	b_1	b_2	$\ldots$	$\ldots$	b_{n-1}	
4	b_{n-1}	b_{n-2}	b_{n-3}	$\ldots$	$\ldots$	b_0	
5	c_0	c_1	c_2	$\ldots$	c_{n-2}		
6	c_{n-2}	c_{n-3}	c_{n-4}	$\ldots$	c_0		
$\ldots$	$\ldots$	$\ldots$	$\ldots$				
$2n-5$	u_0	u_1	u_2	u_3			
$2n-4$	u_3	u_2	u_1	u_0			
$2n-3$	v_0	v_1	v_2				

The following definitions provide the $b_k, c_k, \ldots, v_k$:

$$b_k = \begin{vmatrix} a_0 & a_{n-k} \\ a_n & a_k \end{vmatrix} = a_0 a_k - a_n a_{n-k} \tag{10.13}$$

$$c_k = \begin{vmatrix} b_0 & b_{n-k-1} \\ b_{n-1} & b_k \end{vmatrix} = b_0 b_k - b_{n-1} b_{n-k-1} \tag{10.14}$$

$$v_0 = \begin{vmatrix} u_0 & u_3 \\ u_3 & u_0 \end{vmatrix}; \quad v_1 = \begin{vmatrix} u_0 & u_2 \\ u_3 & u_1 \end{vmatrix}; \quad v_2 = \begin{vmatrix} u_0 & u_1 \\ u_3 & u_2 \end{vmatrix} \tag{10.15}$$

5. Test whether $|b_0| > |b_{n-1}|$. If $|b_0| \leq |b_{n-1}|$, the system is unstable.
6. Test whether $|c_0| > |c_{n-2}|$. If $|c_0| \leq |c_{n-2}|$, the system is unstable.
7. Continue these tests for all remaining coefficients. For the last three lines of the Jury array, $|u_0| > |u_3|$ and $|v_0| > |v_2|$ is required for stability.

10.5 Jury Arrays for Low-Order Systems

Special cases for low-order polynomials are given in this section. Unlike continuous systems, even first- and second-order time-discrete systems can exhibit instability. For first-order systems with $q(z) = a_1 z + a_0$, we know that the single root exists at $-a_0/a_1$. The Jury test does not apply to this case.

A second-order polynomial $q(z) = a_2 z^2 + a_1 z + a_0$ has only one line in its Jury array, and the Jury test in Section 10.4 ends after Item 3. Listed explicitly as an example, the Jury test for the second-order system contains the following steps:

1. Compute $q(z)$ for $z = 1$. Here, $q(1) = a_0 + a_1 + a_2$, and $q(1) > 0$ is required for stability.
2. Compute $q(z)$ for $z = -1$, that is, $q(-1) = a_0 - a_1 + a_2$. Since n is an even number, we require $q(-1) > 0$ for stability.
3. If $|a_0| < a_2$, and the two conditions above are met, the system is stable.

For a third-order system with the polynomial $q(z) = a_3 z^3 + a_2 z^2 + a_1 z + a_0$, the Jury array has three lines. First, we test the requirements $q(1) > 0$ and $q(-1) < 0$, then we can construct the Jury array as:

$q(z) = a_3z^3 + a_2z^2 + a_1z + a_0$				
1	a_0	a_1	a_2	a_3
2	a_3	a_2	a_1	a_0
3	b_0		b_2	

We use Eq. (10.15) to obtain $b_0 = a_0^2 - a_3^2$ and $b_2 = a_0a_2 - a_1a_3$. For stability, we require the following four conditions to be met (note that we do not need to compute b_1):

1. $a_0 + a_1 + a_2 + a_3 > 0$.
2. $a_0 - a_1 + a_2 - a_3 < 0$.
3. $|a_0| < a_3$.
4. $|b_0| > |b_2|$.

To examine a fourth-order system, we begin with the first three requirements, $q(1) > 0$ and $q(-1) > 0$ and $|a_0| < a_4$. If all three requirements are met, we continue with the Jury array for a fourth-order system:

$q(z) = a_4z^4 + a_3z^3 + a_2z^2 + a_1z + a_0$					
1	a_0	a_1	a_2	a_3	a_4
2	a_4	a_3	a_2	a_1	a_0
3	b_0	b_1	b_2	b_3	
4	b_3	b_2	b_1	b_0	
5	c_0		c_2		

The only two remaining requirements are:

1. $|b_0| > |b_3|$ with $b_0 = a_0^2 - a_4^2$ and $b_3 = a_0a_3 - a_1a_4$. For the next step, we also need $b_1 = a_0a_1 - a_3a_4$ and $b_2 = a_0a_2 - a_2a_4$
2. $|c_0| > |c_2|$ with $c_0 = b_0^2 - b_3^2$ and $c_2 = b_0b_2 - b_1b_3$.

10.6 Example Applications

Let us consider several examples for continuous and for time-discrete systems. First, a fourth-order system has the characteristic polynomial $q(s) = s^4 + 8s^3 + 24s^2 + 32s + 16$, and its Routh array below has only positive coefficients in the first column. There are no roots of $q(s)$ in the right half-plane. We can verify this observation by numerically finding the roots (e.g., with Scilab): all four roots lie at (-2) in the left half-plane.

$q(s) = s^4 + 8s^3 + 24s^2 + 32s + 16$			
s^4	1	24	16
s^3	8	32	0
s^2	160/8	16	
s^1	512/20	0	
s^0	16		

For the next two examples, consider the feedback loop in Figure 10.1. A stable process of third order has the transfer function

$$G(s) = \frac{1}{s^3 + 3s^2 + s + 1} \tag{10.16}$$

and is part of a P control loop with gain $k_p = 5$. We first examine the stability of the control loop with P control only (i.e., the shaded differential compensator omitted). The closed-loop transfer function is

$$H(s) = \frac{Y(s)}{X(s)} = \frac{5}{s^3 + 3s^2 + s + 6} \tag{10.17}$$

To examine whether the characteristic polynomial $q(s) = s^3 + 3s^2 + s + 6$ has roots in the right half-plane, we use the special case of a third-order system above, and the only coefficient of concern is b_2, which we obtain as $b_2 = -1$. One sign change occurs from a_2 to b_2, and the second from b_2 to a_0, and this indicates a complex conjugate pole pair in the right-half plane. Consequently, there are two roots with positive real component. In fact, the explicit roots are at -3.26 and $+0.129 \pm 1.351j$. The complex conjugate pole pair lies in the right half-plane. The closed-loop system is therefore unstable and exhibits an exponentially increasing oscillation.

To improve stability, we now introduce a phase-lead compensator (shaded block in Figure 10.1) with an adjustable parameter k_D. We can use the Routh-Hurwitz scheme to determine the value range for k_D that is needed to make the closed-loop system stable. With the compensator, the closed-loop transfer function becomes

$$H(s) = \frac{Y(s)}{X(s)} = \frac{k_D s + 5}{s^3 + 3s^2 + (k_D + 1)s + 6} \tag{10.18}$$

Once again, we use the special third-order case for the characteristic polynomial of the transfer function in Eq. (10.18),

$q(s) = s^3 + 3s^2 + (k_D + 1)s + 6$		
s^3	1	$k_D + 1$
s^2	3	6
s^1	b_2	0
s^0	6	

where only b_2 is a concern and compute b_2 from Eq. (10.8). For stability, we require

$$b_2 = \frac{3(k_D + 1) - 6}{3} > 0 \tag{10.19}$$

and obtain $k_D > 1$ as a requirement for stability and $k_D = 1$ for the marginally stable case. For $k_D = 1$, we can determine the explicit roots of $q(s) = 0$ and find a real-valued pole of the closed-loop system at -3 and a complex conjugate pole pair on the imaginary axis at $\pm\sqrt{2}j$, which confirms the marginally stable case. This third example

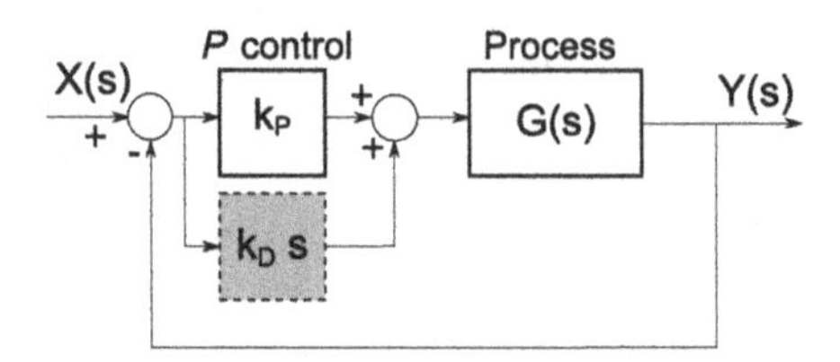

Figure 10.1 Feedback system with a process $G(s)$, a P controller with gain $k_p = 5$ and unit feedback path. The shaded block is an optional D-component with the transfer function $k_D s$.

demonstrates how the Routh-Hurwitz scheme can be used to determine the value range for an unknown parameter in which a system is stable.

For an example time-discrete system, let us examine the positioner control example from Section 9.6. A closed-loop transfer function with *G(z)* from Eq. (9.34) and *PID* control with variable k_p and fixed integral and differential component is

$$\frac{X(z)}{X_{set}(z)} = \frac{k_p(0.4662z^2 - 0.4335z + 0.0265)}{z^3 + z^2(-1.7788 + 0.4662k_p) + z(0.7788 - 0.4335k_p) + 0.0265k_p} \tag{10.20}$$

For this third-order system, the Jury test provides us with the requirements:

1. $q(1) > 0$, therefore $k_p > 0$. This requirement is always met.
2. $q(-1) < 0$, therefore $k_p < 3.5567/0.9292$ or $k_p < 3.85$. This requirement limits the maximum value for k_p.
3. $|a_0| < a_3$, in this case $0.0265k_p < 1$ limits the range for k_p to $k_p < 37.7$, which is included in the previous requirement.
4. $|b_0| > |b_2|$, which leads to $|(0.0265k_p)^2 - 1| > |0.0124k_p^2 + 0.3864k_p - 0.7788|$. This requirement resolves to $k_p < 4.07$, which is included in the second requirement.

This example illustrates how the Jury test can be used, similar to the Routh-Hurwitz scheme, to determine the stable range for a parameter. However, the computation can get fairly complex and may lead to multiple regions of stability along discontinuous polynomial surfaces.

11 Frequency-Domain Analysis and Design Methods

Abstract

Analysis of a system's frequency response is a powerful tool to predict the behavior of a control system. In fact, frequency-domain analysis allows to conclude closed-loop stability from the open-loop frequency response. When the input to a linear system is a sinusoid, its output is a sinusoid of the same frequency, but with different amplitude and phase. Amplitude and phase as functions of the frequency can be plotted in Nyquist- and Bode-diagrams. Particularly in Bode diagrams, the concept of *gain* and *phase margin* give rise to the notion of relative stability, that is, how much variation of process parameters is allowable before a feedback control system turns unstable. Similar insights are gained from the Nyquist criterion, which allows to conclude from the loop gain to the presence of closed-loop poles in the right half-plane.

Up to this point, we have predominantly examined the response of feedback control systems to step changes at the input. We have used the step response to characterize a system and to determine the system coefficients. Frequency-domain methods are based on sinusoidal signals at equilibrium. By examining the response of a linear system to sinusoidal signals, we obtain an additional powerful method for system analysis and design. Moreover, frequency-domain methods are particularly well-suited to examine the concept of relative stability.

11.1 Frequency Response of LTI Systems

The steady-state response of any linear, time-invariant system to a sinusoidal signal is another sinusoidal signal with a different amplitude and phase, but with the same frequency as the input signal (Figure 11.1). The amplitude modulation m and the phase shift φ are generally functions of the frequency. We can interpret this behavior as a multiplication of the sinusoidal input signal by a frequency-dependent complex number $m \cdot e^{j\varphi}$. A system can be fully characterized by analyzing its frequency response, specifically, the frequency-dependent output magnitude and phase shift for all possible frequencies. Furthermore, the frequency response of an open-loop system allows us to predict the stability of the corresponding closed-loop system.

To obtain the frequency response in practice, measurements can be taken for a limited range of frequencies if some *a-priori* knowledge of the system response is available. Devices are available that perform this task automatically. A *frequency plot* of the system, often also called *Nyquist plot*, would then show real and imaginary part of the complex number $m \cdot e^{j\varphi}$ for the examined frequency range.

Linear Feedback Controls. http://dx.doi.org/10.1016/B978-0-12-405875-0.00011-5
© 2013 Elsevier Inc. All rights reserved.

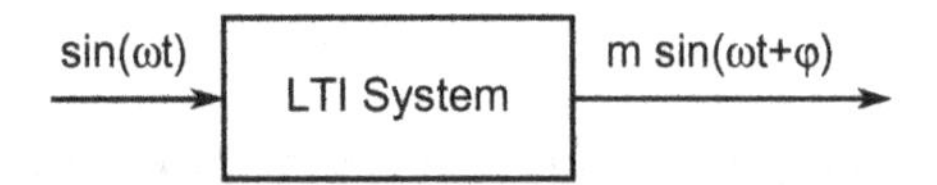

Figure 11.1 If the input of a linear, time-invariant system is excited with a continuous sinusoidal signal, the output is again sinusoidal, but with a modulated amplitude m and a phase shift φ with respect to the input signal. For a sinusoidal signal, the LTI system behaves as if it multiplies the input signal with a complex number $m \cdot e^{j\varphi}$.

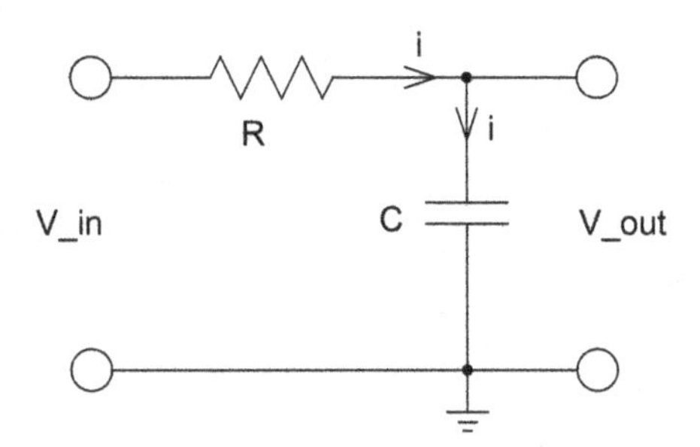

Figure 11.2 *R-C* lowpass as an example of a first order system. The capacitor acts as a complex impedance $Z_C = 1/j\omega C$ in a voltage divider. The output voltage and phase angle are frequency-dependent.

Let us introduce the *R-C* lowpass filter as a simple example (Figure 11.2). The capacitor has a complex impedance $Z_C = 1/j\omega C$ which allows us to use the voltage divider rule:

$$\frac{V_{out}}{V_{in}} = \frac{Z_C}{Z_C + R} = \frac{1}{1 + j\omega RC} = \left.\frac{1}{\tau s + 1}\right|_{\sigma=0} \tag{11.1}$$

with $\tau = RC$. Note that we can use s instead of $j\omega$ if we assume that the real part of s is zero. If we compare this equation to the equation of the waterbath and the DC motor, we realize that these systems have similar character and a similar step response. The reason for the similar behavior lies in the presence of a single energy storage element in each of these systems: The capacitor stores charges (integrated current), the waterbath stores heat (the integrated heating power), and the motor stores angular velocity (the integrated torque).

It is possible to apply a sinusoidal input voltage $V_{in}(s) = V_0\omega/(s^2 + \omega^2)$ and compute the output voltage in the Laplace-domain by using partial fraction expansion (see Section 3.5.1), but it is more practical for steady-state sinusoids $V_0 e^{j\omega t}$ to use the complex impedance and compute the output voltage in the time domain:

$$V_{out}(t) = V_0 e^{j\omega t} \frac{1}{1 + j\omega RC} \tag{11.2}$$

Equation (11.2) essentially confirms that we are multiplying the sinusoidal input signal with a complex number $me^{j\varphi}$. To obtain the magnitude m and phase φ of the transfer function, we apply the rules in Eq. (11.3) for a complex number c:

$$m = \sqrt{\Re(c)^2 + \Im(c)^2}, \quad \tan\varphi = \frac{\Im(c)}{\Re(c)} \tag{11.3}$$

We now expand the fraction in Eq. (11.2) with the complex conjugate of the denominator and separate the real and imaginary parts (Eq. (11.4)), which gives us the closed-term solutions for m and φ:

$$m \cdot e^{j\varphi} = \frac{1 - j\omega RC}{1 + \omega^2 (RC)^2} = \frac{1}{1 + \omega^2 (RC)^2} - j\frac{\omega RC}{1 + \omega^2 (RC)^2} \tag{11.4}$$

$$m = \frac{1}{\sqrt{1 + \omega^2 (RC)^2}}, \quad \tan\varphi = -\omega RC \tag{11.5}$$

For very low frequencies, the imaginary part in Eq. (11.4) vanishes and the real part tends toward unity. For very large values of ω, on the other hand, both the real and imaginary parts approach zero. In between, real and imaginary parts describe a semicircle in the Nyquist plot (Figure 11.3).

Some information about the transfer function cannot be seen in Nyquist plots. One example is the location of poles (the pole at $s = -1/RC$ is highlighted by a dotted arrow in Figure 11.3). A more relevant graphical representation of the frequency response are Bode plots, which are covered further below.

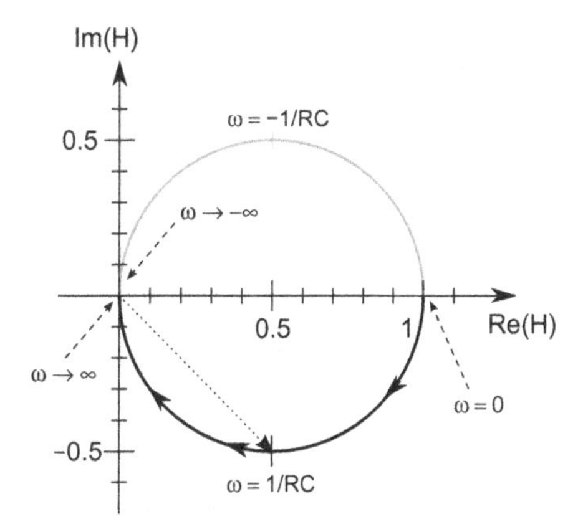

Figure 11.3 Nyquist plot of the RC-lowpass. For very low frequencies, the transfer function is close to unity. When $\omega = 1/RC$, we can see from Eq. (11.4) that $\Re(H) = 0.5$ and $\Im(H) = -0.5$. This point is indicated by the dotted arrow. When $\omega \to \infty$, both real and imaginary parts of $H(s)$ approach zero. The curve can be constructed for negative frequencies as well, and we obtain the shaded frequency response as ω goes from 0 to $-\infty$. Note that this is not an s-plane plot. Rather, we are plotting the real and imaginary parts of a transfer function $H(j\omega)$ as it changes with increasing frequency.

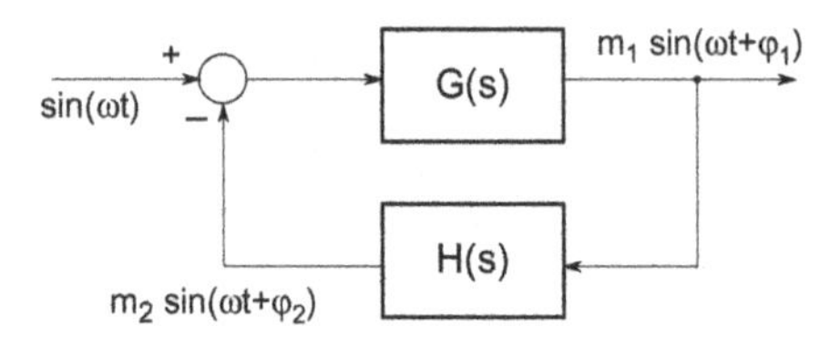

Figure 11.4 In a closed-loop feedback system, a sinusoidal input is propagated to the output and further to the feedback node. When the open-loop phase shift φ_2 reaches 180°, the phase shift corresponds to a negative sign because $\sin(\omega t + \pi) = -\sin(\omega t)$, and the negative feedback loop turns into a positive feedback loop.

11.2 Frequency Response and Stability

Consider a simple closed-loop feedback system as shown in Figure 11.4. A sinusoidal input signal propagates to the output with modulation and phase shift, but it also propagates to the feedback node with a different modulation and phase shift. A critical frequency is reached when the combined phase shift becomes 180°, because $\sin\omega t + \pi = -\sin\omega t$. This sign inversion causes the feedback loop to turn into a positive feedback loop. The system is unstable when at this point the open-loop modulation m_2 is greater than unity. In such a case, a sinusoidal input is not even necessary. According to the Fourier theorem, any periodic signal can be written as a superposition of harmonic oscillations of different frequency. Even the most miniscule fluctuation at the input feeds back through the loop, and the harmonic at which $\varphi_2 = \pi$ is *added* to the input signal with a new amplitude m_2. If $m_2 > 1$, the signal is amplified and leads to an exponentially increasing oscillation: instability ensues. **From the phase/magnitude behavior of the open-loop system, we can determine stability of the closed-loop system.** Even more, we can determine how much gain reserve we have before a system becomes unstable (i.e., relative stability).

We make two observations. First, for the closed-loop system to become unstable, the loop gain needs to be greater than unity when the phase shift reaches 180° (oscillation condition). A first-order system asymptotically reaches 90° phase shift, and a second-order system, 180°. First- and second-order systems can therefore never reach the oscillation condition, and first- and second-order feedback control systems with stable subsystems are always stable. We already made this observation with the Routh-Hurwitz stability criterion for the second-order special case.

Second, in oscillators and signal generators, oscillations are desired. Many oscillators are based on the principle of a feedback system that has a total phase shift greater than 180°, and a gain at the critical phase shift slightly greater than unity. Some examples are presented in Chapter 14.

11.3 Bode Plots

Bode plots or Bode diagrams are plots of the magnitude and phase of a system's response as a function of frequency. In other words, we plot the complex number

$m \cdot e^{j\varphi}$ (Figure 11.1) as a function of the frequency. The Bode diagram contains two graphs: magnitude over frequency and phase angle over frequency. In a Bode diagram, the magnitude and the frequency are plotted in a logarithmic scale, whereas the phase angle is plotted in a linear scale.

First, let us define the logarithmic magnitude A in decibels (dB):

$$A = 20\ \mathrm{dB} \cdot \log_{10} m \tag{11.6}$$

With this definition, the Bode diagram becomes the plot of A and φ, both in linear scale, over the frequency in a logarithmic scale. Furthermore, the decibel scale allows us to more conveniently describe asymptotic behavior: For example, if the magnitude m of a system decays with the inverse of the frequency ($m \propto \omega^{-1}$), we can express this behavior as m decreasing by 20 dB per decade (i.e., increase by a factor of 10) of the frequency. Note that in some contexts the octave (i.e., doubling of the frequency) is used. A change of 20 dB per decade corresponds to 6 dB per octave.

It is worth-while examining the asymptotic behavior of the functions in Eq. (11.5). For low frequencies $\omega \ll 1/RC$, we obtain $m \approx 1$ and $\varphi \approx 0$. Consequently, $A \approx 0$ dB. For high frequencies $\omega \gg 1/RC$, $m \approx 1/\omega RC$ and $\varphi \approx -90°$. In the high-frequency case, A drops by 20 dB/decade. Only when $\omega \approx 1/RC$, a deviation from the asymptotes is expected. By using Eq. (11.5), we can determine the actual values of m and φ at the cutoff frequency $\omega_0 = 1/RC$, and we obtain $m = 1/\sqrt{2}$ with a corresponding $A = -3\mathrm{dB}$ and $\varphi = -45°$. Note that the transfer function (Eq. (11.1)) has a pole at $s = -1/RC$. This is not a coincidence. Figure 11.5 shows the Bode diagram of the R-C lowpass with the asymptotes. Here, the frequency axis is drawn in Hertz instead of the inverse seconds used for the angular frequency ω. Therefore, it is scaled by a factor of 2π.

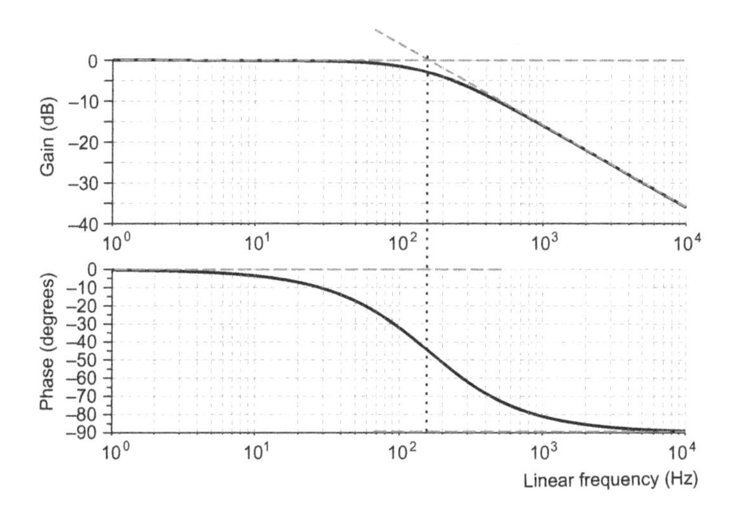

Figure 11.5 Bode diagram of the R-C lowpass with $R = 10\ \mathrm{k\Omega}$ and $C = 0.1\mu\mathrm{F}$. The cutoff frequency is $f_0 = 1/(2\pi RC) = 159$ Hz, indicated by the vertical dotted line. The asymptotes are drawn as gray dashed lines. The magnitude asymptote for $\omega \gg 1/RC$ has a slope of -20 dB/decade.

11.4 Definition of Phase and Gain Margin

It follows from the frequency-response behavior that first- and second-order systems are always stable in a feedback configuration that has only proportional components. A third-order system can cause a phase shift of up to 270° and is therefore the lowest-order system where P control may cause instability. Let us consider the example in Figure 11.6. All time-dependent functions of the controller have either been merged with the process ($G(s)$) or with the sensor ($H(s)$), and the gain factor k is the only remaining variable parameter. Let us further assume the loop gain $L(s)$ to be

$$L(s) = k \cdot G(s) \cdot H(s) = \frac{k}{s^3 + 52s^2 + 780s + 3600} \quad (11.7)$$

The function $L(s)$ is the *open-loop* transfer function, yet we can determine *closed-loop* stability. We choose arbitrarily $k = 1$ and draw the Bode diagram of the function in Eq. (11.7), which is shown in Figure 11.7. In a Bode diagram, phase and gain margin are defined as:

- The *gain margin* is the reciprocal of the loop gain where the phase crosses $-180°$.
- The *phase margin* is the difference between $-180°$ and the phase angle that the system exhibits at the frequency where its gain is exactly 1 (0 dB).

In the example of Eq. (11.7) and Figure 11.7, the gain margin is 92 dB or approximately 40,000. A phase margin is not defined, because the loop gain at $k = 1$ never reaches unity. From the gain margin, we know that any $k < 40,000$ keeps the closed-loop system stable, whereas $k > 40,000$ leads to an unstable system. We now know the highest possible value of k that we can choose.

The phase margin, on the other hand, tells us how much phase shift (for example a dead-time delay or fluctuations in one pole) the system can tolerate before becoming unstable. The phase margin becomes more relevant when a system is designed close to the limit of stability and the impact of additional phase shifts need to be examined.

Let us consider the system with the open-loop transfer function in Eq. (11.8):

$$L(s) = G(s) \cdot H(s) = \frac{5}{s+5} \cdot \frac{\omega^2}{s^2 + 2\zeta\omega s + \omega^2} \quad (11.8)$$

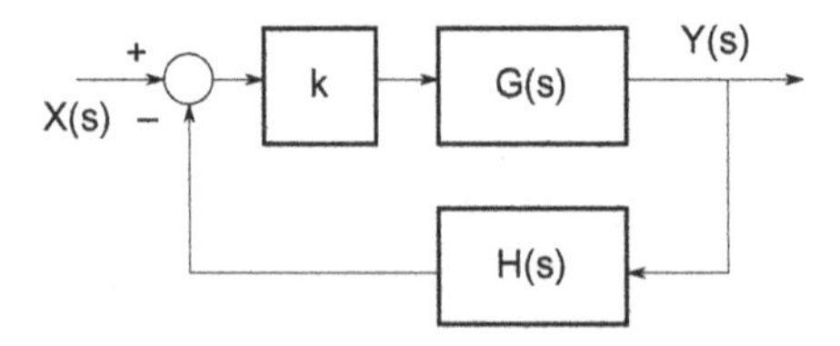

Figure 11.6 An example feedback system where all controller functions are either merged with the process $G(s)$ or the sensor $H(s)$, isolating the real-valued gain factor k. A Bode diagram can answer the question how large k can be chosen without causing system instability.

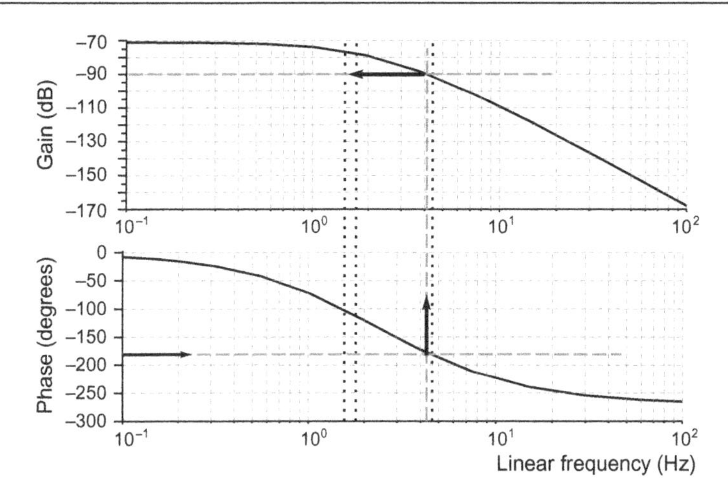

Figure 11.7 Bode diagram for the open-loop transfer function in Eq. (11.7) with $k = 1$. Dotted lines indicate the frequencies at the three poles. To obtain the gain margin, we use the phase part of the diagram to determine the frequency where the phase shift reaches $-180°$. From there, we determine the gain at that frequency, which in this case is -92 dB (dashed lines). The gain margin, therefore, is 92 dB. The phase margin is undefined, because the magnitude never reaches 0 dB.

Depending on ζ, a complex conjugate pole pair close to the imaginary axis can cause oscillatory behavior. For example, when $\omega = 10\ \mathrm{s}^{-1}$ and $\zeta = 0.12$, a complex pole pair exists at $p_{1,2} = -1.2 \pm 10j$. The open-loop system is stable, but the Bode diagram (Figure 11.8) reveals that the magnitude exceeds 0 dB at the critical phase angle of 180°. Without calculating the closed-loop transfer function, we know that a loop with unit gain feedback would turn unstable. The open-loop system has a negative gain margin.

11.5 Construction of Bode Diagrams

Bode diagrams can be computed (for example with Scilab) or sketched with the help of its asymptotes. The code for Scilab comprises (1) the definition of s, (2) the definition of a linear system with the transfer function $H(s)$, and (3) plotting the Bode diagram:

```
s = poly (0, 's');
H = 10/(s^3+5*s^2+6*s+10);
lsys = syslin ('c', H);
bode (lsys);
```

Furthermore, Scilab offers convenient functions to draw a Bode diagram complete with the gain and phase margins and to numerically compute gain and phase margins: the functions `show_margins`, `g_margin`, and `p_margin`, respectively. These functions are called in the exact same way as `bode`, i.e., with the linear system as argument.

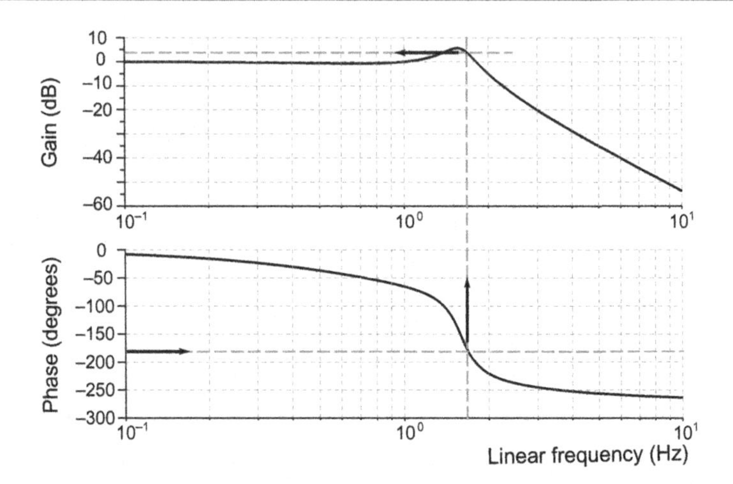

Figure 11.8 Bode diagram for a third-order system with a resonance peak caused by a complex conjugate pole pair. The Bode diagram clearly reveals that the resonance peak exceeds 0 dB at the critical phase angle of 180°. Although the open-loop transfer function is stable, a closed-loop system with no additional gain becomes unstable.

To manually sketch a Bode diagram from asymptotes, follow these steps:

1. Compute the poles and zeros of the transfer function.
2. On a sheet of paper, draw the axes for the logarithmic magnitude (A in dB, linear scale—note that this is equivalent to the magnitude m drawn in logarithmic scale) over the frequency ω, logarithmic scale. Underneath, draw the axes for the phase angle (linear scale) over the same frequency ω. Make sure that the frequency range of the abscissa exceeds the inverse of the time constant of all poles and zeros. For example, if you have poles at $s = -1$ and $s = -5$, the associated angular frequencies are 1 and 5 s^{-1} with frequencies $f = \omega/2\pi$. A good range in this example would be $\omega_{min} = 0.1\ \mathrm{s}^{-1}$ and $\omega_{max} = 100\ \mathrm{s}^{-1}$.
3. Does the transfer function have poles or zeros in the origin?

   - If not, its magnitude begins at low frequencies with a horizontal asymptote. The asymptote lies vertically at $A = 20 \log_{10} H(s = 0)$. The phase angle begins with a horizontal asymptote at $\varphi = 0$.
   - If there are n poles in the origin, the asymptote for A has a slope of $n \cdot (-20)$ dB/decade, and the vertical position is undefined at this point. The phase angle has a horizontal asymptote with a phase angle of $n \cdot (-90)°$.
   - If there are m zeros in the origin, the asymptote for A has a slope of $m \cdot 20$ dB/decade, and the vertical position is undefined at this point. The phase angle has a horizontal asymptote with a phase angle of $m \cdot 90°$.
   - If there are no horizontal asymptotes which could be used to determine their vertical position, a point on the curve can be determined, for example, the magnitude at $\omega = \omega_c$ or any other convenient frequency.

4. Now move along the negative real axis of the s-plane. Every time, you encounter a pole on the real axis at σ_p, the asymptote of A bends downwards by 20 dB/decade at the angular frequency of σ_p. Every time you encounter a zero, the asymptote bends upwards by 20 dB/decade. At the same point, a pole causes the asymptote of the phase angle to drop by 90°, and a zero raises the phase asymptote by 90°. Phase asymptotes are always horizontal.
5. If you encounter a complex conjugate pole pair, the asymptote of A bends downwards by 40 dB/decade and the phase angle decreases by 180°. However, complex conjugate pole pairs can be associated with a resonance peak that will be illustrated further below. A higher resonance peak is also accompanied by a steeper transition of the phase angle from the present asymptote to the new value that is lower by 180°. Complex zero pairs haver a similar, but opposite effect: With low damping, a complex conjugate zero pair causes a distinct notch in the frequency response A and a steep transition of the phase angle to a new, 180° higher, value.
6. Once you finished drawing the asymptotes, the actual course of $A(\omega)$ and $\varphi(\omega)$ runs smoothly between the asymptotes.

11.6 Frequency Response of a Second-Order System

We will now consider the example of a second-order system to illustrate resonance effects. Figure 11.9 shows a commonly used second-order active lowpass filter. The specific arrangement of the passive components around the operational amplifier is known as *Sallen-Key topology*. The op-amp is configured as a voltage follower, that is, the output voltage V_{out} is identical to the positive input voltage of the op-amp.

The voltage divider with R_2 and C_2 provides the relationship between V_x and V_{out}:

$$\frac{V_{out}(s)}{V_x(s)} = \frac{1}{1 + sR_2C_2} \tag{11.9}$$

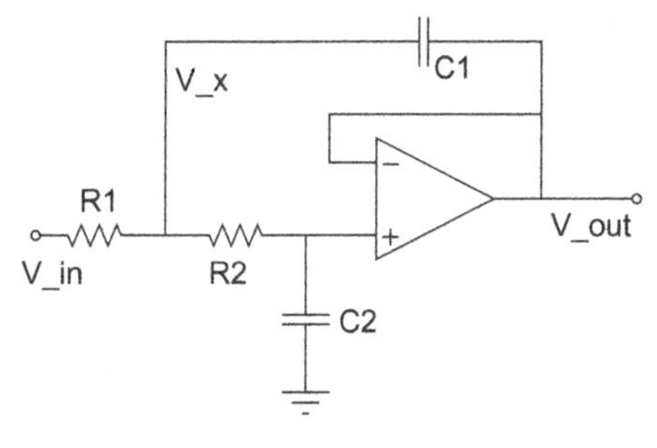

Figure 11.9 Second-order lowpass filter in Sallen-Key topology. If we use Z_1 and Z_2 for the complex impedances of C_1 and C_2, respectively, we can see that the complex voltage divider R_2, Z_2 is uncoupled by the op-amp, which is configured as a voltage follower. Feedback of the output voltage into the filter network yields a second-order system. For the determination of the transfer function, a voltage V_x is defined at the node of R_1, R_2, and C_1 toward ground.

The node rule provides the relationship between V_x, V_{in}, and V_{out}:

$$\frac{V_{in}(s) - V_x(s)}{R_1} = \frac{V_x(s) - V_{out}(s)}{1/(sC_1)} + \frac{V_x(s) - V_{out}(s)}{R_2} \tag{11.10}$$

Elimination of $V_x(s)$ in Eq. (11.10) with Eq. (11.9) yields, after some arithmetic manipulation, the transfer function of the circuit:

$$\frac{V_{out}(s)}{V_{in}(s)} = \frac{1}{s^2 C_1 C_2 R_1 R_2 + sC_2(R_1 + R_2) + 1} \tag{11.11}$$

It is now convenient to define the cutoff frequency ω_c and the damping coefficient ζ as

$$\begin{aligned} \omega_c &= \frac{1}{\sqrt{R_1 C_1 R_2 C_2}} \\ \zeta &= \frac{1}{2\omega_c C_1}\left(\frac{1}{R_1} + \frac{1}{R_2}\right) \end{aligned} \tag{11.12}$$

which leads to the simplified transfer function

$$\frac{V_{out}(s)}{V_{in}(s)} = \frac{\omega_c^2}{s^2 + 2\zeta\omega_c s + \omega_c^2} \tag{11.13}$$

The transfer function of the filter in Figure 11.9 is equivalent to the system function of a spring-mass-damper system. In fact, with a suitable choice of C_1 and C_2, the damping ratio and the center frequency can be adjusted independently. The roots of the transfer function, Eq. (11.13), are

$$p_{1,2} = -\omega_c\left(\zeta \pm \sqrt{\zeta^2 - 1}\right) \tag{11.14}$$

Very much like the spring-mass-damper system, a highly overdamped system ($\zeta \gg 1$) has one pole very close to the origin and one pole near $-2\zeta\omega_c$. As ζ decreases, the two poles move toward a common location, and critical damping is achieved when $\zeta = 1$ (double real-valued pole at $-\zeta\omega_c$; this occurs when $C_1 = C_2$ and $R_1 = R_2$).

Often, the Q-factor ("quality" coefficient) is used in place of the damping coefficient,

$$Q = |H(s = j\omega_c)| = \frac{1}{2\zeta} \tag{11.15}$$

which better describes the resonance peak in the frequency response. More precisely, Q is the magnitude of the frequency response at the cutoff frequency. For the critically damped case, we obtain $Q = 0.5$. Further undamping (i.e., reduction of ζ and the corresponding increase in Q) leads to two more interesting cases:

1. When $\zeta \approx 0.86$ or $Q \approx 0.58$, the group delay is approximately constant (in other words, the phase shift is proportional to the frequency in the passband of the filter). This case is particularly interesting in audio applications, because it is the lowpass with the optimal step response. A filter with constant group delay is called *Bessel filter*.

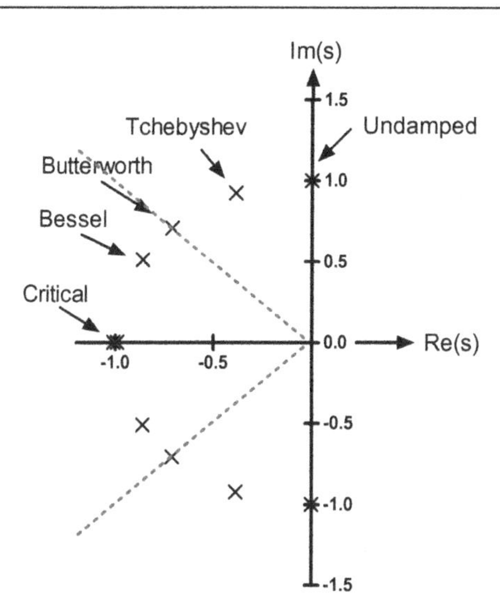

Figure 11.10 Location of the complex poles of special filter types along the root locus of an underdamped second-order system. Critical damping leads to a real-valued double pole. This is the lowest damping that does not to overshoot in the step response. A Bessel filter is characterized by a constant group delay. The Butterworth filter exhibits an overshoot as step response, but the step response approaches the final value from above, i.e., without subsequent undershoot. Tchebyshev filters exhibit a resonance peak; higher-order Tchebyshev filters show a distinct passband ripple (in this example 3 dB). These filters respond to a step input with multiple-oscillation overshoot.

2. When $\zeta = 1/\sqrt{2}$ or $Q = 1/\sqrt{2}$, the step response overshoots, but does not undershoot again—the step response reaches the final value from above. This filter has the steepest transition from the passband to the stopband without ripples in the passband. A filter with the damping factor $\zeta = 1/\sqrt{2}$ is called *Butterworth filter*.[1]

Further reduction of ζ causes longer oscillations in the step response, and the passband shows ripples. In this range, we find the family of *Tchebyshev filters*, which are defined by the amount of passband ripple. For example, a Tchebyshev filter with 3 dB passband ripple uses $\zeta = 0.385$ ($Q = 1.3$). Finally, when $\zeta = 0$, undamped oscillations occur, and the roots lie on the imaginary axis. A plot of the poles in the complex s-plane for a normalized frequency $\omega_c = 1$ is shown in Figure 11.10.

The frequency response of the four filters (critically damped, Bessel, Butterworth, Tchebyshev) is shown in Figure 11.11. At the resonance frequency ω_c, the critically damped filter attenuates the signal by 3 dB, and the attenuation is less for the Bessel and Butterworth filters. The Tchebyshev filter shows a small resonance peak of +3 dB. Further undamping (i.e., $\zeta < 0.385$) further increases the resonance peak, which actu-

[1] Higher-order Butterworth filters have equally-spaced poles on a semicircle in the left half-plane. However, detailed filter theory is beyond the scope of this book.

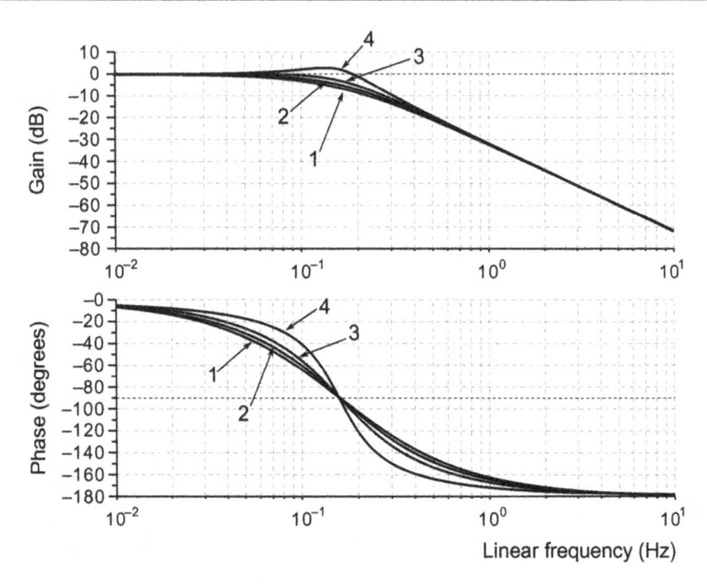

Figure 11.11 Bode diagrams corresponding to the poles of second-order systems shown in Figure 11.10. Asymptotically, all filters behave identical, but differences exist near the resonance frequency $\omega_C = 1\ \text{s}^{-1}$. **1:** Critically damped filter, **2:** Bessel filter, **3:** Butterworth filter, and **4:** Tchebyshev filter with 3 dB resonance peak. A moderate overshoot of the step response allows a steeper transition from the passband to the stopband.

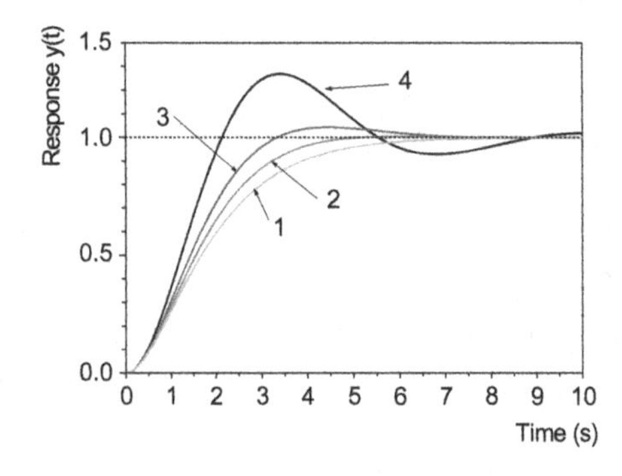

Figure 11.12 Step responses of the four filters discussed. The critically damped filter (**1**) has no overshoot. Bessel (**2**) and Butterworth (**3**) filters have overshoot and approach the final value from above. The 3 dB Tchebyshev filter (**4**) shows multiple oscillations in its step response.

ally reaches infinity for $\zeta = 0$. For completeness, the step responses of these filters are shown in Figure 11.12.

A resonance peak has a strong effect on filter design and on the design of feedback control systems. When the dynamic response is critical, it is often a good idea to design a controller that makes the closed-loop system of second or higher order. For example, a first-order system with $G(s) = b/(s+a)$ (please refer to Figure 11.6) can be converted into a second-order system when the controller is a *PID* controller, i.e., a controller with the transfer function

$$H_{FB}(s) = k\left(1 + k_D s + \frac{k_I}{s}\right) \tag{11.16}$$

The closed-loop transfer function is now

$$\frac{Y(s)}{X(s)} = \frac{kbs}{(1 + kbk_D)s^2 + (a + kb)s + kbk_I} \tag{11.17}$$

and suitable selection of k, k_I and k_D determine the dynamic response. Such a controller allows tuning of the rise and settling times.

To provide some rules of thumb, the maximum magnitude m_{peak} in the Bode diagram can be calculated for $\zeta < 1/\sqrt{2}$:

$$m_{peak} = \frac{1}{2\zeta\sqrt{1-\zeta^2}} \tag{11.18}$$

Note that the resonance frequency shifts when $\zeta < 1/\sqrt{2}$. Here, the resonance frequency ω_0 can be calculated from the natural, undamped resonance ω_n through

$$\omega_0 = \omega_n\sqrt{1-\zeta^2} \tag{11.19}$$

11.7 Frequency Response of Digital Filters

In Chapters 4 and 9 we have introduced an interpretation of time-discrete control systems as digital filters. Both time-discrete feedback controls and digital filters are described by their z-transform transfer functions. If a time-discrete system with the transfer function $H(z)$ receives a sinusoidal input sequence $x_k = \sin(\omega kT)$, the output signal is also a discrete approximation of a sinusoid. The frequency response characterizes the behavior of a system $H(z)$ somewhat similar to the frequency response of a continuous system, but some crucial differences must be noted.

In Chapter 4, the interpretation of the transfer function $H(z)$ on the unit circle as the discrete Fourier transform of its impulse response was presented. The Fourier transform of the system's impulse response sequence is identical to its frequency response. In analogy to continuous systems where the frequency response $H(\omega)$ was obtained from the transfer function $H(s)$ when $s = j\omega$, we can obtain the discrete-system frequency-response $H(\omega)$ by traversing the unit circle with $z = e^{j\omega}$. The first major difference to continuous systems is that the frequency spectrum is periodic. The point $z = -1 = e^{j\pi}$ on the unit circle corresponds to the Nyquist frequency $\omega_N = \pi/T$. Increasing the frequency above π/T yields the frequency response for negative frequencies ($\omega - 2\pi/T$). For frequencies above the sampling frequency $\omega_s = 2\pi/T$, the spectrum repeats itself periodically, because

$$e^{j\omega T} = e^{j(\omega T + 2k\pi)} = e^{j(\omega + 2k\pi/T)T} = e^{j(\omega + k\omega_s)T} \tag{11.20}$$

where k is any positive or negative integer number.

We will now examine a time-discrete approximation of the RC-lowpass (Figure 11.2) with a cutoff frequency $\omega_c = 1/RC \ll \omega_s$. The filter function is obtained either through pole mapping (i.e., the s-plane pole at $-\omega_c$ maps to a z-plane pole at $+e^{-\omega_c T}$) or by matching the step response (Table B.2). Both approaches lead to the infinite-impulse response filter function $H(z)$ and the corresponding finite-difference equation given in Eq. (11.21):

$$\begin{aligned} H(z) &= \left(1 - e^{-\omega_c T}\right) \cdot \frac{z}{z - e^{-\omega_c T}} \\ y_k &= y_{k-1} \cdot e^{-\omega_c T} + x_k \left(1 - e^{-\omega_c T}\right) \end{aligned} \tag{11.21}$$

The brute-force approach (and also the most impractical one) is to obtain a function of frequency by traversing the unit circle. With $z = e^{j\omega T}$, we can rewrite $H(z)$ as

$$H(\omega) = \left(1 - e^{-\omega_c T}\right) \cdot \frac{1}{1 - e^{-\omega_c T} e^{-j\omega T}} \tag{11.22}$$

To simplify the following section, let us define $a = e^{-\omega_c T}$. Furthermore, $e^{-j\omega T} = \cos \omega T - j \sin \omega T$, which also invites the definition of a normalized frequency $\Omega = \omega T$. Together, this provides

$$H(\omega) = \left(1 - a\right) \cdot \frac{1}{1 - a \cos \Omega + aj \sin \Omega} \tag{11.23}$$

Following the same path laid out in Eqs. (11.4) and (11.5), we expand the fraction with the complex conjugate of the denominator and separate real and imaginary parts:

$$\begin{aligned} H(\Omega) = \quad & (1 - a) \frac{1 - a \cos \Omega}{(1 - a \cos \Omega)^2 + (a \sin \Omega)^2} \\ & - j(1 - a) \frac{a \sin \Omega}{(1 - a \cos \Omega)^2 + (a \sin \Omega)^2} \end{aligned} \tag{11.24}$$

By using Eq. (11.3), we obtain magnitude and phase as a function of Ω:

$$m = \frac{1 - a}{\sqrt{(1 - a \cos \Omega)^2 + (a \sin \Omega)^2}}, \quad \tan \varphi = -\frac{a \sin \Omega}{1 - a \cos \Omega} \tag{11.25}$$

The periodicity of this function with increasing Ω is obvious. Furthermore, $\Omega = 2\pi$ is the sampling frequency and $\Omega = \pi$ is the Nyquist frequency. Lastly, for any ω that is very small compared to the Nyquist frequency ($\Omega \approx 0$), the approximations $\cos \Omega \approx 1$ and $\sin \Omega \approx \Omega$ apply, and magnitude and phase resemble more the continuous lowpass filter in Eq. (11.5):

$$m \approx \frac{1 - a}{\sqrt{(1 - a)^2 + (a\Omega)^2}}, \quad \tan \varphi \approx -\frac{a\Omega}{1 - a} \tag{11.26}$$

Clearly, this approach rapidly leads to very complex algebra. A somewhat simpler path to the frequency response is the w-transform (Section 4.4). The w-transform is a suitable method to obtain gain and phase margins, possible resonance peaks, and the phase and gain crossover frequencies. Substituting $z = (1+w)/(1-w)$ in Eq. (11.21) yields the frequency response in the w-plane:

$$H(w) = A \cdot \frac{1+w}{w+A} \tag{11.27}$$

where A is shorthand for $(1-a)/(1+a) = (1-e^{-\omega_c T})/(1+e^{-\omega_c T})$. Bode diagrams for $H(w)$ can now be constructed as described for continuous systems in Section 11.5. In the w-plane Bode diagram, the usual frequency-response criteria can be determined, such as phase and gain margin, passband ripple, or resonance effects. However, the imaginary part of w (i.e., the frequency in the w-transformed system) is not linearly related to the frequency ω in $z = e^{\omega T}$. Rather, if $w = \mu + j\nu$, any relevant frequency (e.g., the frequency where the gain passes through 0 dB or the resonance frequency) needs to be transformed back into the z-domain frequency ω with the relationship

$$\omega = \frac{2}{T} \tan^{-1} \nu \tag{11.28}$$

Note that Eq. (11.28) does not reflect the periodicity of the frequency response. Since the inverse tangent returns values in the range from $-\pi$ to π, the inverse-transformed ω will always remain below the sampling frequency.

For time-discrete systems, the most practical way to obtain the frequency response is the use of computer tools. Similar to the Bode diagram of a time-continuous system, Scilab can plot the Bode diagram of a time-discrete system. The sequence of commands is almost identical, but the variable z is used instead of s. For the example above ($T = 0.1$ ms, $\omega_c = 1000$ s^{-1}), the Bode diagram can be obtained with:

```
z = poly (0, 'z');
T = 1e-4; omgc = 1000; a = exp (-omgc*T);
H = (1-a)*z/(z-a);
zsys = syslin (T, H);
bode (zsys);
```

The use of the sampling time T in the definition of the discrete system in `syslin` is crucial to distinguish this linear system from a continuous system where `'c'` indicates the continuous system. The resulting plot is shown in Figure 11.13.

A second example is a time-discrete system that is created from a time-continuous transfer function with the help of the bilinear transform. Our desired second-order system has the Laplace-domain transfer function

$$H(s) = \frac{\omega_c^2}{s^2 + 2\zeta\omega_c s + \omega_c^2} \tag{11.29}$$

By using the bilinear transform and substituting $s = (2/T)(z-1)/(z+1)$, and further defining $\Omega = \omega_C T$, we arrive at a z-domain transfer function,

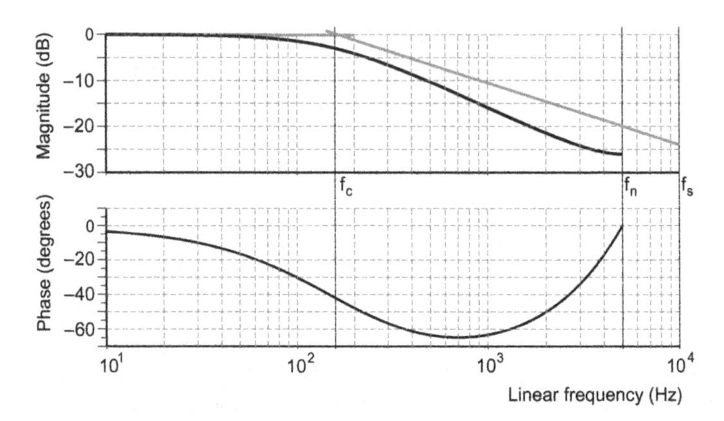

Figure 11.13 Bode diagram of the time-discrete approximation of the RC lowpass. The solid gray lines in the magnitude plot indicate the asymptotes of the continuous system. Also indicated are the cutoff frequency $f_c = \omega_c/2\pi$, the Nyquist frequency $f_N = 1/2T$, and the sampling frequency $f_S = 1/T$. At low frequencies, the digital filter shows a behavior similar to the continuous counterpart, but the behavior diverges as the frequency approaches the Nyquist frequency.

$$H(z) = \frac{\Omega^2(z^2 + 2z + 1)}{(4 + \Omega^2 + 4\zeta\Omega)z^2 + (2\Omega^2 - 8)z + 4 + \Omega^2 - 4\zeta\Omega} \tag{11.30}$$

For a Butterworth filter with a cutoff frequency of 10 Hz, we need $\omega_c = 20\pi\ \text{s}^{-1}$ and $\zeta = 0.71$. With a sampling rate $T = 10$ ms, we obtain $\Omega = 0.61$. The resulting Bode diagram for Eq. (11.30) is shown in Figure 11.14. Once again, the analog filter and its digital counterpart have somewhat similar behavior for low frequencies, but the behavior diverges when the input frequency approaches the Nyquist frequency.

11.8 The Nyquist Stability Criterion

The Nyquist stability criterion is a frequency-domain method to determine the number of unstable closed-loop poles from the loop gain $L(s)$. We begin with a closed-loop system with the transfer function $H(s)$ that is a fraction of two polynomials of s:

$$H(s) = \frac{p(s)}{q(s)} = \frac{p(s)}{1 + L(s)} = \frac{p(s)}{1 + \dfrac{n(s)}{d(s)}} \tag{11.31}$$

We are concerned with the roots of the denominator polynomial, $q(s) = 0$. To explain the Nyquist criterion, let us define a *contour mapping* process that is similar to creating a Nyquist plot (see Figure 11.3) with one crucial difference: The Nyquist plot in Figure 11.3 was obtained by mapping the imaginary axis of the s-plane into the $H(s)$-plane. For the Nyquist criterion, we need to ensure a clockwise, closed path in the right half of the s-plane. A closed contour can be achieved as shown in Figure 11.15 by connecting three sections: A straight line along the positive imaginary axis from the origin to

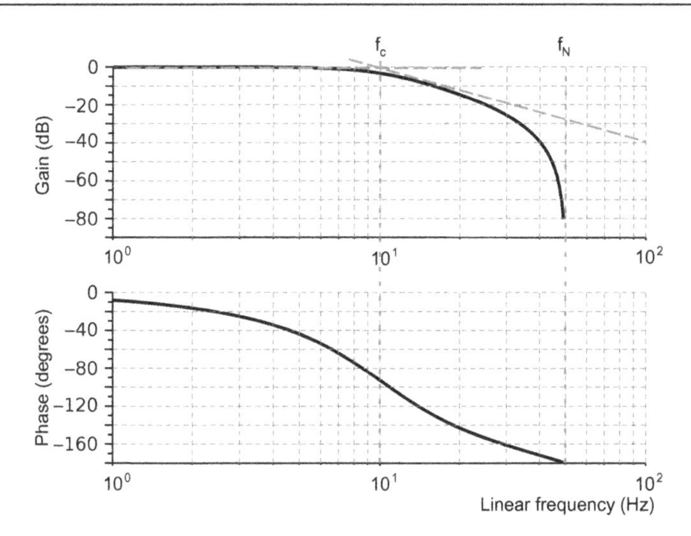

Figure 11.14 Bode diagram of a digital filter that was created from a time-continuous Butterworth filter through the bilinear transform. Dotted lines indicate the filter cutoff frequency $f_C = 10\text{Hz}$ and the Nyquist frequency $f_N = 50\text{Hz}$. The dashed lines represent the magnitude asymptotes of the continuous system. At low frequencies, the filter behaves similar to its analog counterpart, but the stopband magnitude drops off much faster in the digital filter as the input frequency approaches the Nyquist frequency.

$s = jr$, a semicircle in the right half-plane from $s = jr$ through $s = r$ to $s = -jr$, and finally a straight line from $s = -jr$ back to the origin. To encircle the entire right half-plane, we need $r \to \infty$. Let us call this contour Γ.

For a given fraction of polynomials $F(s)$, the Nyquist contour in the s-plane can be mapped to the $F(s)$-plane by plotting $\Im(F(s))$ over $\Re(F(s))$ as s follows the Nyquist contour. The key to the Nyquist stability criterion is Cauchy's theorem that stipulates that *if the contour* Γ *encircles N poles and M zeros of* $F(s)$*, the mapped contour in the* $F(s)$*-plane encircles the origin of the* $F(s)$*-plane* $M - N$ *times.* It is important to note that the contour Γ must not pass through any poles or zeros of $F(s)$. This requirement leads to modifications of the contour when poles or zeros of $F(s)$ exist on the imaginary axis, and examples further below illustrate how functions $F(s)$ with poles or zeros on the imaginary axis are handled.

Up to this point, we have examined the mapping of a function $F(s)$ that corresponds to the characteristic polynomial of the closed-loop system $q(s)$. Since $q(s) = 1 + L(s)$, the mapped contours in the $q(s)$- and $L(s)$-planes are related by a simple coordinate shift where the origin of the $q(s)$-plane is shifted to the point $(-1, 0)$ of the $L(s)$-plane.

Moreover, let us assume that $L(s)$ is in turn a fraction of polynomials, that is, $L(s) = n(s)/d(s)$ as defined in Eq. (11.31). In this case, $L(s)$ has M zeros for $n(s) = 0$ and N poles for $d(s) = 0$. Cauchy's theorem tells us that the mapped contour Γ encircles the origin of the $q(s)$-plane $Z = M - N$ times. Correspondingly, the mapped contour Γ in the $L(s)$-plane encircles the point $(-1, 0)$ $Z = M - N$ times in the clockwise

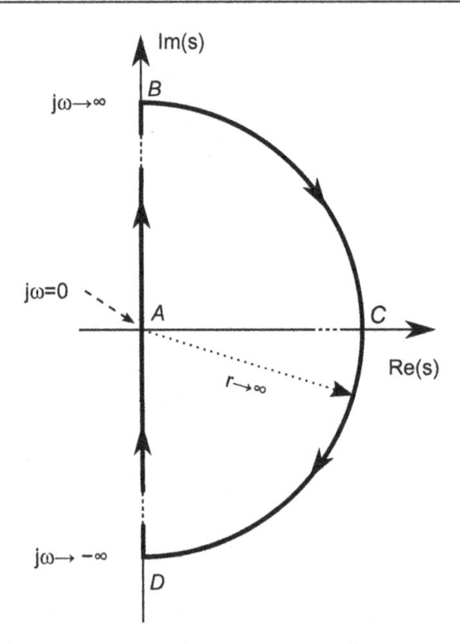

Figure 11.15 The Nyquist contour, that is, a closed contour in the s-plane that encircles the entire right half-plane. The contour starts in the origin (A) with $j\omega = 0$ and follows the positive imaginary axis to the point labeled B, where $s = \lim_{r\to\infty} jr$. A semicircle with $r \to \infty$ crosses the real axis at C and connects the point B with the opposite point D on the negative imaginary axis with $s = \lim_{r\to\infty} -jr$. Finally, the contour is closed by traversing the negative imaginary axis from D to the origin A.

direction. The number of *unstable* poles P of the closed-loop function is the number of encircled zeros $M = Z + N$. We can therefore formulate the Nyquist stability criterion:

- If the loop gain function $L(s)$ has no poles (that is, $N = 0$), the closed-loop system is stable when the mapped contour Γ in the $L(s)$-plane does not encircle the point $(-1, 0)$. Often, the denominator function $q(s)$ is available as a simple polynomial of the form $q(s) = a_0 + a_1 s + a_2 s^2 + \cdots$, and the loop gain function in the form of $L(s) = (a_0 - 1) + a_1 s + a_2 s^2 + \cdots$ has no poles.
- If the loop gain function $L(s)$ has N poles, the closed-loop system is stable when the mapped contour Γ in the $L(s)$-plane encircles the point $(-1, 0)$ N times in the *counterclockwise* direction. The counterclockwise direction is crucial, because we require $M = Z + N = 0$ for stability, but Z counts clockwise encirclements with a positive sign. Thus, N counterclockwise encirclements ($N = -Z$) yield $M = 0$. Note that a possible clockwise and a counterclockwise encirclement cancel each other out.

How can a plot of the mapped contour be obtained? Often, it is sufficient to examine a number of discrete points along the contour Γ. Clearly, the part of the contour that follows the positive imaginary axis is identical to the frequency response $F(j\omega)$. Its negative counterpart (that is, the part of the contour that follows the negative imaginary axis) creates a complex conjugate mirror image of the frequency response, as we have seen briefly in Figure 11.3.

Let us examine an example that follows the general form of feedback loop shown in Figure 11.6. In the first example, we assume that $G(s)$ has two real-valued, stable poles, and that $H(s) = 1$. The loop gain is

$$L(s) = k\frac{1}{(s+a)(s+b)} \tag{11.32}$$

When a and b are known, for example, $a = 2$ and $b = 5$, we can compute several points of $L(s)$ along the contour Γ (Table 11.1) and plot those points. For the imaginary axis, we can make use of the symmetry $L(-j\omega) = L^*(j\omega)$. Furthermore, the entire semicircle BCD in Figure 11.15 maps to the origin, because $|s| \to \infty$. With these points, a sketch of the Nyquist plot of $L(s)$ and thus the mapped Nyquist contour in the $L(s)$-plane can be obtained (Figure 11.16). It can be seen that the mapped contour never touches or encircles the point $(-1, 0)$, irrespective of k. The closed-loop system is therefore stable for all values of k.

To provide an example where closed-loop instability can occur, let us review the third-order system described earlier (Eq. (11.7)). We examine the Nyquist plot for $k = 30,000$, that is,

$$L(s) = \frac{30,000}{s^3 + 52s^2 + 780s + 3600} \tag{11.33}$$

The mapped contour shows a small second loop on the left side of the origin (Figure 11.17). When k is small enough that this loop does not touch the critical point $(-1, 0)$, the closed-loop system is stable. However, the contour expands linearly with k, and a value for k exists where the point $(-1, 0)$ lies inside the small loop. In this case, the critical point is encircled twice, which indicates the presence of a complex conjugate pole pair in the right half-plane. The distance of the intersection point to the critical point is directly related to the gain margin and provides information about the

Table 11.1 Some computed points along the contour Γ for the system in Eq. (11.32).

ω	$L(j\omega)$	$L(-j\omega) = L^*(j\omega)$
0	$0.1k$	$0.1k$
0.05	$k(0.1 - 0.0035j)$	$k(0.1 + 0.0035j)$
0.1	$k(0.1 - 0.07j)$	$k(0.1 + 0.007j)$
0.2	$k(0.1 - 0.014j)$	$k(0.1 + 0.014j)$
0.5	$k(0.09 - 0.033j)$	$k(0.09 + 0.033j)$
1.0	$k(0.07 - 0.054j)$	$k(0.07 + 0.054j)$
2.0	$k(0.026 - 0.060j)$	$k(0.026 + 0.060j)$
5.0	$k(-0.01 - 0.024j)$	$k(-0.01 + 0.024j)$
10.0	$k(-0.007 - 0.0054j)$	$k(-0.007 + 0.0054j)$
∞	0	0

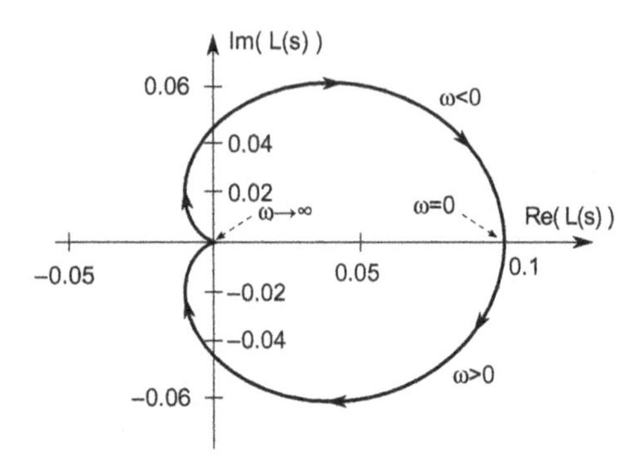

Figure 11.16 Mapping of the Nyquist contour in the $L(s)$-plane with Eq. (11.32) and $k = 1$. The lower branch corresponds to the positive imaginary axis ($\omega > 0$) from points A to B in Figure 11.15, starting for $\omega = 0$ on the positive real axis. The mapping of the semicircle BCD lies in a single point in the origin. The negative imaginary axis ($\omega < 0$, points D to A) is mapped to the upper branch. The critical point $(-1, 0)$ is not encircled. The contour scales with k, but no value for k exists where the contour touches $(-1, 0)$.

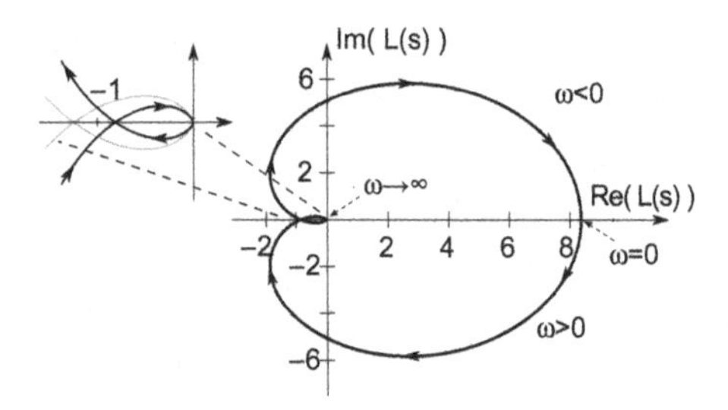

Figure 11.17 Mapping of the Nyquist contour in the $L(s)$-plane with Eq. (11.33). A small loop (magnified in inset) emerges for higher frequencies between the origin and the critical point $(-1, 0)$. As shown, the system is stable, because the critical point is not encircled. However, the curve scales with the gain k, and a value for k exists where the small loop extends beyond the point $(-1, 0)$ (gray curve in inset). With higher values of k, therefore, the critical point is encircled twice, indicating the presence of two unstable poles in the closed-loop system.

relative stability of the system. In this example, the loops intersect at $\sigma = -0.78$, and a 28% increase of k is possible before the closed-loop system becomes unstable.

When the loop gain has poles or zeros on the imaginary axis, the Nyquist contour Γ needs to be modified to make an infinitely small "detour" around the pole. A good example is the positioner system introduced in Chapter 9. The integrating leadscrew creates a pole in the origin (*cf.* Eq. (9.3)), and the loop gain with P control can be written in simplified form as

$$L(s) = \frac{k}{s(s + a)} \tag{11.34}$$

The marginally stable pole in the origin is counted as part of the left half-plane when the Nyquist contour makes a small, semicircular detour around the pole as shown

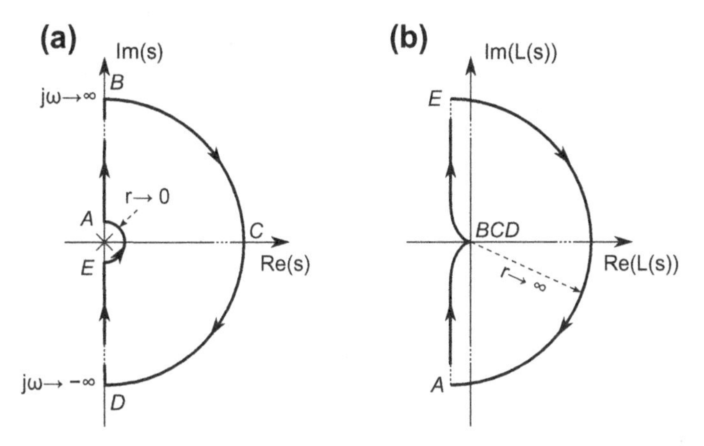

Figure 11.18 Nyquist mapping of a system with the loop gain $L(s) = k/s(s+a)$. (a) The pole in the origin requires a small semicircular and counterclockwise detour *E-A* with radius $\epsilon \to 0$. (b) The ascending mapped branch *A-B* lies in the negative half-plane at $\sigma = -k/a^2$ and curves near the origin such that the tangent in the origin is at $-180°$. The negative branch *D-E* is its complex conjugate counterpart, and the infinite semicircle *BCD* maps into the origin. Lastly, the detour semicircle *E-A* maps into a clockwise semicircle of infinite radius.

in Figure 11.18a. We begin the mapped contour with point *A*, that is, at a very small distance $s = j\epsilon$ above the origin. We obtain from Eq. (11.34) and with $\epsilon \ll a$,

$$L(j\epsilon) = \frac{k}{j\epsilon(j\epsilon + a)} = -\frac{k}{\epsilon^2 + a^2} - j\frac{ak}{\epsilon^3 + \epsilon a^2} \approx -\frac{k}{a^2} - j\frac{k}{\epsilon a} \tag{11.35}$$

Point *B* is mapped to the origin, and the contour approaches the origin at an angle of $-180°$. Similarly, points *C* and *D* map to the origin. The section from *D* to *E* is the complex conjugate of the section *AB* and thus ends at $-k/a^2 + jk/\epsilon a$. Lastly, the detour semicircle with a very small radius $r = \epsilon$ can be described as $\epsilon e^{j\varphi}$ with φ running counterclockwise from $-\pi/2$ to $\pi/2$. The mapping can be obtained through

$$L(\epsilon e^{j\varphi}) = \frac{ke^{-j\varphi}}{\epsilon(\epsilon e^{j\varphi} + a)} \approx -\frac{ke^{-j\varphi}}{a\epsilon} \tag{11.36}$$

This segment of the mapped contour therefore describes a clockwise semicircle of infinite radius. Once again, we see that the critical point $(-1, 0)$ is not encircled, and the closed-loop system is always stable. The main goal of this example, however, was to demonstrate how to exclude poles that lie *on the Nyquist contour.*

11.8.1 The Nyquist Stability Criterion for Time-Discrete Systems

Time-discrete systems are stable when no poles lie outside the unit circle in the z-domain. The Nyquist stability criterion can therefore be adapted to time-discrete systems by defining a z-domain contour that encircles the entire z-plane with the exception of the area inside of (and including) the unit circle. The definition $z = e^{sT}$ implies a mapping between the s- and z-planes. The upper half of the unit circle in the z-plane is the mapped positive imaginary axis of the s-plane from $\omega = 0$ to the Nyquist frequency $\omega_N = \pi/T$, that is, $z = e^{j\omega T}$ with $0 \le \omega < \omega_N$. The unit circle can either be completed by continuing with $\omega_N \le \omega < 2\omega_N$, or the complex conjugate property can be used with the negative frequency range $-\omega_N \le \omega < 0$.

The unit circle needs to be traversed *counterclockwise*, because its inside is excluded from the plane. With two straight lines along the positive real axis and an infinitely large circle, the contour is completed. The z-domain Nyquist contour is shown in Figure 11.19. If poles of the z-domain loop gain occur on the unit circle, a small detour, much like the one shown in Figure 11.18a, needs to be added to the path.

To illustrate how the z-domain Nyquist contour is mapped, let us consider the feedback control system shown in Figure 11.20. The process approximates the motor/positioner behavior with an integrating pole at $z = 1$, and the transfer function is $G(z) = z(1 - e^{-aT})/[(z - 1)(z - e^{-aT})]$. The controller is a simple P controller, but we assume that the microcontroller causes a one-period delay. Thus, the controller transfer function is $H(z) = k_p z^{-1}$. Combined, the loop gain becomes

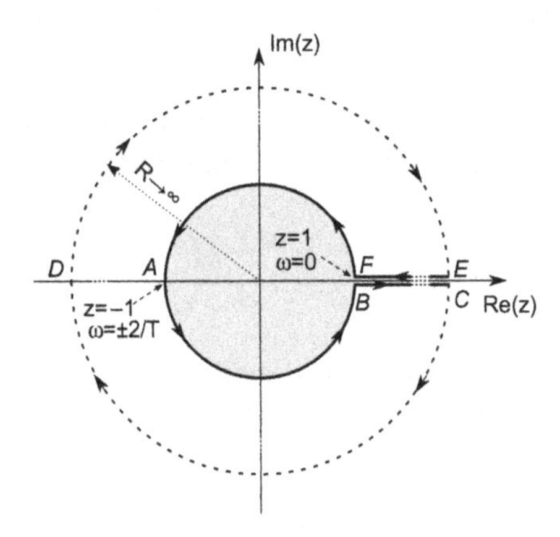

Figure 11.19 Nyquist contour for the z-plane. The contour encloses the entire z-plane with the exception of the stable region $|z| < 1$ (shaded). Starting at point A, the lower semicircle is represented by $z = e^{j\varphi}$ with $-\pi < \varphi < 0$. The path from B to C can be described as $z = r - j\epsilon$ with r increasing from 1 to ∞. The dashed semicircle CD is obtained from $z = Re^{j\varphi}$ with $R \to \infty$ and $0 > \varphi > -\pi$. The other half of the contour is obtained from inverting the sign of the frequencies.

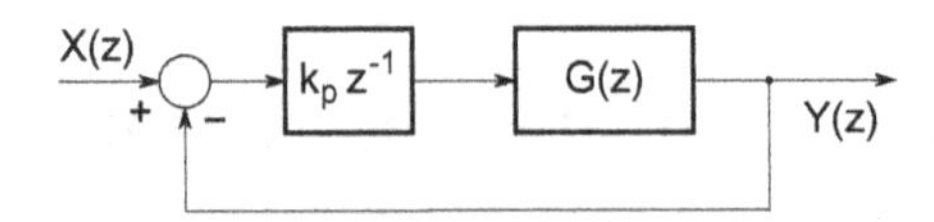

Figure 11.20 Example feedback control system with a time-discrete process $G(z)$ and a controller that delays the output by one sampling period and can be described as $H(z) = k_p z^{-1}$.

$$L(z) = k_p z^{-1} \frac{z(1 - e^{-aT})}{(z - 1)(z - e^{-aT})} = k \frac{1}{(z - 1)(z - e^{-aT})} \tag{11.37}$$

where $k = k_p(1 - e^{-aT})$ is a combined gain coefficient. Furthermore, let us assume a sampling period $T = 10$ ms and a time constant of the process of 25 ms, leading to $a = 40\ \text{s}^{-1}$ and $e^{-aT} = 0.67$. The goal is to construct the mapping of the Nyquist contour in Figure 11.19 into the $L(z)$-plane, which is explained in the following steps for $k = 1$:

- Semicircle from *A* counterclockwise to *B* (frequency from $-\omega_N$ to 0, excluding the pole at $z = +1$): By using $z = e^{-j\pi} = -1$, we know that the contour begins at $L(-1) = +0.3$ at an angle of $-90°$ with the real axis. A second prominent point is $z = e^{-j\pi/2} = -j$, for which we obtain $L(-j) = -0.114 - 0.58j$. The contour curls around the origin and finally follows a vertical asymptote at -10.7 toward the mapped point *B* at $(-10.7 + j\infty)$.
- Detour around the pole at $z = 1$. We can describe this path as a quarter circle $\lim_{r\to 0}(1 + re^{j\varphi})$ with φ varying from $-90°$ to $0°$. The mapped quarter circle becomes a clockwise quarter circle with infinite radius in the $L(z)$-plane. It continues the contour from $(-10.7 + j\infty)$ almost to the real axis.
- Straight line parallel to the real axis (point *B* to *C*): This line can be described as $z = R - j\epsilon$ with R increasing from just right of $z = +1$ to ∞, and with ϵ being a very small quantity that keeps the line below the real axis. This section of the contour maps to a curve above the real axis of the $L(z)$-plane that connects the mapped quarter circle to the origin.
- Outer circle *CDE*: The circle in the z-plane with $R \to \infty$ maps to the origin of the $L(z)$-plane.
- Remaining segments *EF* and *FA*: These segments are the complex conjugates of the segments above.

The resulting mapped contour is shown in Figure 11.21, and we can see that the critical point $(-1,0)$ is encircled twice. This indicates the presence of two poles of the closed-loop system outside of the unit circle and thus an unstable system. Since the contour scales with k, we can see that a reduction of k by a factor of 3 is needed to stabilize the system. With the definition $k = k_p(1-e^{-aT})$ and the requirement $k < 1/3$, we obtain $k_p < 1$ for stability. In fact, examination of the closed-loop transfer function reveals a complex conjugate pole pair at $p_{1,2} = 0.835 \pm 0.99j$ for $k_p = 3$ and at $0.835 \pm 0.55j$ for $k_p = 1$. The second pole pair lies exactly on the unit circle.

11.8.2 Nyquist Stability in Scilab

Scilab offers a `nyquist` function to plot the frequency response of a s- or z-domain function. In the simplest form, we can generate a Nyquist plot with the following steps (*cf.* Eq. (11.33)):

```
s = poly (0,'s')
loopgain = 30000/(s^3+52*s^2+780*s+3600)
lsys = syslin ('c', loopgain)
nyquist (lsys)
```

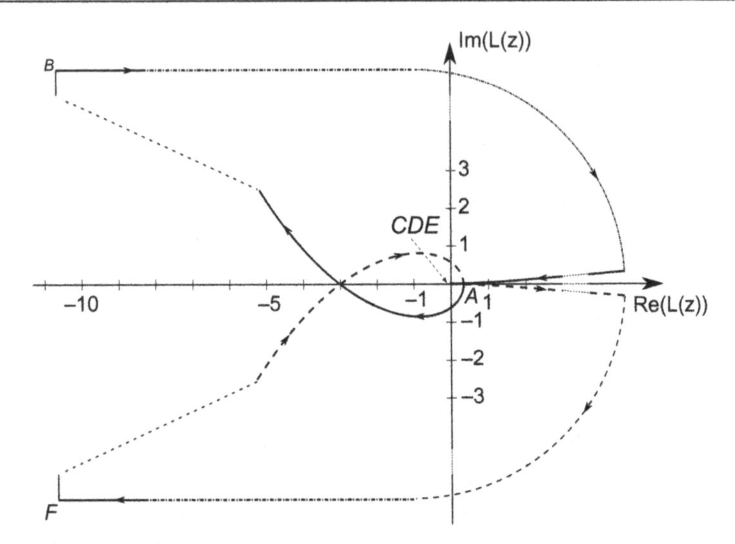

Figure 11.21 Nyquist plot of the z-domain contour for the system in Eq. (11.37). The contour begins at point A at $z = 0.3$, where the branch emerges at an angle of $-90°$. The branch then curves upward and moves toward the mapped point B at $z = -10.7 + j\infty$. The mapped straight line BC, excluding the pole at $z = 1$, circles around the first quadrant, but converges toward the origin as $z \to \infty$. The circle CDE maps into the origin. For positive frequencies, the curve is mirrored along the imaginary axis. The critical point at $(-1, 0)$ is encircled twice, which indicates the presence of two poles of the closed-loop system outside the unit circle.

These steps produce a plot similar to Figure 11.17. Moreover, the function `show_margins` allows to plot a Nyquist plot with gain and phase margins when it is called with an optional argument that indicates a Nyquist plot:

```
...
lsys = syslin ('c', loopgain)
show_margins (lsys, 'nyquist')
```

However, the results of the `nyquist` and `show_margins` functions are difficult to interpret when poles of the loop gain function lie on the Nyquist contour, or even when the Nyquist plot covers a large area. Sometimes, it is sufficient to limit the frequency range in the call to the `nyquist` function. Scilab offers a special function to examine the Nyquist plot inside a given rectangle in the $L(s)$-plane. For example, we can examine the small loop of the system in Eq. (11.33). We can see that the small loop lies inside a box from -1 to 0 on the real axis and from -0.5 to 0.5 on the imaginary axis. The function `nyquistfrequencybounds` provides us with the frequency range that limits the Nyquist plot to the bounding box:

```
...
lsys = syslin ('c', loopgain)
bbox = [-1 -0.5; 0 0.5]
[fmin fmax] = nyquistfrequencybounds (lsys, bbox)
nyquist (lsys, fmin, fmax)
```

In the presence of poles on the Nyquist contour, the mapped contour can be obtained step-by-step by using conventional complex arithmetic. To obtain the negative unit circle in the z-domain and its mapping (Eq. (11.37)), for example, discrete values along the circle can be produced (Scilab notation for π is `%pi` and for the imaginary unit j, `%i`):

```
omega_vector = linspace (-%pi,0.01, 200)
z_mapping = exp (%i*omega_vector)
```

The vector `z_mapping` contains 200 discrete values along the lower unit semi-circle, excluding $z = 1$. `z_mapping` can now be used to compute the corresponding points of $L(z)$, and the segment plotted with `plot(real(L),imag(L))`. This is a convenient method to obtain discrete values along the mapped contour, for example in Table 11.1 or in the step-by step construction of Figure 11.21.

12 The Root Locus Method

Abstract

The root locus method is arguably the most powerful tool in the design engineer's toolbox to design a feedback control system that meets the design specifications. The significance of s-plane roots for a system's dynamic response was highlighted in Chapter 9. The root locus method allows to determine the traces of the roots in the s-plane as any one process parameter (for example the controller gain) is varied. Conversely, the root locus method allows to determine the specific value of a process parameter for a desired root location, and with it, for a desired dynamic response.

The root locus method is a graphical method to obtain the location of the roots of the characteristic equation (the poles of the system) when one parameter is varied, usually from 0 to ∞. With this method, we can analyze a system by observing the pole locations at specific values of the parameter under observation. This allows us to draw conclusions on the stability and dynamic response of the system. We can also determine the value of the parameter when a pole (or poles) are desired to be in a specific location of the s-plane. The root locus method is applied to the open-loop system, and the analysis is based on the loop gain rather than the closed-loop transfer function. The root locus method is also valid for z-domain transfer functions, but the interpretation of the pole locations is fundamentally different.

Consider the simple feedback system in Figure 12.1. $G(s)$ is a linear system and can be described as the fraction of the numerator and denominator polynomials,

$$G(s) = \frac{p(s)}{q(s)} \tag{12.1}$$

The closed-loop transfer function of the feedback system in Figure 12.1 is

$$H(s) = \frac{Y(s)}{X(s)} = \frac{k\frac{p(s)}{q(s)}}{1 + k\frac{p(s)}{q(s)}} \tag{12.2}$$

To obtain the poles of $H(s)$, we set

$$1 + L(s) = 1 + k\frac{p(s)}{q(s)} = 0 \tag{12.3}$$

We can move the additive 1 to the right-hand side. However, we know that the fraction $p(s)/q(s)$ is a complex number, so we can expand the -1 on the right-hand

Linear Feedback Controls. http://dx.doi.org/10.1016/B978-0-12-405875-0.00012-7
© 2013 Elsevier Inc. All rights reserved.

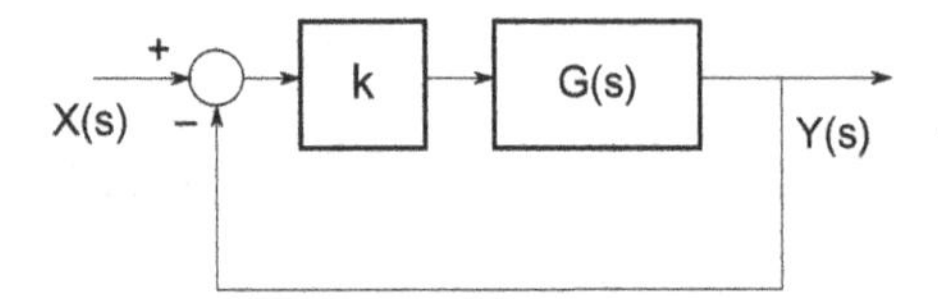

Figure 12.1 Feedback system with controller gain k and unit feedback path.

side as $-1+0j$:

$$k\frac{p(s)}{q(s)} = -1+0j = 1\cdot e^{j\pi} \tag{12.4}$$

This representation may seem counterintuitive at first, but it provides us with two conditions for the equality: Any point s_1 in the complex s-plane is a root locus if both the magnitude condition (Eq. (12.5)) and the phase condition (Eq. (12.6)) apply:

$$|k|\cdot\frac{|p(s_1)|}{|q(s_1)|} = 1 \Leftrightarrow |k| = \frac{|q(s_1)|}{|p(s_1)|} = \left|\frac{q(s_1)}{p(s_1)}\right| \tag{12.5}$$

$$\arg\left(\frac{p(s_1)}{q(s_1)}\right) = (2l+1)\pi, \quad l \in \tag{12.6}$$

These two conditions lead to a number of rules that specify whether a point on the s-plane is a root locus or not. More specifically, these two conditions allow to derive construction rules for root locus plots. We will now describe these rules step-by-step.

12.1 Graphical Construction of Root Locus Plots

The root locus plot shows the location of the poles as one parameter is varied from 0 to ∞, such as, for example, a P controller gain k_p. From a root locus sketch, we can immediately see whether varying the parameter in question can lead to instability, and we can follow changes of the dynamic response of the closed-loop system. Moreover, we can graphically determine the design value of this parameter when a pole or pole pair is desired to be in a specific location.

12.1.1 Prepare the Characteristic Equation in Root Locus Form

Let us assume that a system has a known transfer function with the characteristic polynomial

$$q(s) = a_n s^n + a_{n-1}s^{n-1} + \cdots + a_1 s + a_0 \tag{12.7}$$

The parameter for which the system is to be analyzed needs to exist in the coefficients of $q(s)$. Let us call this parameter κ. Our goal is to factor out κ so that $q(s) = 0$ can be rewritten as

$$q(s) = 1 + \kappa \cdot F(s) = 0 \tag{12.8}$$

Note that this is the same equation as Eq. (12.3). The new function $F(s)$ is (as seen in Eq. (12.3)) a fraction of two polynomials of s. We can rewrite the two polynomials in product form, thus revealing the poles and zeros of $F(s)$,

$$1 + \kappa F(s) = 1 + \kappa \frac{\prod_{m=1}^{M}(s + z_m)}{\prod_{n=1}^{N}(s + p_n)} = 0 \tag{12.9}$$

If there are no zeros ($M = 0$), the product of the $(s + z_m)$ needs to be set to unity. Equation (12.9) can be rewritten as

$$\prod_{n=1}^{N}(s + p_n) + \kappa \cdot \prod_{m=1}^{M}(s + z_m) = 0 \tag{12.10}$$

We can instantly see that the traces of the root locus emerge from the poles of $F(s)$ when $\kappa = 0$ and that the traces end in the zeros of $F(s)$ as $\kappa \to \infty$.

Example 1. A system has the characteristic polynomial $q(s) = s^3 + 3s^2 + Ks + 5$. It is usually possible to divide the equation $q(s) = 0$ by all terms of $q(s)$ which do not contain κ (or, in this case, K). In this example, we want to factor out K and find $F(s)$:

$$\begin{aligned} & s^3 + 3s^2 + Ks + 5 = 0 \quad \Rightarrow \\ & \frac{s^3 + 3s^2 + Ks + 5}{s^3 + 3s^2 + 5} = 0 \quad \Rightarrow \\ & 1 + K \cdot \frac{s}{s^3 + 3s^2 + 5} = 0 \end{aligned} \tag{12.11}$$

where $F(s)$ emerges as

$$F(s) = \frac{s}{s^3 + 3s^2 + 5} \tag{12.12}$$

We can see that $F(s)$ has one zero ($M = 1$) and three poles ($N = 3$). We also recognize that the denominator of $F(s)$ is identical to the characteristic equation $q(s)$ for $K = 0$ and that the numerator of $F(s)$ is identical to $q(s)$ for $K \to \infty$. Note that:

- We should always get $N \geq M$.
- The poles and zeros of $F(s)$ are *not* the poles and zeros of the closed-loop transfer function $H(s)$ in Eq. (12.2).
- However, the poles and zeros of $F(s)$ are identical to the poles and zeros of the closed-loop transfer function when $K = 0$, i.e., $F(s)$ is the open-loop transfer function.

We now determine the zeros and poles of $F(s)$. In this example, we get $z_1 = 0$ and $p_1 = -3.426$, $p_{2,3} = +0.213 \pm 1.19j$. Note that in this example $F(s)$ has poles in the right half-plane, which makes $H(s)$ unstable for small K.

12.1.2 Open-Loop Poles and Zeros and Asymptotes

Start the root locus sketch by drawing the real and imaginary axes of the s-plane. Mark all poles of $F(s)$ with an $\times$ symbol and all zeros of $F(s)$ with a $\circ$ symbol (pole-zero plot of $F(s)$). We know that all traces of the root locus begin in the poles and end in the zeros. This immediately brings up the question: what if we have fewer zeros than poles? The answer: In that case, we have branches of the root locus that end at infinity, and these branches follow asymptotes for very large κ.

When $F(s)$ has M zeros and N poles, the root locus plot has $N-M$ asymptotes. Because of Eq. (12.6), each asymptote l subtends the real axis at an angle ϕ_A for which holds

$$\phi_{A,l} = \frac{2l+1}{N-M} \cdot \pi \quad \text{with} \quad l = 0, 1, 2, \ldots, (N-M-1) \tag{12.13}$$

Furthermore, all asymptotes meet in one point σ_A, which behaves like a "center of gravity" of the root locus curve. For reasons of symmetry, this point must lie on the real axis. We can calculate σ_A through

$$\sigma_A = \frac{1}{N-M} \cdot \left(\sum_{n=1}^{N} p_n - \sum_{m=1}^{M} z_m \right) \tag{12.14}$$

For the third step in the root locus sketch we therefore determine σ_A and add the asymptotes to the s-plane sketch we started above. Note that Eq. (12.13) limits the possible configuration of asymptotes to well-defined cases, of which some are sketched in Figure 12.2. Furthermore,

- σ_A is not necessarily a root locus!
- All asymptotes must show mirror symmetry along the real axis, because complex poles only occur as conjugate pairs.
- The configurations in Figure 12.2 are valid only for negative feedback systems. For positive feedback, Eq. (12.13) needs to be modified for multiples of 2π. In this case, the asymptotes in Figure 12.2 would be rotated by $\pi/(N-M)$.

12.1.3 Branches of the Root Locus on the Real Axis

Each section of the real axis that is to the left of an odd number of zeros plus poles is a root locus. Complex conjugate pole pairs count—naturally—as two poles. As fourth step of drawing your root locus diagram, draw all branches of the root locus on the real axis.

Example 2. We use $F(s)$ in Eq. (12.12). No root locus exists to the right of the rightmost pole/zero. Moving along the real axis from right to left, we eventually cross a double pole pair at $\Re(s) = 0.213$. Now, we are to the left of two poles. This is an even number, and there is still no root locus on the real axis. Next, we reach the zero in the origin, to the left of which the sum of poles and zeros is three. This is an odd number, and a branch of the root locus exists between the zero and the final pole at -3.426. To the left of that pole, the sum of poles and zeros is four. This is an even number,

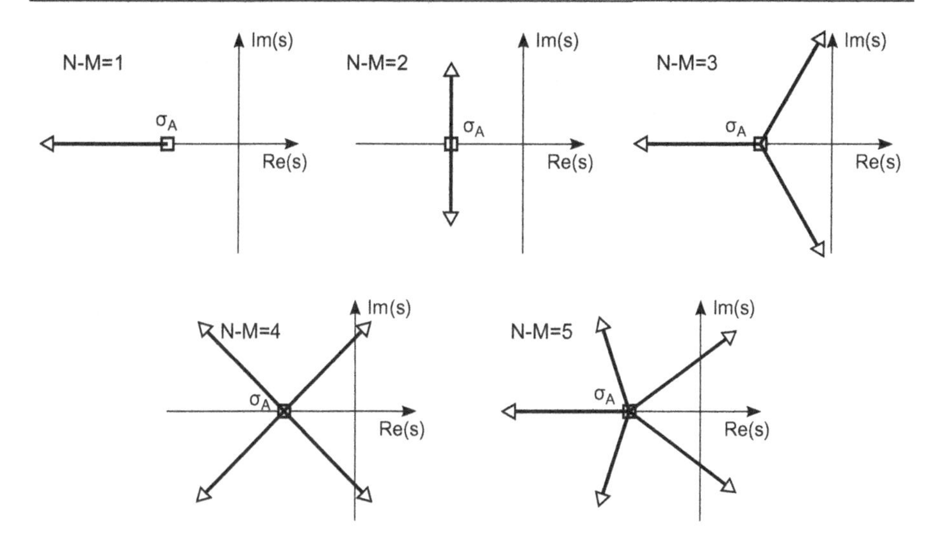

Figure 12.2 Asymptote configuration for $N - M = 1$ through 5. All asymptotes meet in the centroid σ_A, which always lies on the real axis due to symmetry reasons. All asymptote branches are symmetrical along the real axis. The angles of the asymptotes with the real axis are given by Eq. (12.13). Note that these configurations are valid only for negative feedback systems (shown in Figure 12.1).

and the pole therefore ends the root locus branch. This example is consistent with the asymptote configuration, Figure 12.2 for $N - M = 2$, which has no asymptotes on the real axis.

12.1.4 Branchoff Points

In some cases, poles move toward each other as κ increases, meet (double pole of $H(s)$), and branch off into the imaginary plane. These points are called branchoff points σ_B and are solutions of Eq. (12.15):

$$\sum_{n=1}^{N} \frac{1}{\sigma_B - p_n} = \sum_{m=1}^{M} \frac{1}{\sigma_B - z_m} \tag{12.15}$$

All branchoff points σ_B lie on the real axis due to symmetry reasons. Note that there are generally more solutions of Eq. (12.15) than actual branchoff points. Only real-valued solutions of Eq. (12.15) that meet the condition under Section 12.1.3 are true branchoff points, because no root locus exists to the left of an even number of poles and zeros. As the last step in drawing your root locus diagram, indicate all valid σ_B. From the σ_B, sketch the complex conjugate branches as they approach their respective asymptotes. The root locus diagram is now completed.

Example 3. The characteristic equation of a system has the root locus form of

$$1 + \kappa \frac{1}{s^3 + 5s^2 + 6s} = 0 \tag{12.16}$$

The open-loop poles (i.e., the poles of $F(s)$ are in the origin and at $p_1 = -2$ and $p_2 = -3$. There are three asymptotes that meet (rather, begin) at $\sigma_a = -5/3$. Next, we determine the branchoff points. Since we have no zeros, Eq. (12.15) becomes for this specific case

$$\frac{1}{\sigma_B} + \frac{1}{\sigma_B + 2} + \frac{1}{\sigma_B + 3} = 0 \tag{12.17}$$

By multiplying the entire equation with $\sigma_B(\sigma_B + 2)(\sigma_B + 3)$, we obtain

$$(\sigma_B + 2)(\sigma_B + 3) + \sigma_B(\sigma_B + 3) + \sigma_B(\sigma_B + 2) = 0 \tag{12.18}$$

which can be rearranged to

$$\sigma_B^2 + (10/3)\sigma_B + 2 = 0 \tag{12.19}$$

We find two solution candidates for σ_B, at -0.785 and -2.549. We know that the second solution lies to the left of an even number of poles and is not a valid root locus. Our branchoff point therefore lies at $\sigma_B = -0.785$.

12.1.5 Departure Angles for Complex Poles

We can determine the angle at which a root locus branch exits a complex pole. Similarly, we can determine the angle at which a root locus branch arrives at a complex zero. The recipe for estimating the departure angle is based on the angle criterion, Eq. (12.6). If we rewrite $F(s)$ in product form (Eq. (12.9)), the phase angles become additive. The phase angle for any point s_0 in the s-plane is the sum of all phase angle contributions from the zeros minus the sum of all phase angle contributions from the poles, obtained by setting $s = s_0$ in Eq. (12.9). When s_0 is a root locus, the phase angles must add up to 180°. To obtain the departure angle of a pole p_k, we assume that we examine a point s_0 in the s-plane at a very small distance from the pole. For s_0 to be a root locus, it must meet the angle condition, and we can write

$$\alpha + \sum_{n=1, n \neq k}^{N} \arg\left(\overline{p_n p_k}\right) - \sum_{m=1}^{M} \arg\left(\overline{z_m p_k}\right) = \pi \tag{12.20}$$

where α is the unknown departure angle (hypothetically, the angle between pole p_k and the infinitesimally close point s_0), the first sum contains all angles from the poles to the pole p_k, and the second sum contains all angles from the zeros to the pole p_k.

Graphically, we can draw connecting lines from all poles to the pole p_k and determine the angles of these lines with the real axis. Similarly, we draw connecting lines from all zeros to the pole p_k. The unknown departure angle α for pole p_k is 180° plus all angles of the connecting lines from the zeros minus all angles from the connecting lines from the poles.

Example 4. Consider Figure 12.3 (inspired by Figure 12.8). We want to determine the departure angle of the pole at $p_k = -1+3j$. From the double zero, two connecting lines

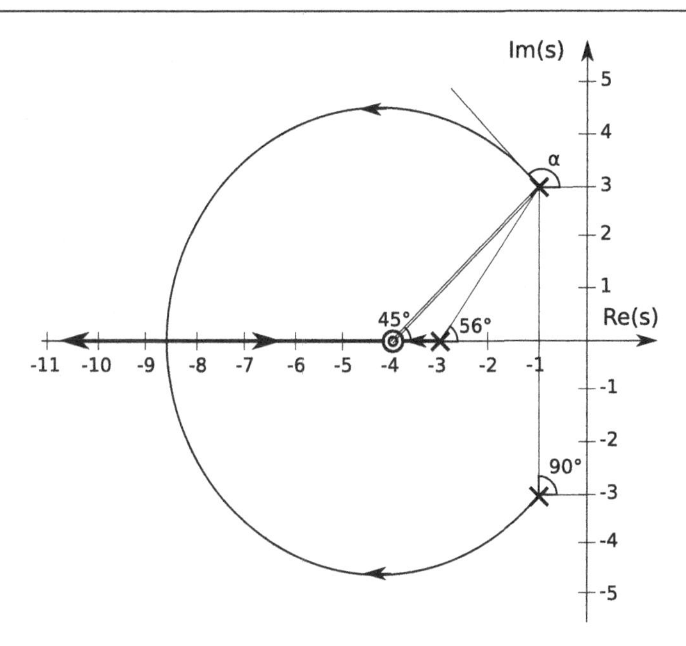

Figure 12.3 Calculation of the departure angle in a root locus sketch. In this example, the departure angle α from the pole at $p_k = -1 + 3j$ is sought. Connecting lines from each zero and each pole to p_k are drawn and their angles measured. The sum of angles from other poles to p_k is $90° + 56°$. The double zero connects to p_k with two lines at 45°. The final departure angle is 124°.

exist at an angle of 45°. The connecting line from its complex conjugate counterpart is at 90°, and the connecting line from the real-valued pole at -3 meets the real axis at an angle of 56°. For the angles,

$$\alpha + 90° + 56° - 45° - 45° = 180° \tag{12.21}$$

holds, and we obtain the departure angle as $\alpha = 124°$.

12.2 Root Locus Diagrams in Scilab

If our fraction of polynomials $F(s)$ is completely numerical, we can use Scilab to plot a root locus curve. Follow these steps after you have determined $F(s)$—we use Eq. (12.12) as an example:

```
s = poly (0,'s')                //Define the complex variable s
num = s                         //Define the numerator of F(s)
den = s^3 + 3 * s^2 + 5         //Define the denominator of F(s)
sys = syslin ('c',num/den)      //Create a linear system from F(s)
evans (sys)                     //Draw the root locus curve
```

In this example, num, den, and sys are arbitrary variable names. Note that you need to use $F(s)$ and never $H(s)$ in the evans function. evans simply assumes that the system you provide has the parameter of interest already factored out and meets Eq. (12.3). The evans function can restrict the gain, for example,

```
evans (sys,15)                //Draw the root locus up to k = 15
```

draws the root locus only from $\kappa = 0$ to $\kappa = 15$. This feature is useful when the most relevant features exist for small κ, and for larger κ the root locus merely follows its asymptotes.

It is sometimes helpful to (1) print the root locus diagram as provided by Scilab and (2) zoom in into interesting regions and print the magnified section. Note that Scilab can be helpful even if you have symbolic parameters left in $F(s)$. For example, if $F(s)$ contains an unknown parameter α, you can simply assume a reasonable value for α and plot the root locus curve with Scilab. Next, you compare your hand-drawn curve (which can be drawn for the unknown α) to the computer curve and verify that your special-case assumption for α leads to identical results.

Finally, if you have to zoom in into regions of interest to the extent that the discrete step size of the curve plays a role, root locus diagrams created by Scilab may actually contain impossible configurations, for example, a root locus that is not mirror-symmetric along the real axis. In this case, it advisable to redraw the curve with a reduced gain range.

12.3 Design Example: Positioner with *PI* Control

We will return to the positioner example to illustrate how the root locus method can help us in the design process. Referring to Figure 12.4, we recall that the positioner consists of a first-order armature motor with inertia J and friction R_F. The motor constant is k_M. The leadscrew has a pitch of r (this includes the factor of 2π) and acts as an integrator to obtain a position from the rotational speed of the motor. A feedback control system with a feedback path calibrated to unity and a controller with the transfer function $H(s)$ are provided.

We now assume that the controller is a *PI* controller with the transfer function

$$H(s) = k_p + \frac{k_I}{s} = \frac{k_p s + k_I}{s} \tag{12.22}$$

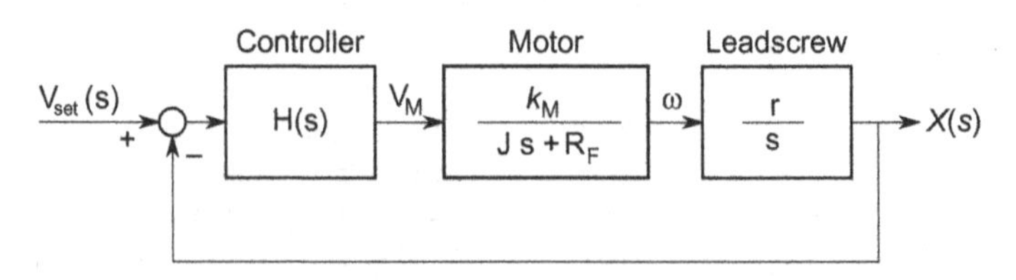

Figure 12.4 Positioner (motor with leadscrew) and feedback control.

which leads to the closed-loop transfer function

$$\frac{X(s)}{V_{set}(s)} = \frac{k_M r(k_p s + k_I)}{Js^3 + R_F s^2 + k_M r(k_p s + k_I)} \tag{12.23}$$

We can now define $\alpha = k_M r/J$ and $R_J = R_F/J$. We divide the equation by J and re-interpret k_p and k_I as αk_p and αk_I, respectively. In this case, the transfer function can be simplified to

$$\frac{X(s)}{V_{set}(s)} = \frac{k_p s + k_I}{s^3 + R_J s^2 + k_p s + k_I} \tag{12.24}$$

We will now examine how k_I influences the poles of the closed-loop system when k_p is chosen to provide critical damping in the absence of integral control.[1] Setting $k_I = 0$ reduces the order of the system, and we obtain the poles of the second-order system as

$$-\frac{R_J}{2} \pm \sqrt{\frac{R_J^2}{4} - k_p} \tag{12.25}$$

which provides us with the condition $k_p = R_J^2/4$ for critical damping. Setting the denominator polynomial of Eq. (12.24) to zero provides us with the equation for the poles, and we treat this equation as suggested in Eq. (12.8) to obtain $F(s)$:

$$q(s) = s^3 + R_J s^2 + R_J^2 s/4 + k_I = 0 \tag{12.26}$$

$$1 + k_I \cdot \frac{1}{s^3 + R_J s^2 + R_J^2 s/4} = 0 \tag{12.27}$$

Following the steps described above, we see that $F(s)$ has no zeros and three poles, namely $p_1 = 0$ and a double pole $p_{2/3} = -R_J/2$. The double pole for $k_I = 0$ is consistent with our initial critically damped system. With $M = 0$ and $N = 3$, we have three asymptotes (see Figure 12.2), and the asymptote centroid is at $\sigma_A = -R_J/3$. The entire real axis left of the origin is a root locus. Without knowing numerical values for R_F and α, calculation of σ_B is tedious, but a rough estimate would be $\sigma_B = \sigma_A/2$. We can now sketch the entire root locus plot for k_I (Figure 12.5).

Two branches of the root locus follow the complex conjugate asymptotes into the right half-plane, and we can see that large values of k_I make the system unstable. Furthermore, the system reveals that a new double pole (critical damping) can be achieved, but with two differences to the critically damped P controlled system: (1) A third pole exists, but this is a fast pole with a very short transient component; (2) the new double pole is closer to the origin, and its transient component is slower than that of the original P controlled system. The advantage of the newly introduced integral component, however, is the suppression of steady-state tracking errors.

[1] This choice of k_p is a design choice. In other designs, other criteria for choosing k_p may apply.

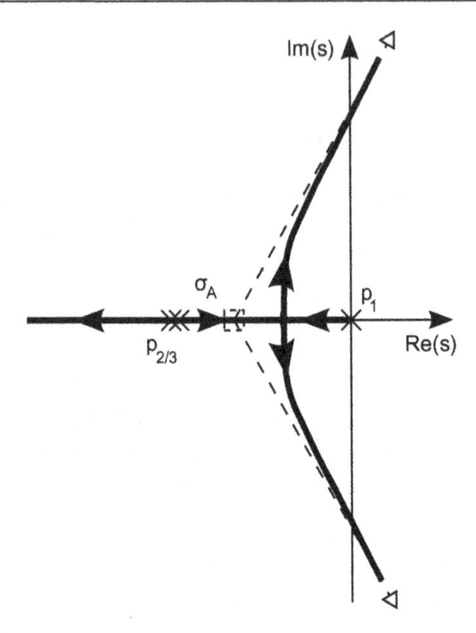

Figure 12.5 Root locus plot for k_I. The poles p_1 and $p_{2/3}$ are the poles for the system when $k_I = 0$. Asymptotes are indicated by dashed lines. Pole p_1 emits a segment to the left, and p_2 emits a segment to the right. These segments meet at the branchoff point and move into the complex plane, where the segments follows the asymptotes. At high k_I, the root locus follows the asymptotes into the right half-plane, and the closed-loop system becomes unstable. The third pole, p_3, emits a segment to the left, which follows the third asymptote along the negative real axis.

It is possible to use the root locus diagram to obtain the values of κ for a desired pole location s_d. We can use the magnitude criterion, Eq. (12.5), which can be rewritten as

$$\kappa = \frac{\prod_{n=1}^{N} |s_d - p_n|}{\prod_{m=1}^{M} |s_d - z_m|} \tag{12.28}$$

where $|s_d - p_n|$ is the length of the vector that connects the desired pole location with the pole p_n of the function $F(s)$ and $|s_d - z_m|$ is the vector that connects the desired pole location with the zero z_m of the function $F(s)$. This equation readily lends itself for a semi-graphical solution, and we will examine this approach in the next examples.

In the first example, we wish to obtain the values of k_I where the system becomes unstable. From the sketch, we estimate that the asymptotes cross into the right half-plane at $s_I = \pm j\sigma_A \cdot \tan 60° \approx 0.58 \cdot j \cdot R_J$. There are three vectors from the poles of $F(s)$ to the point $0.58 \cdot j \cdot R_J$ on the imaginary axis. The first has a length of $0.58 R_J$, and the two vectors from the double pole to the imaginary axis have the length $\sqrt{0.5^2 + 0.58^2} R_J$. The product of the three vector lengths is $0.34 \cdot R_J^3$, and we know that for $k_I = 0.34 \cdot R_J^3$, a complex conjugate pole pair lies on the imaginary axis.

The special case where a system becomes unstable can much better be solved with the Routh-Hurwitz scheme, however, which provides us with $k_I < 0.25 R_J^3$ for stability.

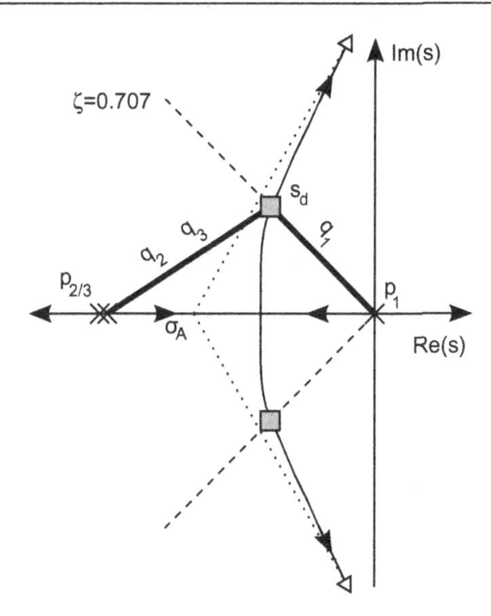

Figure 12.6 We can use the magnitude criterion to graphically determine κ (in this case, k_I), to place poles in a desired location s_d. For example, if we wanted to obtain a complex conjugate pole pair with a damping factor $\zeta = 0.707$, indicated by the gray rectangles, the lengths of the three vectors that connect the "open-loop" poles (i.e., those with $k_I = 0$) with the desired pole location can be used in the magnitude component of Eq. (12.9) to solve for k_I. Note that a third closed-loop pole exists to the left of the open-loop pole pair $p_{2/3}$.

The deviations from the value obtained from the root locus sketch may primarily be attributed to rounding errors and our use of the asymptote rather than the actual root locus trace, which, when raised to the third power, may deviate significantly from the true value.

Another example is the graphical solution of the magnitude component of Eq. (12.9) to place closed-loop poles in a desired location. In Figure 12.6, this process is demonstrated for an underdamped pole pair with $\zeta = 0.707$, that is, a pole pair on the diagonal where $\Re(s) = \Im(s)$. We designate the desired pole location s_d. We can measure the lengths of the vectors q_1, q_2, and q_3 and insert these into (12.9), For this example, the magnitude component of Eq. (12.9) becomes

$$1 + |k_I| \cdot \frac{1}{|(s_d - p_1)| \cdot |(s_d - p_2)| \cdot |(s_d - p_3)|} = 0 \tag{12.29}$$

We can measure the length of the three lines to obtain $q_1 \approx 0.28R_J$ and $q_2 = q_3 \approx 0.36R_J$, and find the corresponding value for $k_I = 0.036R_J^3$. In our third-order system, the choice for k_I will create a third pole to the left of the open-loop pole pair $p_{2/3}$, but this fast pole has a short transient response and therefore a small influence on the overall dynamic response.

12.4 Design Example: Resonance Reduction

In the second example, consider an electromagnetic actuator system. Examples for such actuators are head positioning systems in hard drives or optical drives. A strong multipole permanent magnet and a coil (often referred to as voice coil) provide the motive force, determined by the current through the coil. Typically, the moving element is supported by low-friction bearings, and the mass of the moving magnet combined with the electromagnetic force result in an underdamped second-order system. An example for such a system is shown in Figure 12.7. The process consists of the moving magnet (second-order system) and the voice coil with its power driver. For this example, we assume that the driver and the coil inductance cause a relatively short time-lag of about 1/3 s. It can be seen that the process has a real-valued pole at -3 and a complex conjugate pole pair at $-1 \pm -3j$. The latter is the cause of the resonance behavior. The resonance peak is about 2 dB, causing a 20% overshoot of the magnet on a step input at the power driver.

The design goal is to achieve a step response without overshoot, and to reduce the settling time to better than 1.5 s. For this purpose, feedback control is provided as shown in Figure 12.7. The gain k of the error amplifier can be adjusted to tune the closed-loop system. To achieve the design goal, a compensator needs to be found that creates a closed-loop system with only real-valued poles.

We can see that a system with a single asymptote (see Figure 12.2) would create an ideal configuration that is always stable and where the poles become real-valued for sufficiently high k. To achieve $N - M = 1$, the compensator needs to have two zeros in its transfer function $H_c(s)$, and both zeros would ideally lie to the left of the fast pole at -3. One possible compensator is a second-order highpass filter (i.e., a two-stage phase-lead compensator) with a cutoff frequency of $\omega_0 = 4s^{-1}$. This compensator has the transfer function $H_c(s) = (s/4 + 1)^2$, and the closed-loop transfer function with this new compensator is

$$H(s) = \frac{Y(s)}{R(s)} = \frac{30k}{s^3 + (1.875k + 5)s^2 + (15k + 16)s + 30k + 30} \tag{12.30}$$

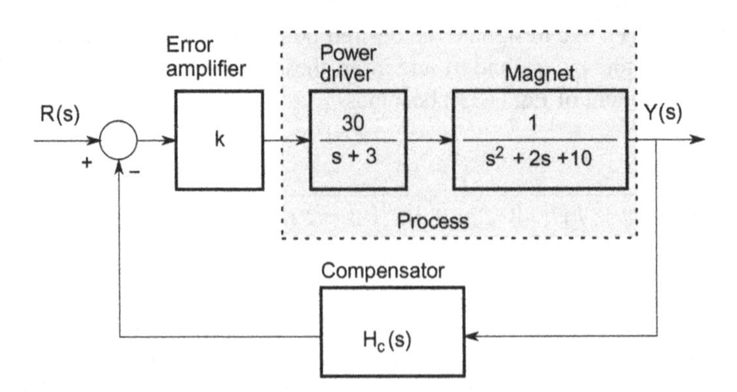

Figure 12.7 Suppression of resonance effects with feedback control.

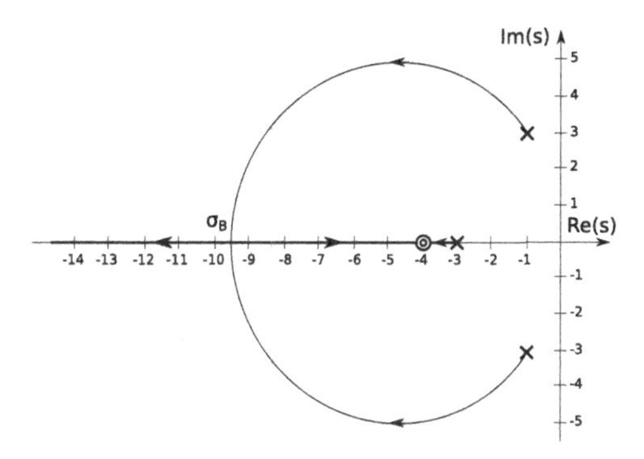

Figure 12.8 Root locus sketch of the closed-loop system in Figure 12.7 for increasing k. The root locus branches begin in the poles of the process and move to the left in all cases. The short branch that emerges from the real-valued pole ends in the double zero at -4. The complex conjugate branches move towards a branchoff point. For even Further increasing k, one pole moves towards the double zero of the compensator, and the other branch follows the asymptote to $-\infty$.

The denominator polynomial, transformed into root locus form, becomes

$$1 + k \cdot \frac{30}{16} \cdot \frac{(s+4)^2}{s^3 + 5s^2 + 16s + 30} = 0 \tag{12.31}$$

where the poles and zeros of the open-loop transfer function $F(s)$ (see Eq. (12.12)) are identical with the poles of the process and the zeros contributed by the compensator. The double zero at -4 attracts the root locus branch from the real-valued pole of the process and at the same time "bends" the complex branches toward the real axis, where they meet at σ_B as shown in Figure 12.8. It is possible (though inconvenient) to calculate the branchoff point σ_B from Eq. (12.15). More conveniently, $\sigma_B \approx -9.52$ can be read from the root locus sketch.

The desired critically damped case requires k to be adjusted such that the double pole forms at $\sigma_B \approx -9.52$. As in the previous example, we can measure the distance of the poles and zeros from the branchoff point and use the magnitude criterion to obtain k:

$$1 + \frac{30}{16}|k| \cdot \frac{|(\sigma_B - z)|^2}{|(\sigma_B - p_1)| \cdot |(\sigma_B - p_2)| \cdot |(\sigma_B - p_3)|} = 0 \tag{12.32}$$

The fraction after k evaluates as 0.0573, and k emerges as $k \approx 9.3$. Using this value for k in Eq. (12.30) provides the actual location of the roots at -3.41 and $9.52 \pm 0.047j$. The step responses of the process without feedback control and the closed-loop system are shown in Figure 12.9. The simulated response shows that both the rise time and the settling time have been shortened. The settling time can further be shortened by reducing k and allowing a minimal overshoot.

The plot in Figure 12.9 also exhibits an attenuated steady-state response that can already be predicted from Eq. (12.30), namely, that the steady-state value of the output is by a factor of $k/(k+1)$ lower than the input. Usually, this steady-state behavior would be unacceptable. One option to improve the steady-state response is to amplify $R(s)$

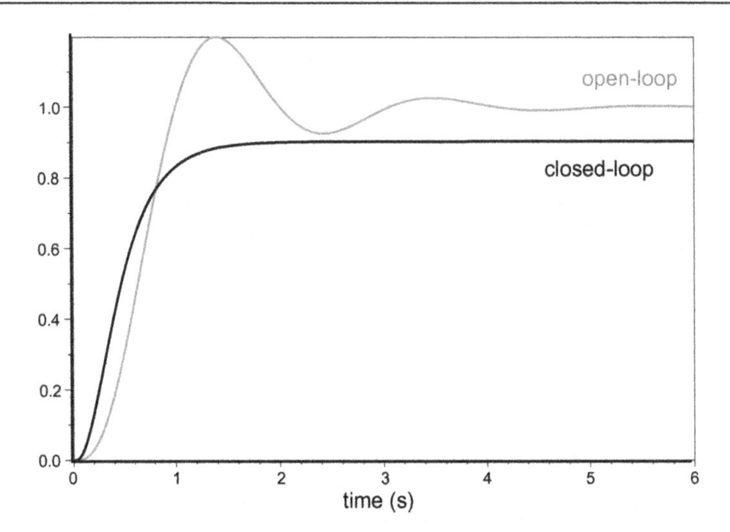

Figure 12.9 Step response of the open-loop and closed-loop systems. The step response of the process alone shows multiple oscillations, whereas the closed-loop system exhibits critically damped behavior without overshoot as expected, and it also shows a faster rise time. Because of the multiple oscillations, the 2% settling time of the process without feedback control is about 3.8 s. With feedback control, the settling time improves to about 1.4 s.

by the steady-state attenuation, that is, $(k + 1)/k$. The dynamic response of the loop would not be affected by the amplification of the input signal, but matching the loop gain and the initial amplification factor can be cumbersome in practice. Alternatively, one compensator zero could be placed in the origin, and the transfer function of the compensator becomes $H_c(s) = k \cdot s(s+4)$. The new compensator has a steady-state gain of zero and therefore does not influence the steady-state behavior of the process. Note that the amplifier gain k, too, needs to be moved into the feedback path as indicated in the transfer function of the compensator. With this compensator, the closed-loop transfer function becomes

$$H(s) = \frac{Y(s)}{R(s)} = \frac{30}{s^3 + (30k + 5)s^2 + (120k + 16)s + 30} \tag{12.33}$$

Note that k does not occur in the zero-order coefficient of the characteristic polynomial. The steady-state gain is now unity, but at the cost of a slower step response: The real-valued root locus branch now ends in the origin, and the dynamic response becomes slower for increasing k. The branchoff point of the alternative compensator is at $\sigma_B \approx -6.35$, requiring a compensator gain $k \approx 0.28$. The slowest real-valued closed-loop pole is at -0.75, and the 2% settling time is more than 6 s. Because of the slow pole, the second system could be undampened by reducing k and allowing the fast poles to become a complex conjugate pair (some of the overshoot would be suppressed by the slow pole). For example, a choice of $k = 0.15$ produces poles at -1.28 and $-4.1 \pm 2.57j$, and the system settles into the 2% tolerance band after a step input after 3.62 s. The settling time is very similar to that of the open-loop process, but the step response does not show any overshoot. The primary goal of reducing the resonance effects has been achieved.

12.5 The Root Locus Method for Time-Discrete Systems

For a time-discrete system, identical rules apply to obtain a root locus plot. Assume that a time-discrete feedback control system has z-domain poles according to the denominator of its transfer function:

$$1 + L(z) = 1 + \kappa \frac{p(z)}{q(z)} = 0 \tag{12.34}$$

Since the construction rules for the root locus do not differ whether $p(s)$ and $q(s)$ are polynomials of s or $p(z)$ and $q(z)$ are polynomials of z, the root locus depends only on the poles and zeros of $L(z)$. However, the *interpretation* of the closed-loop poles in the z-plane differs markedly:

- Closed-loop poles outside the unit circle indicate an unstable system.
- Exponentially decaying responses correspond to poles on the positive branch of the real axis.
- Poles on the negative real axis indicate oscillations with the Nyquist frequency.
- The exponential decay becomes faster as the corresponding pole lies closer to the origin.
- Complex poles occur only as conjugate pairs, but their angle with respect to the positive real axis determines the frequency of their oscillation.
- The location of the poles is partially determined by the sampling period T. Often, an increased (slower) sampling period limits the stable value range of k to lower values.

The *design step* where a specific dynamic response is desired needs to take into account the different interpretation of the pole locations. Other than that, the same techniques of varying a gain factor κ and of placing additional poles or zeros in the z-plane with a digital compensator can be used that were introduced in the previous sections. One advantage of time-discrete systems is that the software realization of compensators allows extreme flexibility for pole/zero placement. Where continuous compensators of higher order assume appreciable complexity, a digital filter of higher order is no different from a low-order digital filter: computation of the output value by weighted addition of the filter memory followed by shifting of the filter memory values into their delayed-cycle positions. In addition, software can react to different operating points and adjust the pole/zero pattern. The design engineer has an enormously increased flexibility when the controller or compensator is realized in software.

To illustrate the differences between the root locus method in continuous and time-discrete systems, let us again consider again DC motor speed control based on the motor model in Chapter 9. The motor angular speed ω is measured by a unit-gain sensor. The controller attempts to match the setpoint speed ω_{set}. The block diagram of the mixed s/z-domain system is given in Figure 12.10. We also provide several motor and controller constants. The controller sampling interval is $T = 5$ ms. The motor time

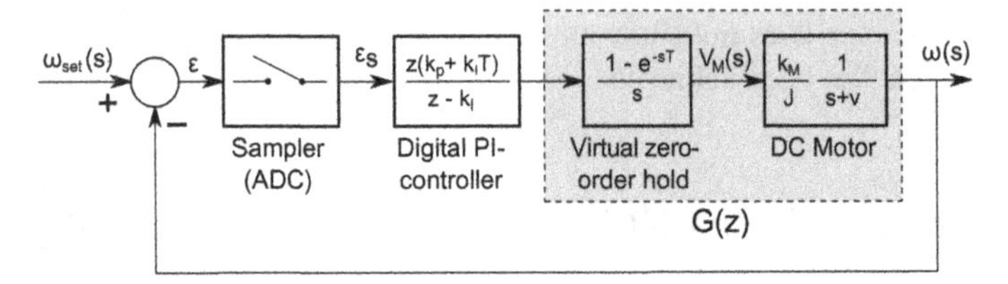

Figure 12.10 Block diagram of a digital motor speed controller with a digital *PI* control element. For the z-domain treatment of the transfer function, the gray-shaded blocks need to be converted into the z-domain.

constant is $v = 20$ ms, and the motor efficiency is $k_M = 0.18$ Nm/V. We used similar values in Section 9.6.

Following Sections 4.3 and 9.6, we begin by computing the process transfer function in the z-domain, $G(z)$. Partial fraction expansion or the equivalence in Table B.5 give

$$G(z) = \frac{k_M}{v \cdot J} \cdot \frac{1 - e^{-T/v}}{z - e^{-T/v}} = \alpha \cdot \frac{1}{z - e^{-T/v}} \tag{12.35}$$

where α is a convenient shorthand, defined as $\alpha = k_M(1 - e^{-T/v})/vJ$. For this system with $e^{-T/v} = 0.78$, the numerical value is $\alpha = 10\ \mathrm{V}^{-1}\ \mathrm{s}^{-1}$. Multiplying the process transfer function $G(z)$ with the controller transfer function of the *PI* controller yields the loop gain $L(z)$

$$L(z) = \alpha \cdot \frac{z(k_p + k_I T)}{(z - k_I)(z - e^{-T/v})} \tag{12.36}$$

For this example, we choose a weak integrator with $k_I = 0.2$ and examine the root locus for k_p. The closed-loop characteristic equation, written in root locus form, becomes

$$1 + \alpha \cdot k_p \cdot \frac{z}{z^2 + z(k_I \alpha T - k_I - e^{-T/v}) + k_I e^{-T/v}} = 0 \tag{12.37}$$

The root locus (Figure 12.11) has two real-valued open-loop poles at $z_1 = 0.87$ and $z_2 = 0.23$, and one zero in the origin. The stability region is inside the unit circle, and k_p needs to be kept small enough that the pole on the asymptote along the negative z-axis remains inside the unit circle, which requires $\alpha k_p < 2.3$ or $k_p < 0.23$. For an identically looking root locus in the s-plane, the requirement would have been a minimum k_p that keeps the poles in the left half-plane.

Moreover, critical damping can be achieved at the first branchoff point ($k_p \approx 0.02$). The corresponding s-plane double pole is at $s = -160$, consistent with the millisecond dynamic response of the motor. The s-plane poles depend on the sampling interval, and a longer sampling interval moves the poles closer to the s-plane origin (slower dynamic response). In the z-plane, a longer sampling interval moves the right pole closer to the origin, the circular root locus trace in the z-plane becomes smaller, and the stable range for k_p diminishes.

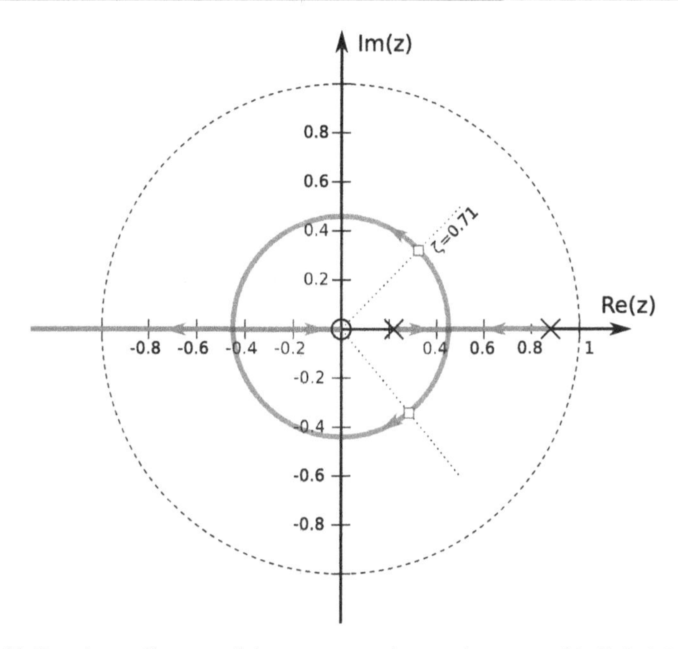

Figure 12.11 Root locus diagram of the motor speed control system with digital *PI* controller. The two open-loop poles are determined by k_I and T, respectively. The dynamic response is exponential for low k_p until critical damping is achieved at approximately $k_p = 0.2$. Two conjugate branches emerge into the complex plane, leading to an oscillatory response with the highest frequency reached when $k_p \approx 0.2$. Further increasing k_p causes one pole to move along the negative real axis, where it eventually leaves the unit circle (dashed), and the system becomes unstable.

From the root locus, it is also possible to obtain a Butterworth-damped pole pair ($\zeta = 0.71, s = -b(1 \pm j)$). Because the z-domain pole pair needs to be on the root locus curve, and because of the relationship $z = e^{-b(1\pm j)T}$, we obtain $b = 160$ for the assumed sampling rate $T = 5$ ms. b also relates to the angular frequency of the oscillatory component: the angle of the pole pair in the z-plane is $bT = 0.8$ rad (46°), and since this angle corresponds to ωT, the oscillations occur at $b/2\pi \approx 25$ Hz. The location of those poles is indicated in Figure 12.11 with the dotted lines that intersect the complex-valued root locus circle.

This example illustrates that the root locus method can be applied similarly for time-discrete systems and for continuous systems. The significance of the poles in the z-plane is different, however, and the prediction of the dynamic response requires moving between s- and z-domains.

13 The *PID* Controller

Abstract

The *PID* controller can be considered the industry standard controller. Because of its widespread use, the *PID* controller is presented in detail in this chapter. The action of the three independent control components (proportional, integral, and derivative) are explained with root locus and frequency-response analyses. Some methods for *PID* controller tuning are covered, and circuits to realize *PID* control are proposed. Furthermore, implementation of *PID* control in a digital, time-discrete system is presented, together with the differences between continuous and time-discrete system responses.

13.1 Intuitive Introduction

The *PID* controller is a very popular type of controller. *PID* controllers are in widespread use, and many control problems can be solved adequately with a *PID* controller. For this reason, we will examine the *PID* controller in detail. *PID* stands for Proportional-Integral-Differential, and the *PID* controller propagates a weighted sum of the input signal, its integral and its first derivative to the output. In the time domain, the output signal can be described as

$$y(t) = k_p \cdot x(t) + k_I \cdot \int x(t)\mathrm{d}t + k_D \cdot \frac{\mathrm{d}x(t)}{\mathrm{d}t} \tag{13.1}$$

under the assumption that the integral component is energy-free at $t = 0$. The constants k_p, k_I, and k_D are adjustable and need to be optimized for the specific type of process and control goal. Often, the *PID* controller coefficients k_p, k_I, and k_D are optimized and fixed for a specific control problem, in which case the *PID* controller is strictly a linear, time-invariant system. In the Laplace domain, the transfer function becomes

$$\frac{Y(s)}{X(s)} = k_p + \frac{k_I}{s} + k_D s = k_D \cdot \frac{(s - z_1)(s - z_2)}{s} \tag{13.2}$$

An alternative form of the transfer function is

$$\frac{Y(s)}{X(s)} = k \cdot \left(1 + \frac{1}{\tau_I s} + \tau_D s\right) \tag{13.3}$$

where k is the overall gain, and τ_I and τ_D are the time constants of the integral and differential parts, respectively. Note that s has units of inverse seconds, and the expression inside the parentheses is therefore unitless. Equation (13.3) often emerges from electronic filter circuits (see Figure 13.8) and is particularly suited for Bode and root locus analysis.

Linear Feedback Controls. http://dx.doi.org/10.1016/B978-0-12-405875-0.00013-9

© 2013 Elsevier Inc. All rights reserved.

We will need to re-examine Eq. (13.2), but for now, let's look at the individual additive components. The proportional component, weighted by k_p, can be seen as the "workhorse." We know that any gain factor (such as k_p) determines the overall loop gain, and a large loop gain leads to low sensitivity, small steady-state error, and good disturbance rejection. Unfortunately, k_p usually has some upper limits, either physical limits or stability and overshoot limits.

The integral component is responsible for eliminating the steady-state error. We know that an integrator can only reach equilibrium if its input is zero. If the input of the *PID* controller is the control deviation ϵ, the integrator component will continue changing the corrective action until $\epsilon = 0$. Since the integrator component contains energy storage, it can lead to undesirable transient responses and even instability. For this reason, k_I is usually kept relatively small.

The derivative component helps providing a rapid transient response. If a transient disturbance or a sudden change of the setpoint occur, the first derivative is huge and causes a correspondingly strong control action. As we will see, the derivative component also acts as a phase-lead component that improves relative stability.

An alternative interpretation is that the *PID* controller uses present, past, and future errors to compute a corrective action. The *P* component is responsible for the present error. The *I* component can be seen as the sum of past errors, and the *D* component can be interpreted as an extrapolation (slope) from present errors to the future. This latter interpretation has the quality of an analogy and cannot be used for a rigorous mathematical analysis of the control problem or for control design.

13.2 Transfer Functions with *PID* Control

A closed-loop control configuration that uses *PID* control can be seen in Figure 13.1. Let us examine the closed-loop transfer function of this system. To simplify the task, we assume a unit-gain sensor, that is, $H_s(s) = 1$.

The closed-loop transfer function is presented in Eq. (13.4):

$$\frac{Y(s)}{X(s)} = \frac{(k_D s^2 + k_p s + k_I)G(s)}{s + (k_D s^2 + k_p s + k_I)G(s)} \tag{13.4}$$

The *PID* controller changes the order of the closed-loop system. If the process has M zeros and N poles, the order O of the closed-loop system is

$$O = \max \begin{cases} N + 1 \\ M + 2 \end{cases} \tag{13.5}$$

In practice, this means that a simple first-order process with no zeros has a second-order closed-loop transfer function and can exhibit overshoot (and, correspondingly, accelerated dynamic response). A second-order process has a third-order closed-loop transfer function, and unstable behavior becomes possible, notably, as we will see, with large k_I.

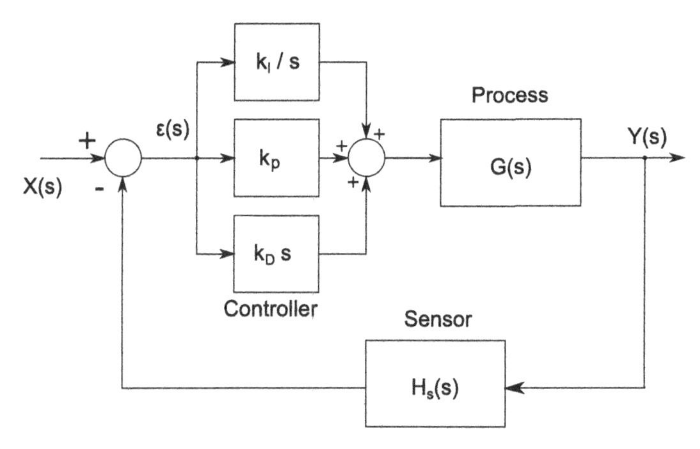

Figure 13.1 Block diagram of a process $G(s)$ with a *PID* controller. The sensor has a transfer function of $H_s(s)$.

13.2.1 PID Control of a First-Order Process

To demonstrate how the pole-zero configuration can be influenced with a *PID* controller, let us examine Figure 13.1 with a first-order system that has the transfer function

$$G(s) = \frac{a}{s+a} \tag{13.6}$$

and for which we obtain the closed-loop function

$$\frac{Y(s)}{X(s)} = \frac{a(k_D s^2 + k_p s + k_I)}{s^2 + as + a(k_D s^2 + k_p s + k_I)} = \frac{a(k_D s^2 + k_p s + k_I)}{(1 + ak_D)s^2 + a(1 + k_p)s + ak_I} \tag{13.7}$$

Let us examine the characteristic equation. First, we divide the transfer function by $(1 + ak_D)$ and compare it to the characteristic equation of a damped second-order system, $s^2 + 2\zeta\omega_n s + \omega_n^2$. We can relate the natural frequency ω_n and the damping coefficient ζ to the coefficients of the *PID* controller through

$$\omega_n = \sqrt{\frac{ak_I}{1 + ak_D}}; \quad \zeta = \frac{a(1 + k_p)}{2\sqrt{ak_I(1 + ak_D)}} \tag{13.8}$$

From Eq. (13.8), we can see that very flexible pole placement in the complex plane is possible. Depending on k_p, k_I, and k_D, we can choose real-valued poles (for example, a critically damped situation), or we can choose complex poles with faster rise and settling times when a small overshoot is acceptable. A first-order system under *PID* control cannot become unstable, although the poles can be placed quite near the imaginary axis with a sufficiently large k_I. In general, controlling a first-order process with a *PID* controller leads to desirable transient and steady-state behavior, and the *PID* controller provides a very high degree of design flexibility.

13.2.2 PID Control of a Second-Order Process

We now turn to a second-order process, again with unit DC gain, with the transfer function

$$G(s) = \frac{\omega_n^2}{s^2 + 2\zeta\omega_n s + \omega_n^2} \tag{13.9}$$

which describes, for example, a spring-mass-damper system, or an electronic circuit with two different kinds of energy storage. Note that the damping factor ζ and the natural resonance frequency ω_n are here properties of the process and not the auxiliary variables that we used in the previous section in Eq. (13.8). The closed-loop transfer function for the second-order process is

$$\frac{Y(s)}{X(s)} = \frac{\omega_n^2(k_D s^2 + k_p s + k_I)}{s^3 + (2\zeta\omega_n + k_D)s^2 + (\omega_n^2 + k_p)s + k_I} \tag{13.10}$$

Closed-term solutions for the roots of the characteristic polynomial are relatively complex. However, we can use the root locus method to get an idea how the individual parameters of the *PID* controller influence the poles of the closed-loop system. The root locus form of the characteristic polynomial for k_I is

$$1 + k_I \cdot \frac{1}{s^3 + (2\zeta\omega_n + k_D)s^2 + \left(\omega_n^2 + k_p\right)s} = 0 \tag{13.11}$$

The root locus diagram has one pole in the origin and either two real-valued poles or a complex conjugate pole pair, depending on k_p and k_D. Furthermore, the root locus diagram has three asymptotes: one along the negative real axis, and two at $\pm 60°$ with the positive real axis. We can therefore qualitatively describe the influence of k_I:

- For a large enough k_I, a pole pair will move into the right half-plane, making the closed-loop system unstable.
- When the closed-loop system has three real-valued poles when $k_I \to 0$, a branchoff point between the pole in the origin and the slower pole exists. Further increasing k_I leads to a dominant complex conjugate pole pair with relatively slow transient response (Figure 13.2a).
- When the closed-loop system has a complex conjugate pole pair when $k_I \to 0$, low values of k_I can improve the dynamic response (Figure 13.2b)

The root locus form of the characteristic polynomial for k_p is

$$1 + k_p \cdot \frac{s}{s^3 + (2\zeta\omega_n + k_D)s^2 + \omega_n^2 s + k_I} = 0 \tag{13.12}$$

With one zero in the origin and three poles that are configured similar to Eq. (13.11), the root locus for k_p has two asymptotes. Unlike k_I, a large value for k_p will not make the closed-loop system unstable. However, large overshoots and oscillations may occur. The three roots of the denominator polynomial in Eq. (13.12) can either be real-valued, or form a complex conjugate pair with one real-valued pole. None of the poles is in the

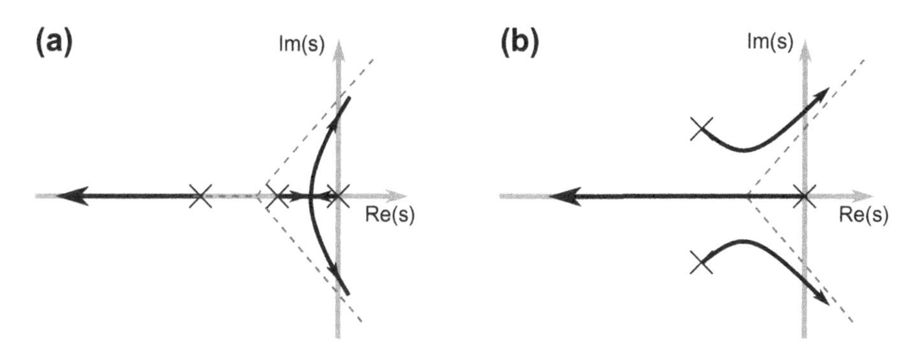

Figure 13.2 Root locus plots for k_I for two sample second-order systems. In (a), the process under *PD* control has three real-valued poles, and increasing k_I first moves the slowest pole out of the origin, but rapidly leads to overshoot and then to instability. In (b), the process under *PD* control has one complex conjugate pole pair. Low values for k_I actually move the poles closer to the real axis, thus improving dynamic response. Further increasing k_I, however, worsens the overshoot and eventually leads to instability.

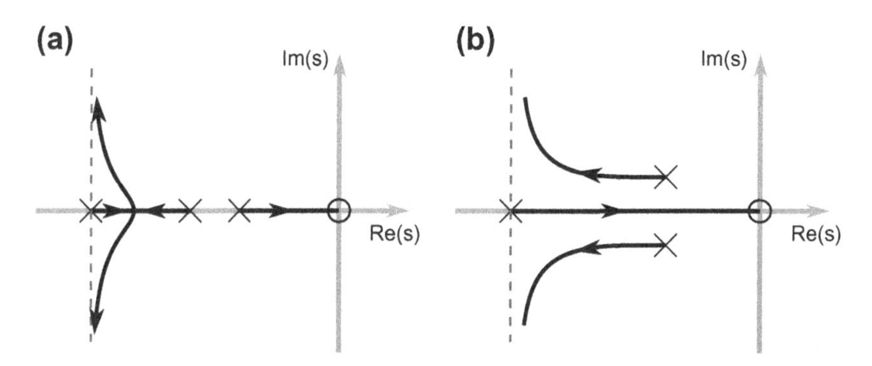

Figure 13.3 Root locus plots for k_p for two sample second-order systems, analogous to Figure 13.2. In (a), the process has three real-valued open-loop poles (when $k_p = 0$), whereas the process in (b) has a complex conjugate open-loop pole pair. In both cases, one closed-loop pole moves toward the zero in the origin with increasing k_p. The other two poles each emit a branch of the root locus; if the pole pair is complex conjugate, the resulting poles remain complex conjugate. If the open-loop system has only real-valued poles, a branchoff point exists.

origin. The root locus curve for two example configurations is shown in Figure 13.3. We can qualitatively describe the influence of k_p:

- For a large enough k_p, one pole will follow the real-valued branch to the zero in the origin. A large value for k_p paradoxically leads to a slow step response and therefore to undesirable dynamic behavior.
- When the closed-loop system has three real-valued poles when $k_p \to 0$, a branchoff point between the two faster poles exists, and a critically damped case can be achieved. Further increasing k_p leads to a complex conjugate pole pair with an

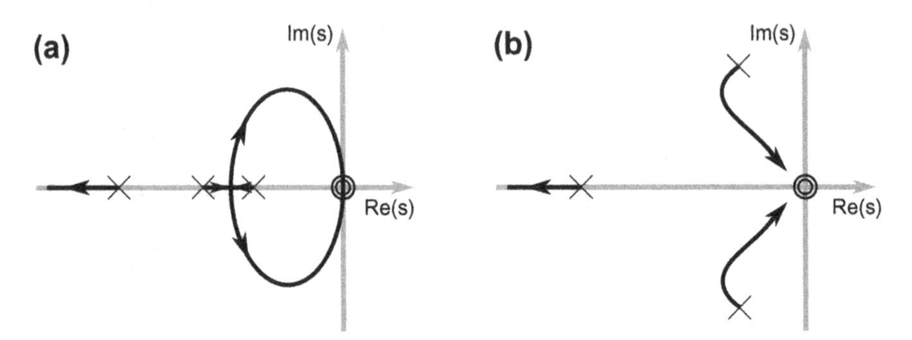

Figure 13.4 Root locus plots for k_D for the same second-order systems in Figure 13.3. Here, a double-zero in the origin exists. In both cases, one branch moves to the left along the only asymptote and creates a fast pole that becomes less and less dominant for larger k_D. However, two branches move toward the origin with increasing k_D, either from the complex poles (b), or from the real axis with an excursion into the complex plane (a). No branches extend into the right half-plane.

oscillatory response (Figure 13.3a). Increasing k_p reduces the damping and increases the oscillatory frequency.

- When the closed-loop system has a complex conjugate pole pair when $k_p \to 0$, the three poles can be brought in close proximity, with an overall optimal dynamic response (Figure 13.3b).

Lastly, we examine the influence of k_D. The root locus form of the characteristic polynomial for k_D is

$$1 + k_D \cdot \frac{s^2}{s^3 + 2\zeta\omega_n s^2 + (\omega_n^2 + k_p)s + k_I} = 0 \tag{13.13}$$

with an interesting double zero in the origin that "attracts" the root locus for large k_D. The root locus for k_D has only one asymptote, and one pole rapidly becomes non-dominant for larger k_D. Two branches end in the origin, either from the complex pole location or through a branchoff point on the real axis (Figure 13.4). Although a large k_D may lead to an undesirably slow system response, no branches exist in the right half-plane, and the system is always stable for arbitrarily large k_D.

It becomes obvious that higher-order systems under *PID* control gain rapidly in complexity. *PID* control is not necessarily the optimum control for high-order processes. However, it is a workable general-purpose solution that rapidly leads to acceptable control behavior in many, especially lower-order, applications.

13.3 Frequency-Domain Aspects of *PID* Control

The effect of *PID* control can also be understood in terms of the open-loop frequency response and its influence on phase and gain margin of the open-loop system. We recall from Eq. (13.2) that the *PID* controller added a pole in the origin and two zeros in the

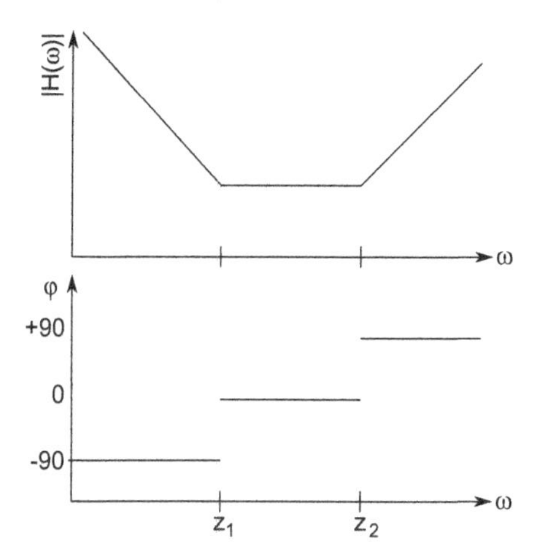

Figure 13.5 Asymptotes of the Bode plot of a *PID* controller. The integrator component causes the gain to decrease with 20 dB/decade for low frequencies. The first zero raises the phase to 0° and causes an asymptotically level gain response. The second zero turns the *PID* controller into a phase lead network for high frequencies with a phase asymptote of +90°.

left half-plane. The frequency response asymptotes of the *PID* controller are shown in Figure 13.5. The integrator component causes the gain to decrease with 20 dB/decade, and gives the *PID* controller an infinite DC gain (i.e., loop gain when $\omega = 0$) that is responsible for the elimination of the steady-state tracking error.

One interpretation of the *D* component is its high-pass character. Higher frequencies have a higher open-loop gain and are therefore attenuated in the closed-loop system. Another interpretation of the *D* component takes into consideration its phase behavior: At high frequencies, the *D* component turns the controller into a phase-lead network with a +90° phase shift. Suitable selection of the overall gain and the position of the second zero can improve stability of a higher-order system, because a suitably tuned *PID* controller increases the gain margin. To provide one example, assume that the second zero of the *PID* controller is placed on the critical frequency of the process where the phase shift of the process reaches −180°. If the first controller zero is sufficiently far away at lower frequencies, the *PID* controller adds +3 dB to the open-loop gain when $k_D = (z_1 z_2)^{-1}$. The phase angle, however, is raised by 45°. This constitutes a very significant increase of the phase margin. At the same time, the critical point where the entire open-loop system crosses −180° will have moved toward higher frequencies. If the order of the system is two or higher, the overall gain still decreases, and gain margin also improves.

13.4 Time-Discrete *PID* Controllers

The discrete *PID* controller was briefly introduced as a building block in Section 4.5, and we present the z-domain transfer function once more in this context:

$$H_{PID}(z) = \frac{z(k_p + k_I T + k_D/T) - k_D/T}{z - k_I} \tag{13.14}$$

The digital *PID* controller contributes one pole at $z = k_I$ and one zero at $z = (k_D/T)/(k_p + k_I T + k_D/T)$. For high sampling rates, the zero exists on the real axis of the z-plane near $z = 1$. Controller realization is possible by digital filter implementation. Equation (13.14) can be simplified with pre-computed coefficients, and both the z-domain transfer function and the digital filter equation are given in Eq. (13.15):

$$\begin{aligned} H_{PID}(z) &= \frac{az - b}{z - k_I} \\ y_k &= k_I y_{k-1} + a x_k - b x_{k-1} \end{aligned} \tag{13.15}$$

with $a = k_p + k_I T + k_D/T$ and $b = k_D/T$. A microcontroller would invoke a *PID* function in regular intervals (for example, in an interrupt service routine under control of a timer). A pseudocode example follows, under the assumption that the interval between interrupts occurs at constant intervals of T. In the pseudocode example, `Y1` is used to store y_{k-1} and `X1` is used to store x_{k-1}. Outside of the interrupt code, the software can perform additional tasks, such as updating the registers `A` and `B` when changes of k_p, k_I, or k_D occur.

Algorithm 13.1. Interrupt service routine (ISR) for a time-discrete *PID* controller.

1 *Interrupt*;
2 $X \leftarrow$ result from ADC;
3 start new ADC cycle;
4 $Y \leftarrow KI \cdot Y1 + A \cdot X - B \cdot X1$;
5 **if** $|X| < \epsilon_{max}$ **then**
6 $Y1 \leftarrow Y$;
7 **else**
8 $Y1 \leftarrow 0$;
9 **end**
10 $X1 \leftarrow X$;
11 output Y to analog port or PWM register;
12 return from interrupt;

- Line 2: This step is valid if the ADC can provide some form of signed value under the assumption that ϵ is the ADC input. If the ADC serves to measure the sensor or the output variable, a step where the ADC value Z is subtracted from the setpoint R, namely $X \leftarrow R - Z$, is necessary. In this case, the setpoint is maintained in software.
- Line 3: It is a good idea to start a new conversion cycle early, because the ADC can work on the conversion in parallel to the calculations, thus reducing the wait time until the microcontroller is ready for the next interrupt.
- Line 4: The system constants A and B are computed beforehand. With suitable scaling factors, these operations can be performed very fast with integer arithmetic. This operation needs to consider the sign as X and $Y1$ may be positive or negative.
- Lines 5–9: This step prevents integral windup. The integral register ($Y1$) is reset to zero if the control deviation X exceeds a predetermined value, but is updated when the control deviation lies within the specified tolerance band.

In the z-domain transfer function in Eq. (13.15), an additional factor of z^{-1} may become necessary. This factor reflects the combined delay caused by the analog-to-digital conversion (ADC) and the execution time of the interrupt service function. Depending on the microcontroller, conversion time is highly variable. Specialized sub-microsecond converters are available. However, many common converters take between 10 and 30 μs for a conversion. The computation steps inside the *PID* algorithm can take a similar time to complete. *PID* execution time is dependent on the microcontroller and on the implementation. An implementation of the *PID* algorithm with fast integer arithmetic on a low-cost 8-bit microcontroller might take about 20 μs. By sequentially executing ADC and *PID* computation, a sampling interval of $T = 100$ μs is possible that leaves a moderate amount of processing time for other tasks. Alternatively, ADC and *PID* computation could run in parallel as shown in Figure 13.6. As a consequence, T can be reduced and the dynamic behavior of the controller improved. However, parallel execution of ADC and *PID* computation leads to a one-cycle delay: the *PID* algorithm relies on the conversion result of the previous cycle, that is, to compute y_k, the conversion x_{k-1} is used. This time lag needs to be considered with an additional z^{-1} in the transfer function.

In the practical realization of digital controls, delays such as those shown in Figure 13.6 need to be carefully considered, because the associated phase lag reduces the system's relative stability. Strictly, Figure 13.6 shows an additional lag of roughly $T/2$ for the time between the start of the *PID* computation and the application of Y to the output. Any system that strictly relies on a specific sampling time would be designed such that Y is applied at the start of the subsequent interval, creating a total time lag of $2T$.

Compared to analog *PID* control systems, the software implementation allows additional "tricks" that improve controller performance. The *PID* control algorithm shown (Algorithm 13.1), for example, includes a simple means to prevent integral windup (see Section 13.6.1). A more rigorous approach would be to switch between *P* control and

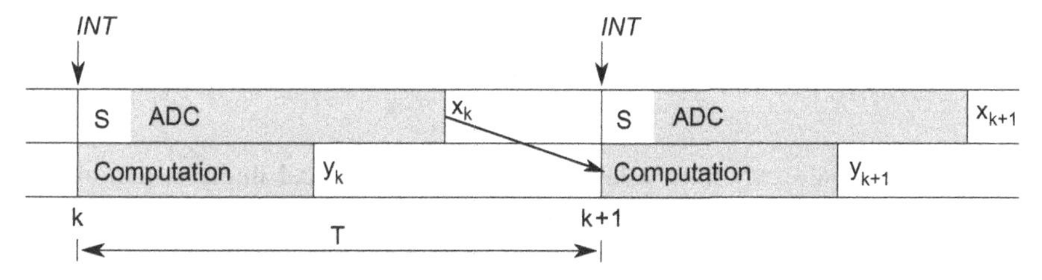

Figure 13.6 Timing diagram of a possible microcontroller realization of *PID* control. An interrupt (INT), occurring at time intervals of T, starts the analog-to-digital conversion (ADC). Depending on the microcontroller type, conversion time may range from several microseconds to tens of microseconds. Sampling (*S*) takes place at the beginning of the ADC cycle. The *PID* algorithm itself also takes tens of microseconds, depending on its implementation. ADC and *PID* computation can be run in parallel to reduce T. In that case, however, the *PID* algorithm relies on the conversion result from the previous cycle.

full *PID* control, depending on the control deviation. This method is described in more detail in Section 14.1.

Another interesting alternative advertises itself for many types of mechanical actuators, such as levers, step motors, valves or servos, that control the process. Rather than controlling the *absolute position* of the actuator, the *PID* controller can control the relative position. This approach, often referred to as *velocity control*, uses the integrator characteristic of the actuator and provides the necessary *change* in the actuator's position to move the process back to the control optimum. Such a system has a higher robustness against disturbances and even a controller failure. If we denote the actuator position as a, the controller's corrective action becomes $c = k_A \cdot \dot{a}$, where k_A is the actuator's proportionality constant. A velocity *PID* algorithm could be formulated as

$$\begin{aligned} \dot{a} &\approx \frac{c_k - c_{k-1}}{k_A T} \\ &= \frac{k_I}{2}\left(x_k + x_{k-1}\right) + \frac{k_p}{T}\left(x_k - x_{k-1}\right) + \frac{k_D}{T^2}\left(x_k - 2x_{k-1} + x_{k-2}\right) \end{aligned} \quad (13.16)$$

The input x to the controller is used in its first derivative for the P component and in its second derivative for the D component. This reflects taking the first derivative on both sides of Eq. (13.1). The integral component uses the more robust trapezoidal rule.

Multiplying Eq. (13.16) with $k_A T$ and moving the past output c_{k-1} to the right-hand side gives the digital filter equation

$$\begin{aligned} c_k = c_{k-1} &+ x_k k_A \left(\frac{k_I T}{2} + k_p + \frac{k_D}{T}\right) \\ &+ x_{k-1} k_A \left(\frac{k_I T}{2} - k_p - \frac{2k_D}{T}\right) + x_{k-2} k_A \frac{k_D}{T} \end{aligned} \quad (13.17)$$

Similar to Eq. (13.15) we can summarize the coefficients of the input terms x_k, x_{k-1}, and x_{k-2} as a_0, a_1, and a_2, respectively, and obtain the simplified digital filter equation together with its z-domain transfer function

$$\begin{aligned} c_k &= c_{k-1} + a_0 x_k + a_1 x_{k-1} + a_2 x_{k-2} \\ H_{PID}(z) &= \frac{a_0 z^2 + a_1 z + a_2}{z(z-1)} \end{aligned} \quad (13.18)$$

Once again, the exact time delays caused by the practical implementation may require that a modified transfer function is considered.

Lastly, the importance of input filtering needs to be stressed again. In Chapter 4, we mentioned that any frequency component in the input signal that is higher than $1/2T$ would be folded back into the frequency range below $1/2T$. These high-frequency components may be caused by sharp transients or by measurement noise. To prevent high-frequency components from reaching the controller's input, the digital controller must be preceded by an *analog* lowpass filter, such as the one proposed in Figure 11.9. This filter introduces two additional s-plane poles. A typical filter configuration would place the cutoff frequency near the digital filter's Nyquist frequency, that is, $\omega_c = \pi/T$. Although a first-order lowpass (usually a RC circuit) is often sufficient, the high-frequency suppression of a second-order active filter is superior. Often, the active filter can be combined with additional gain and signal-level adjustment in one stage, and the relative additional effort for the second-order lowpass becomes less. Typically, Butterworth configurations would be used, and the analog filter contributes the transfer function

$$H_{LP}(s) = \frac{\omega_c^2}{s^2 + 2\zeta\omega_c s + \omega_c^2} \tag{13.19}$$

to the loop gain function, where $\omega_c = \pi/T$. The damping coefficient ζ determines the steepness of the transition at ω_c. For a critically damped filter, $\zeta = 1$, and for a Butterworth filter, $\zeta = 0.71$.

When the process dynamics are very slow compared to the sampling rate (and, by association, to the Nyquist frequency), the filter poles can be omitted as nondominant poles. However, in cases such as the motor control example in Section 12.5, where the process time constants are in the same order of magnitude as the sampling rate, the lowpass poles have a major influence on the closed-loop response. It is possible, to some extent, to implement additional phase-lead components in the digital controller and partly compensate for the phase lag introduced by the anti-aliasing filter.

13.5 *PID* Controller Tuning

The *PID* controller has three independent parameters, which can either be interpreted as the weights of the proportional, integral, and differential component, or as the overall gain and the location of the open-loop zeros (Eq. (13.2)). Two general approaches exist. If the transfer function of the process $G(s)$ is known, frequency-domain or root locus methods can be applied to achieve the desired dynamic response (note that the steady-state response is always unity because of the integral term). If the transfer function of the process is not known, heuristics or algorithmic approaches are needed, because the optimization of the three-dimensional parameter space is usually not practical. Since the design of a controller for known $G(s)$ is a straightforward process, we will focus on *PID* tuning for unknown $G(s)$.

The methods described below usually assume *PID* control of a second-order system, i.e., a closed-loop system that exhibits a dominant complex conjugate pole pair, and may have one or more real-valued, non-dominant poles. For higher-order systems with multiple complex conjugate pole pairs (oscillations at multiple frequencies), the methods are not suitable.

13.5.1 Iterative Adjustment with an Oscilloscope

The iterative adjustment is particularly suitable for electronic or micro-mechanical systems that have response times in the order of microseconds to seconds. If we have a sufficiently fast process, a square-wave input to the closed-loop system simulates a repeated step function. It can therefore be displayed on the oscilloscope, and changes in the dynamic response can be rapidly observed as the parameters are adjusted. This method typically applies to optimizing the disturbance rejection of high-gain processes that try to keep the output at a fixed setpoint. One example is the load regulation of switching power supplies. Unlike linear voltage regulators, where a very high closed-loop gain can be applied without causing unstable behavior, more complex regulators require a well-designed loop compensation to prevent out-of-control situations or overshoots. A primary-side overshoot in a switching power supply can easily reach hundreds of volts!

The square-wave signal is typically applied as a disturbance. We begin with $k_I = 0$ and $k_D = 0$, i.e., with a pure P controller. We adjust k_p to obtain a strong oscillatory response. We aim for an angle of the dominant complex conjugate pole pair of 75° from the negative real axis, which corresponds to a damping factor of approximately $\zeta = 0.25$. In the second step, we increase k_D until the oscillations disappear and the step response is optimal. The dominant pole pair is still complex, but ζ is now close to unity. In the third step, we monitor the control deviation, i.e., the difference between the output and input signals. We now increase k_I until the control deviation rapidly vanishes after the step change (the edge of the square-wave signal). As a consequence, overshoots will occur. A good compromise is a single overshoot near $\zeta = 45°$. We can now return to adjusting either k_p or k_D to suppress the overshoot. Repeated adjustment of k_I and k_D leads to the desired step response.

To demonstrate the method let us assume a second-order process with a poor step response (a real-valued pole close to the imaginary axis causes a two-second decay time). The transfer function of the process is

$$G(s) = \frac{1}{(s + 0.5)(s + 2)} \tag{13.20}$$

We attempt to use PID control to improve the step response. A suitable initial value for k_p is 200, and the square-wave disturbance causes an oscillatory response (Figure 13.7a). Adding a D component with $k_D \approx k_p/10$ yields an almost ideal step response. However, the disturbance is not fully rejected (Figure 13.7a)—in the idealized case, the output would be a flat line. Enabling the I component improves disturbance rejection, although the integrator reacts relatively slow. For this reason, a large k_I needs to be selected. In Figure 13.7c, $k_I = 800$ was chosen, and the disturbance causes a single "bump" at the output, which is suppressed after 0.5 s, that is, 1/4 of the process time constant. Increasing k_p and slightly decreasing k_I reduces the amplitude of the "bumps" at the expense of settling time (Figure 13.7d), in this example $k_D = 50$ and $k_I = 700$. The parameters can now be balanced until a suitable compromise between overshoots and settling time is reached. Note that the large values for k_D, and k_I reflect the large magnification of the process output signal, which is already strongly attenuated

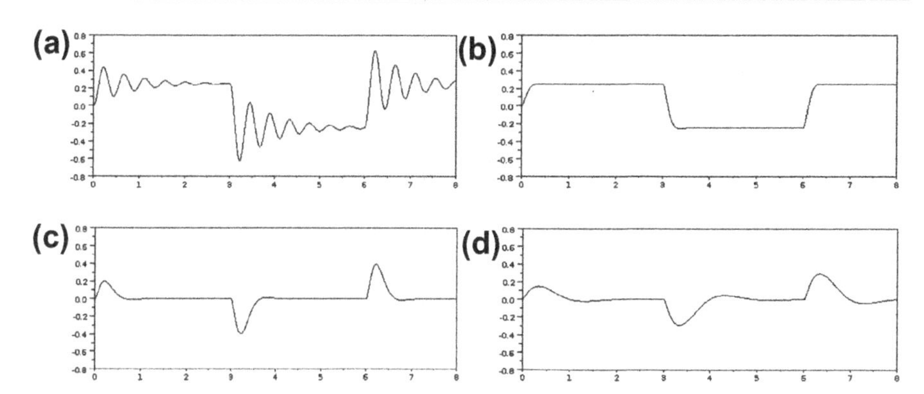

Figure 13.7 Tuning of a *PID* control. The process suffers from a square-wave disturbance with a period of 6 s. Pure *P* control with a large k_p leads to a strong oscillatory overshoot (a). Adding a *D* component ($k_p = 200$, $k_D = 20$) dampens the oscillations, but a steady-state error remains (b). The *I* component ($k_I = 800$) suppresses the steady-state error, and only a small transient "bump" remains (c). k_D and k_I can now be balanced to achieve a compromise between height and duration of the transient error (d).

by the *P* component. However, as this example shows, the tendency of the closed-loop system to oscillate with large k_p can be effectively reduced by the *I* and *D* components. The large values for the *PID* coefficients also indicate that the closed-loop process is widely dominated by the *PID* controller itself. With these large values, the closed-loop transfer function becomes approximately

$$H(s) \approx \frac{s}{s^3 + k_D s^2 + k_p s + k_I} \tag{13.21}$$

and we can see that the pole locations only depend on the choice of the *PID* controller parameters.

13.5.2 Ziegler-Nichols Tuning Method

Ziegler and Nichols proposed this method in the 1940s. Once again, we begin with a pure *P* controller with $k_D = k_I = 0$ and adjust k_p until the output oscillates almost undamped (pole pair very close to the imaginary axis). It is necessary to measure the period of these oscillations. We define the gain at which these oscillations occur as k_U and the corresponding period T_U. We now choose the final values of $k_p = 0.6k_U$, $k_D = 1.2k_U/T_U$, and $k_I = 0.075k_U T_U$.

The Ziegler-Nichols method prescribes a fairly high gain k_p, very much like the iterative method in the previous section. The dominant pole pair can be brought closer to the real axis by using a much lower k_p. Values that bring the closed-loop system near the critically damped case are $k_p = 0.2k_U$, $k_D = 0.4k_U/T_U$, and $k_I = 0.067k_U T_U$.

Another disadvantage of the Ziegler-Nichols method is the need to increase k_p to the limits of stability. In many systems, eliciting an almost undamped response is not advisable, because minor process variations can lead to an unstable system and therefore to damage to the process.

13.5.3 Cohen-Coon Tuning Method

The Cohen-Coon method assumes a first-order process that may have an additional dead-time delay. The method can be applied to higher-order processes with only real-valued poles and one dominant pole i.e., a process that approximately behaves similar to a first-order process. The Cohen-Coon method differs from the previous methods in that it can be based on measurement of the open-loop system. More precisely, the step response of the process is measured and the steady-state output rise $\Delta y = y(\infty) - y(0)$ is determined. Furthermore, the times are measured when 50% of the steady-state output τ_{50} and when 63% of the steady-state output τ_{63} are reached. From these values, the system time constant τ and the dead-time t_0 can be determined, and formulas exist for the recommended settings of k_p, k_I, and k_D for different levels of overshoot.

The limitation to first-order processes restricts the applicability of the Cohen-Coon method. *PID* control of a first-order process can be optimized in a straightforward manner. However, when the process exhibits a dead-time, the Cohen-Coon method leads much faster to acceptable control results than other tuning methods.

13.6 Variations and Alternatives of *PID* Control

13.6.1 Integral Windup

The integral component poses a challenge whenever large setpoint changes occur, because the error is integrated throughout the transient period as the output follows the setpoint. This effect is called *integral windup*. As the output reaches the setpoint, the integral component needs to be "discharged," which causes overshoot. Much more aggressive tuning is possible if integral windup is prevented. Figure 13.8 shows a simple circuit that realizes a *PID* controller. Without the Zener diodes, the transfer function is

$$H(s) = -\frac{R_2}{R_1}\left(1 + sC_1R_1 + \frac{1}{sC_2R_2}\right) \tag{13.22}$$

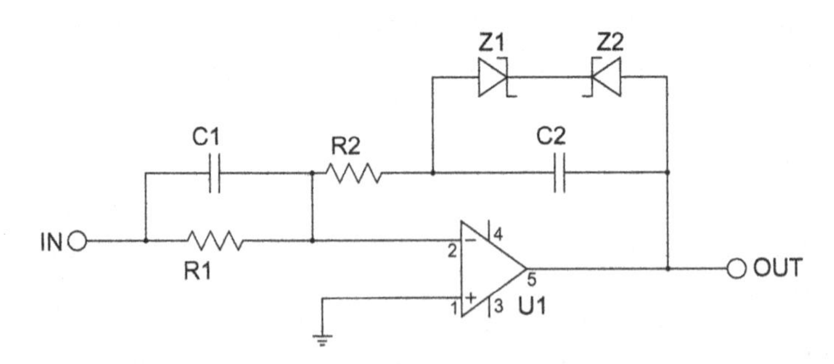

Figure 13.8 Realization of a *PID* controller with one op-amp and a few passive components. The integrating capacitor, C2, is clamped with zener diodes to prevent integral windup.

The Zener diodes introduce a nonlinearity that prevents any voltage above the Zener voltage to build up in the integration capacitor C2. Integral windup is therefore clamped, but not fully prevented. With small values of the control error, normal *PID* operation occurs.

A similar effect can be achieved in software, when the integral component is used only for low values of the control error. The algorithm would set the integral register to zero if ϵ exceeds a threshold ϵ_{max} in any direction, but operate normally for $|\epsilon| < \epsilon_{max}$.

13.6.2 Nonlinear Processes

If the process is nonlinear, the optimum choice of k_p, k_I, and k_D depend on the operating point, and changes of the setpoint over a wide range may cause an unacceptable loop gain and even unstable behavior. The *PID* controller could be tuned conservatively, thus maintaining stablility throughout the operating envelope. The downside of this approach is usually a poor dynamic response. In a software implementation, tables of k_p, k_I, and k_D for different operating point bands can be maintained, which keep the *PID* controller operating close to the optimum irrespective of the operating point.

Nonlinear systems are subject of its own branch of control theory. The mathematical foundation to describe and design nonlinear feedback controls is similar to linear feedback controls, but many mathematical operations that build on linearity are no longer valid. Frequently, frequency-domain methods are used where the system is split into a linear part and a nonlinear, frequency-dependent gain component. Numerous graphical approaches exist where the simple curve of a linear system turns into a family of curves that reflect the system response at different operating points. In addition, computer simulation has become an important tool. However, the simulation of nonlinear systems is severely limited, since it can lead to chaotic behavior and unpredictability of the feedback system.

For a process with pronounced nonlinearity over the operating range, the *PID* controller is generally not an optimum control system, and more advanced control methods need to be employed.

13.6.3 Pole Cancellation

It is possible to design electronic circuits that provide a real-valued pole and a real-valued zero, or two complex conjugate pole-zero pairs (Figure 13.9 provides an example). The zero can be placed at the location of a process pole and therefore cancel the pole in the overall transfer function. As a consequence, a dominant process pole can be replaced by a non-dominant pole that originates from the compensator. In many cases, the cancellation of a dominant slow pole can dramatically improve the transient response. However, the canceling zero needs to be placed very accurately. If process fluctuations cause the pole to move, pole cancellation fails, and the resulting pole-zero pair can cause highly undesirable closed-loop behavior. Adaptive approaches exist that allow the zero to "follow" the pole.

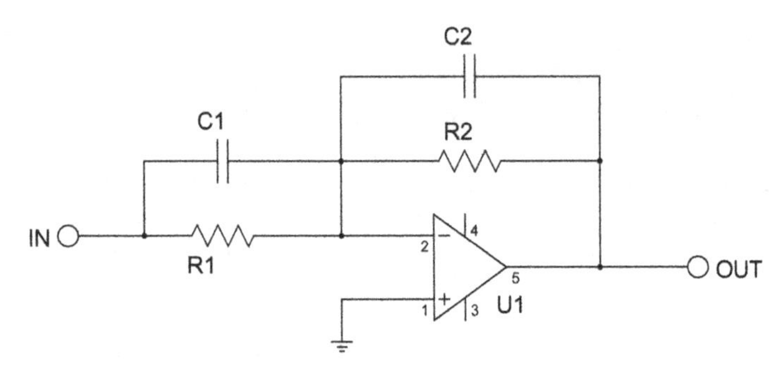

Figure 13.9 Schematic diagram of a lead/lag compensator. This circuit has a transfer function of $-(R_2/R_1) \cdot (1 + R_1C_1s)/(1 + R_2C_2s)$ and thus provides one pole and one zero that can be placed anywhere on the negative real axis. When the zero is closer to the origin than the pole, this circuit is referred to as lead compensator, otherwise as lag compensator.

13.7 Conclusion

The *PID* controller is arguably the most widely used controller in industry. Its popularity stems from its easy implementation in both hardware and software and the flexible adjustment of its coefficients. Many control situations, particularly for first- and second-order processes can be solved satisfactorily with *PID* control. On the other hand, higher-order systems pose a challenge for the universal *PID* controller. In addition, pronounced nonlinearities of the process require adaptive behavior where the *PID* coefficients are adjusted with changes of the operating point. In many cases, compensation networks or controllers with a highly specific transfer function provide better transient behavior than *PID* control.

14 Design Examples

Abstract

In this chapter, the link between theory and practical application is established. With seven different representative examples, the technical realization and optimization of analog and digital controllers is presented. In some cases, technical challenges are highlighted that are not evident from the mathematical model.

14.1 Precision Temperature Control

Temperature control is one of the major examples in this book (Chapters 5 and 6). Additional considerations are necessary to achieve steady-state control with a precision of 0.01 °C or better. High-precision temperature control is important in biochemical and enzymatic processes, spectroscopy, liquid chromatography, and laser stabilization, to name a few examples.

Temperature control with the main goal to achieve high precision in the steady state can be conveniently realized with a digital controller and a *PI* control algorithm. In fact, analog *PI* control is impractical due to the long time constants in the process that would require a large integrating capacitor. Furthermore, due to the long equilibration times of the process, a digital controller can be treated as an s-domain continuous system. High-precision temperature controllers typically use active heating/cooling elements, predominantly the Peltier element. A Peltier element causes heat flow between its end plates with two advantageous properties: heat flow is proportional to the applied electrical current, and the direction of heat flow depends on the polarity of the current. Unlike the resistive heater that was used in the waterbath example, a Peltier element can realize the sign of the heating power (positive: heating, negative: cooling) in a linear system. A representative control system is shown in Figure 14.1.

A first-order approximation of the system, as presented in Chapter 5, is usually sufficient to describe the process, unless a poor thermal junction between the heating/cooling element and the actual process (e.g., laser, chromatographic column, spectroscopic cuvette) creates an additional first-order lag. The sensor itself is also a time-lag system, and in a poor design, the combined process with its sensor would have three distinct real-valued poles. Some semiconductor precision sensors, such as the LM335, have time constants of several seconds. To treat this pole as nondominant, the process time constant needs to be in the range of several minutes. It is usually a good idea to measure the open-loop step response and perform nonlinear regression to determine whether the response is sufficiently well described with a single-exponential association, or whether a bi- or tri-exponential association describes the response significantly better. The process pole configuration is important to determine acceptable initial values for

Linear Feedback Controls. http://dx.doi.org/10.1016/B978-0-12-405875-0.00014-0

© 2013 Elsevier Inc. All rights reserved.

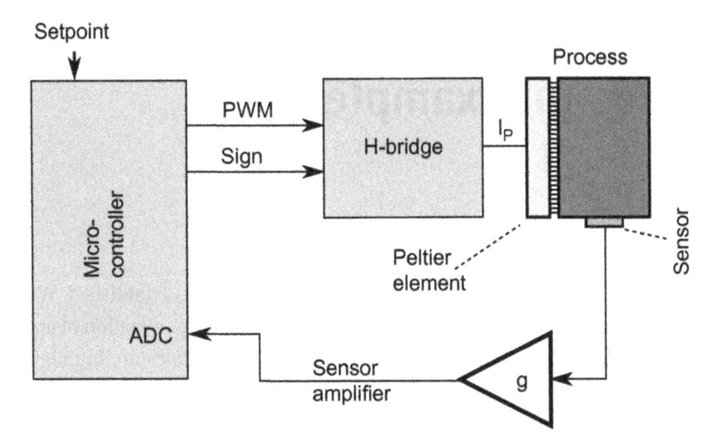

Figure 14.1 Schematic diagram of a precision temperature controller. The control algorithm, including calculation of the control deviation, is realized in software. The microcontroller outputs a pulse-width modulated (PWM) signal for the corrective action (i.e., the Peltier current). An H-bridge serves as the power driver and combines the PWM signal and the sign into the Peltier current I_P. A sensor, placed elsewhere on the process, provides a temperature-dependent voltage or current that needs to be amplified before being evaluated by the microcontroller's ADC. The sensor amplifier also needs to contain a bandlimiting lowpass.

k_p and k_I. Furthermore, if the process cannot be adequately described with a single-exponential step response, the *PI* controlled system will have three or more poles, and unstable behavior is possible.

For precision temperature control the choice of sensor and amplifier is of crucial importance. Semiconductor precision sensors have many desirable properties, but if their time constant is too long, thermocouples need to be used. Thermocouples produce a temperature-proportional voltage in the microvolt range, requiring a large sensor gain g. High-gain amplifiers introduce noise, and analog lowpass filtering is required to remove the noise from the sensor signal.

The combination of temperature range and precision dictates the requirements for the ADC. For a design example, let us assume that we want to achieve any temperature between 0 °C and 100 °C with a precision of 0.01 °C. The output voltage of the LM335 sensor ranges from 2.73 to 3.73 V. With a suitable gain $g = 5$ and an offset correction, the sensor voltage can be translated into the range 0–5 V for the ADC. The combined sensor/amplifier gain is 50 mV/°C, and an increase of 0.01 °C causes a voltage change of 0.5 mV, which is 1/10,000 of the ADC voltage span. A converter of at least 14 bits is required. In a 14 bit converter, the least significant bit changes with every 0.3 mV, leading to discrete temperature steps of 0.006 °C. If a higher accuracy is needed, an ADC with 15 or 16 bits is preferable. The least significant bit of a 16 bit ADC changes with 0.0015 °C, and a precision of 1/650 of a degree is theoretically feasible for the given gain and temperature range.

The Peltier element can be treated as a linear element, but it has different gain values for heating and cooling. Most of the electrical power used to drive a Peltier element is dissipated as heat. When the Peltier element is operating in heating mode, the dissipated electrical power is added to heat the process. Conversely, when the Peltier element is operating in cooling mode, the dissipated electrical power reduces its cooling efficiency. Many Peltier systems use circulating water to carry away the dissipated heat. However, the loop gain is still different in heating and cooling modes. Here, the software implementation becomes advantageous, because a simple loop gain correction can be realized in the control algorithm.

The flexibility of software control can be further exploited by preventing integral windup. If the transient-response overshoot that is associated with a strong integral component could be reduced, the *PI* controller can be much more aggressively tuned. The software solution is to operate the controller in *P* control mode when the control deviation exceeds a pre-defined tolerance threshold, and switch to *PI* control only when the control deviation falls inside the tolerance band. Thus, the integrator starts operating only when the control deviation is small, and integral windup remains limited.

An example pole configuration is shown in Figure 14.2. In the example, a process time constant of approximately 60 s is assumed, and a sensor time constant of 10 s. *P* control (Figure 14.2a) can be configured in a straightforward manner, for example, by placing the closed-loop poles on the dotted lines in the s-plane. A low loop gain is required for this configuration, and both disturbance sensitivity and steady-state tracking error are unacceptably high. In *PI* control mode (Figure 14.2b), the integrator pole in the origin leads to a significantly slower dynamic response, unless a zero is placed between the two process poles. Placement of the zero requires a fairly large $k_I \approx 2.5k_p$, and integral windup would lead to an unacceptably large overshoot (Figure 14.3a). The combination of conservatively tuned *P* control followed by *PI* control with large loop gain and large k_I reduces the overshoot dramatically (Figure 14.3b). The

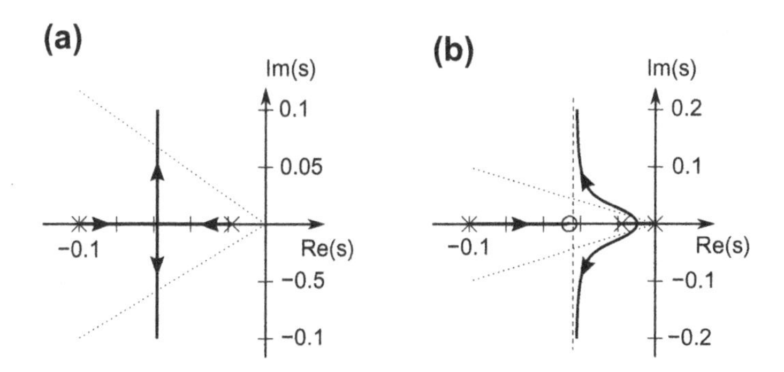

Figure 14.2 Root locus diagrams for the precision temperature control system in *P* control mode (a) and *PI* control mode (b). *P* control has a relatively slow dynamic response, and a Butterworth-damped complex conjugate pole pair (intersection with the dotted line) provides an adequate step response. Once the system switches to *PI* control mode, the integrator pole in the origin and a suitably positioned zero are added. The zero pulls the asymptotes toward the left, ensuring a faster dynamic response. However, $k_I \approx 2.5k_p$ is required. With such large values of k_I, integral windup would lead to an unacceptable overshoot, but by switching to *PI* control only for very low $|\epsilon|$, integral windup is constrained.

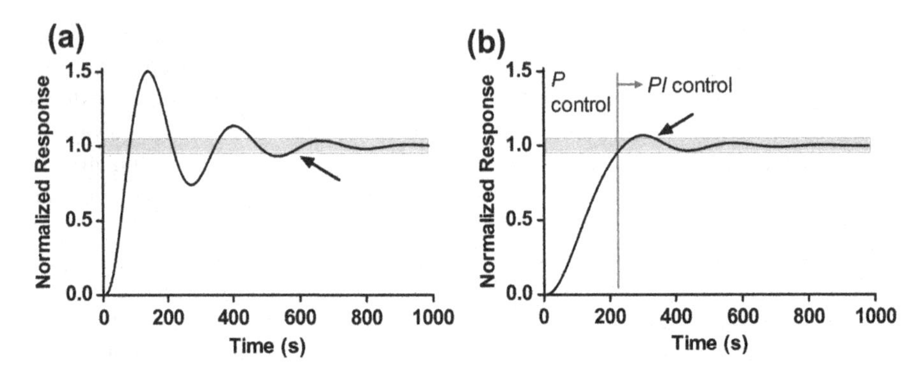

Figure 14.3 Comparison of the dynamic response of an aggressively tuned *PI* control system (a) and a system that approaches the tolerance band under *P* control, then switches to *PI* control (b). The 5% tolerance band is indicated by the gray rectangle. Arrows indicate the settling time into the 5% tolerance band. Although the initial rise time is much shorter with *PI* control and a large k_p, the overshoot of more than 50% prevents fast settling. With about 1/3 of its k_p-value, the initial rate is slower. When the tolerance band is reached after 220 s and the system switches to *PI* control with large k_p- and k_I-values, the momentum in the process is low enough that little overshoot occurs. Further lowering the initial k_p reduces the overshoot even more at the expense of a longer settling time.

values for k_p and k_I are the same in Figure 14.3a and b when *PI* control is active. The step response in Figure 14.3b begins with $k_I = 0$ and k_p set to about 1/3 of its value for *PI* control.

14.2 Fast-Tracking Temperature Control

The control goal for fast-tracking temperature control is to obtain a fast step response combined with somewhat lower precision, and it complements the slow precision temperature control in the previous example. Fast-tracking temperature control is necessary, for example, for the polymerase chain reaction (PCR). PCR is used to amplify short DNA sections, and it requires repeated cycles of DNA denaturation at 98 °C (20–30 s) followed by an annealing step of 20–40 s in the range of 50–65 °C and the elongation step at 72 °C for several minutes. This cycle is repeated many times, and the amount of DNA is doubled at each cycle. Ideally, the system ends with a hold cycle at low temperatures (4–10 °C) to conserve the reaction product.

PCR devices are small. Notably, recent integrated PCR devices in a lab-on-a-chip environment have very low volumes of fluid, and the container walls are thin. The process time constant is therefore very short. With a highly integrated sensor, the process is adequately described as a single-pole time-lag system. A classical approach with *PID* control is appropriate for this control task. When both zeros of the *PID* controller are placed to the left of the process pole (it is possible to form a double zero), two complex conjugate branches exist. Pole placement with the *PID* controller is relatively flexible,

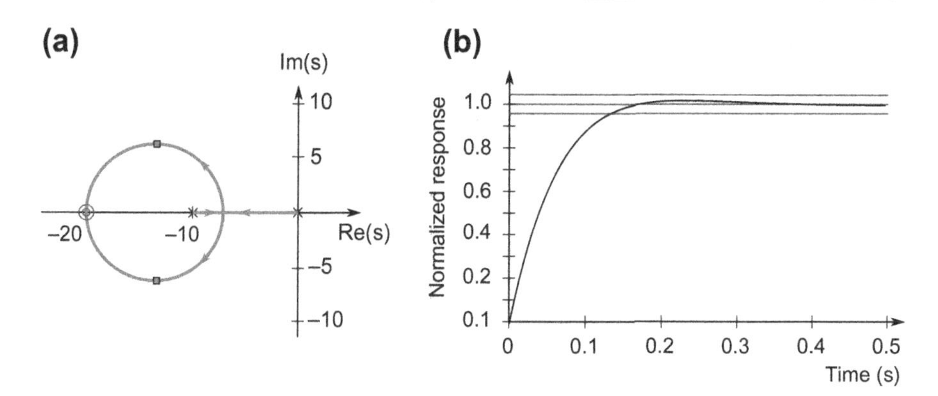

Figure 14.4 Root locus diagram (a) and step response (b) of a first-order *PID* controlled system with a process pole at $s = -10\ \mathrm{s}^{-1}$. The *PID* controller places a double zero to the left of the process pole that attracts the root locus branches. A desirable closed-loop pole location is indicated by the complex conjugate squares on the root locus. The corresponding step response shows approximately 4% overshoot and has a settling time into the 5% tolerance band sightly more than 100 ms. The apparent time constant is approximately 50 ms and therefore faster than the process time constant.

and underdamped responses can be achieved that balance moderate overshoot of a few percent with a closed-loop time constant that is faster than that of the process—despite the integrator pole in the origin.

An example for classical *PID* control is shown in Figure 14.4. We assume a process time constant of 100 ms (pole at $s = -10$). The *PID* controller places a double zero at $s = -20$, which provides the constraints $k_D = k_p/40$ and $k_I = 10k_p$. By increasing k_p (and by association of k_I and k_D), the pole pair can be moved into the complex plane until a suitable step response has been found.

If a digital control system is used in this example, it is still possible to treat the controller as continuous system when the sampling frequency is in the kHz range. This sampling frequency is achievable with basic microcontrollers and fast integer arithmetic. For even faster process time constants (lab-on-a-chip), the sampler pole can no longer be neglected, and z-domain treatment is necessary.

If the process is relatively slow, and either the Peltier element or the sensor provide a second pole, different considerations apply. With conventional continuous feedback control, it may be impossible to meet the design goal of a fast step response (requires large loop gain) and low overshoot (requires small loop gain). The software trick to switch between *P* control and *PI* control does not apply, because it leads to a slower dynamic response. However, it is possible to use active "braking" of the process to reduce overshoot.

Ideally, the control system is somewhat predictive: a digital phase-lead filter can be used to approximately cancel the second process pole. Exact pole cancellation is difficult with traditional linear control systems, but an approximation of the point where

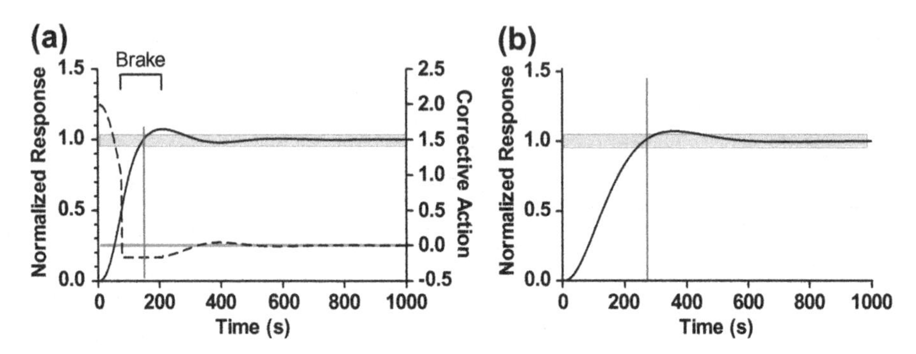

Figure 14.5 Step responses of an algorithmically braked second-order process under *PID* control. The algorithm predicts approximately when the first time-lag component enters its overshoot region and applies a negative corrective action until the output value stops rising (a). This algorithm can only be applied to specifically compensate for large overshoot caused by step changes of the setpoint. It allows to choose a much larger k_p than would be possible with a conventionally controlled system. The system in (b) is tuned to deliver the same overshoot in the step response, but its much lower k_p causes its rise time to be almost twice as long as the system in (a).

the step response of the first process pole crosses into the overshoot region is sufficient. At this point, the software algorithm applies a constant corrective action with an inverted sign until the first derivative of the output signal reaches zero, after which control is returned to the *PID* controller.

The step response of such a system is shown in Figure 14.5. Due to the braking process, a much higher loop gain is possible, and the rise time is correspondingly shortened. It needs to be emphasized, however, that this control algorithm is only valid for large setpoint changes. In the software implementation, therefore, a step change (such as moving from one temperature to the next in the PCR example) would enable the application of the brake. Once the brake has been applied, it is disabled and control returned to the *PID* controller until the next step change. In the example, the rise time is almost halved with the algorithmic control when compared to conventional linear control with the same overshoot (Figure 14.5 a and b). With the same k_p, overshoot of more than 50% is seen (Figure 14.3a).

14.3 Motor Speed and Position Control

Control of motor speed and its integral—position—is one of the most common tasks of feedback controls. A good example that combines both speed and position control are CNC machines, such as the lathe on the cover page. Two fundamentally different approaches exist: step motors and continuous motors. Step motors enjoy high popularity, because they can be operated in an open-loop configuration, and their discrete angular step size makes them attractive for microprocessor applications. Step motors are briefly covered in this section due to their popularity and widespread use as an alternative to controlled continuous motors. Unlike step motors, continuous motors, such as DC motors, require closed-loop feedback to control both speed and position. The advantages of DC motors over step motors are their higher speed, larger torque, their linear behavior, and the absence of resonance effects that characterize step motors at certain speeds.

14.3.1 Open-Loop Control with Step Motors

Step motors are similar to AC synchronous motors in that they follow a rotating magnetic field. Typically, step motors have a large number of magnetic poles compared to AC synchronous motors, and their step angle is accordingly smaller. Electrically, the step motor has two independent, orthogonal windings A and B that need to be alternatingly excited in the sequence $+A$, $+B$, $-A$, $-B$ to move the motor in one direction, and with the reverse sequence to move the motor in the opposite direction (Figure 14.6). The sign $(-A, +A)$ indicates the direction of the current flow through the windings. In this configuration, the step motor requires one H-bridge driver for each winding, and two digital signals per winding to drive each transistor pair of the H-bridge. This configuration is referred to as *bipolar*.

When each winding has a center tap, the configuration is referred to as *unipolar*. Unipolar step motors require only two transistors for each winding, thus reducing the complexity of the driver circuit at the expense of a lower torque. Both the bipolar and the unipolar motors require four digital lines for control. The microcontroller

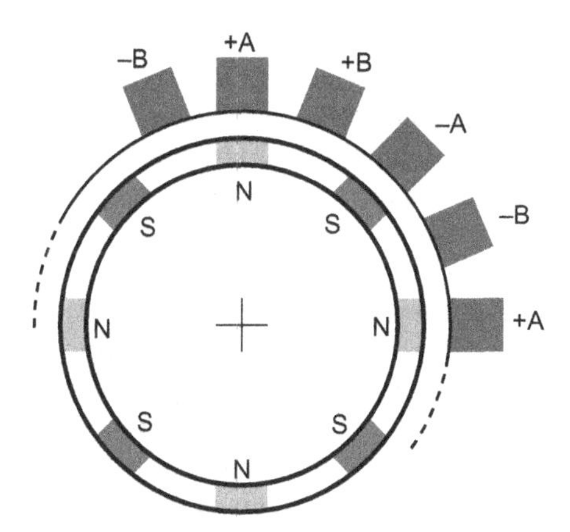

Figure 14.6 Simplified schematic of a step motor. A current in positive direction through the A-windings attracts a magnetic north pole below the stator poles labeled $+A$ and a magnetic south pole below those labeled $-A$. Reversal of the current also reverses the poles. The same principle apples to the B-windings. In the position shown here, a $+A$ current creates a holding torque. De-energizing A and energizing B in the positive direction attracts the poles in a clockwise direction and moves the motor by one step. De-energizing B and energizing A in the reverse direction performs the next step.

Table 14.1 Phase driving sequence for full- and half-step modes in one direction. By cyclically applying the sequence the motor turns in one direction. By applying the sequence in reverse order, the motor turns in the opposite direction. The phase designations Phi 0 through Phi 3 refer to Figure 14.7. AWO stands for All Windings Off, where the holding torque is released.

	Full-step mode				**Half-step mode**			
Step	$+A$ **(Phi 0)**	$+B$ **(Phi 1)**	$-A$ **(Phi 2)**	$-B$ **(Phi 3)**	$+A$ **(Phi 0)**	$+B$ **(Phi 1)**	$-A$ **(Phi 2)**	$-B$ **(Phi 3)**
0	1	0	0	0	1	0	0	0
0a					1	1	0	0
1	0	1	0	0	0	1	0	0
1a					0	1	1	0
2	0	0	1	0	0	0	1	0
2a					0	0	1	1
3	0	0	0	1	0	0	0	1
3a					1	0	0	1
AWO	0	0	0	0	0	0	0	0

applies the digital values in the sequence indicated in Table 14.1. To reverse direction, the sequence is applied in the reverse order. Half-stepping mode makes use of the superposition of the orthogonal magnetic fields to create a resulting field that has a maximum between two adjoining magnetic poles. A typical step motor would advance 1.8° between two full steps and 0.9° between two half-steps. A higher-torque full-step mode uses the steps 0a, 1a, 2a, and 3a in Table 14.1. To complete a full revolution, the full-step sequence needs to be repeated 50 times (200 steps with 1.8° each). If all windings are de-energized, the step motor produces no holding torque.

In the typical open-loop configuration, care must be taken to not exceed the maximum torque and to avoid resonance effects. The resonance is the consequence of an inert mass (the rotor) reacting to a change in the direction of the magnetic force. Depending on the amount of external friction, the rotor can actually overshoot its static position and equilibrate after a few oscillations. Interference between step frequency and resonant frequency can cause the motor to slip and skip steps.

The available torque decreases with increasing step frequency. If the torque demands of the load are larger than the available speed, the motor will also skip steps and possibly stall. This scenario is unacceptable in an open-loop configuration, because the microcontroller relies on each step actually being taken. Careful mechanical design is necessary to dampen resonance effects, and to keep the speed of the motor low enough to prevent stalling.

The diminishing torque at higher speeds is primarily caused by the slow exponential build-up of coil current in voltage-driven circuits. Some motor drivers apply a higher voltage for the first few microseconds after a step to rapidly build up the current (and thus the torque), then drop to the design voltage to prevent overheating. The basic circuit schematic for such a driver is shown in Figure 14.7. Without the kick current feature,

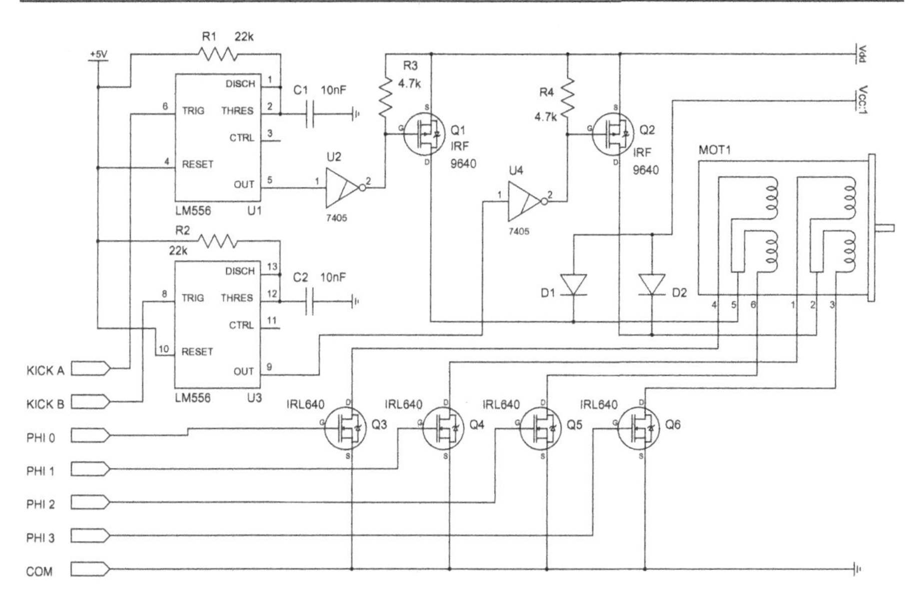

Figure 14.7 Basic circuit schematic for a step motor driver with kick current. Six microcontroller ports are needed, four of them to energize the motor coils according to Table 14.1 and two to activate the kick current. A brief negative pulse at `Kick A` that accompanies steps 0 or 2 triggers U1 for approximately 250 μs and applies the high drive voltage `Vdd` to the motor through Q1. For steps 0 and 3, a similar kick current is applied through `Kick B`, U3, and Q2. The motor's steady-state operating voltage `Vcc:1` is applied through D1 and D2. D1 and D2 are high-current Schottky diodes. Due to the high currents involved, generous bypass capacitors are necessary for all power supply lines (not shown in this diagram).

U1 through U4, Q1, Q2 and their associated passive components can be omitted. Kick current switching is passive, that is, the kick current is applied for a fixed time after a brief trigger pulse. In this design, the time constant is 250 μs, which limits the step rate to 4000 half-steps per second. Even at this rate, the motor is operated beyond its design limits and sufficient time needs to be given to allow the motor to cool. A similar design with a 3.3 V, 2 A motor and a kick voltage of 35 V has been used by the author to achieve step rates up to 10,000 half-steps per second. An advantageous alternative is to use suitable integrated circuits, such as the Sanken SLA7076.

A typical step motor can accelerate to approximately 1–1.5 revolutions per second from a standstill if the external rotational inertia is negligible. For the half-step sequence, this corresponds to a delay of approximately 2 ms between steps. Longer delays are straightforward, and the strength of the step motor lies in very slow, controlled movement. Shorter delays are possible, but the motor needs to be accelerated into the steady-state angular speed. Carefully designed trapezoidal velocity profiles are necessary to allow a step motor to accelerate and decelerate without stalling.Trapezoidal or triangular velocity profiles require advance knowledge of the number of steps to travel, and acceleration/deceleration algorithms can reach appreciable complexity.

When the microcontroller keeps track of the position, a zero position needs to be defined. Often, linear or rotary actuators have end limit switches to stop the motor. Such a switch can be used to move a motor to the reference zero of the position after power-up. The microcontroller then adds or subtracts steps from a position register, thus allowing to drive a step motor in open-loop configuration.

14.3.2 Closed-Loop Control with DC Motors

The DC motor is a linear, voltage- or current-controlled element. The equations provided in Chapter 9 are a good approximation for the practical design of DC motor controllers, and the DC motor can usually be modeled as a first-order system. In this model, the motor windings are considered to have only real-valued impedance. A second-order model would include the inductivity of the motor windings, which adds a second (and faster) real-valued pole to the motor transfer function.

Compared to the step motor driver (Figure 14.7), a DC motor driver is relatively simple. To allow direction reversal, the motor is driven by a H-bridge. Pulse-width modulation of the drive voltage allows to control the mean motor current in a linear fashion with minimal thermal losses in the H-bridge. The principle of an H-bridge driver is shown in Figure 14.8. A microcontroller provides control signals to enable motion on clockwise (CW) or counterclockwise (CCW) direction. A pulse-width signal (PWM) determines the mean drive current and therefore the motor torque. The schematic in Figure 14.8 is provided for reference only. Integrated H-bridge circuits exist, such as the L293 or the more complex UC3176. Integrated drivers typically have shoot-through protection and feature a brake mode where two lower transistors conduct. Many modern microcontrollers (e.g., the Microchip PIC18F4520) offer direct full-bridge drive outputs with integrated PWM, brake, and shoot-through protection. The only discrete components needed in this case are Q1 through Q4.

In rare instances, designs can be found where the motor is driven by a linear power amplifier, such as the OPA548. Similar amplifiers can be used as power drivers for audio applications, and these amplifiers act like operational amplifiers with a high output current. To drive a DC motor with a linear power amplifier, a symmetrical power supply is required to allow direction reversal. In a feedback configuration with a current-sense resistor, linear power amplifiers can provide excellent control of the motor drive current that follows a control voltage. The sign of the control voltage determines the direction of the motor.

The closed-loop configuration of a DC motor requires the use of a sensor. Some servo motor units use a potentiometer to obtain an analog, angle-dependent signal. A possible realization is shown in Figure 14.9a, where the sensor for the angular position Θ is a potentiometer. Many servo systems are controlled with a pulse width signal, and the desired angle depends proportionally on the pulse width τ:

$$\Theta = \frac{3\pi}{2} \cdot \frac{\tau}{T} \tag{14.1}$$

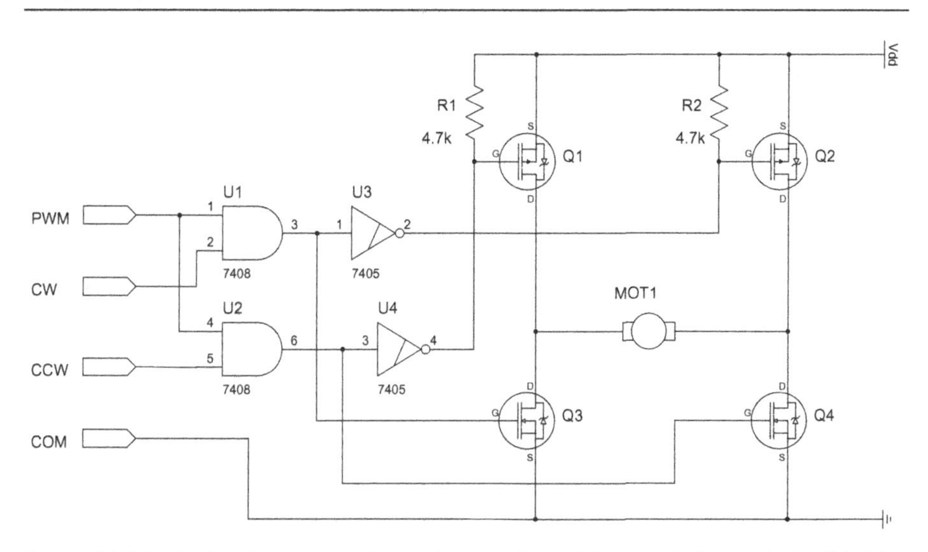

Figure 14.8 Basic circuit schematic for a DC motor driver. The central element is the H-bridge formed by the power transistors Q1 to Q4. U3 and U4 ensure that only diagonal transistor pairs (Q1 and Q4 or Q2 and Q3) are open. Ideally, a direction reversal is accompanied by a brief period where all transistors are off to prevent a shoot-through current through Q1/Q3 or Q2/Q4 that can occur due to switching delays. A positive signal at the `CW` or `CCW` inputs allows the motor to turn in clockwise or counterclockwise direction, respectively, and the mean drive current is influenced by the pulse-width signal `PWM`. Many microcontrollers feature an integrated PWM timer. This basic circuit does not prevent the catastrophic case where the motor power is shorted through Q1/Q3 and Q2/Q4 when both `CW` and `CCW` receive a high signal. Integrated H-bridge drivers exist that replace the discrete circuit shown here.

where T is the period of the pulse-width signal. The factor $3\pi/2$ reflects the maximum angle coverage of the servo arm. The controller can be a purely analog system with an error amplifier and a proportional H-bridge. In compact servo systems, the controller is often a simple error amplifier (*P* control). However, the use of a microcontroller becomes more common. The microcontroller can be used to convert the pulse width into a digital signal, and an on-chip ADC converts the analog potentiometer signal. Use of a microcontroller is attractive, because *PI* or *PID* control algorithms can be implemented without additional space requirements, and the microcontroller can directly generate the PWM signal.

A complete microcontroller-based closed-loop DC control system is shown in Figure 14.9b. The microcontroller drives the motor through a PWM-controlled H-bridge. The position information comes from a quadrature encoder that is attached to the motor shaft. Optical quadrature encoders consist of a slotted disk and two optointerrupters. To provide an example, the encoder disk could have 20 slots, with 9° apertures every 18°. The slot pitch determines the resolution limit of the discrete angular steps. A gear box increases the angular resolution by the gear ratio.

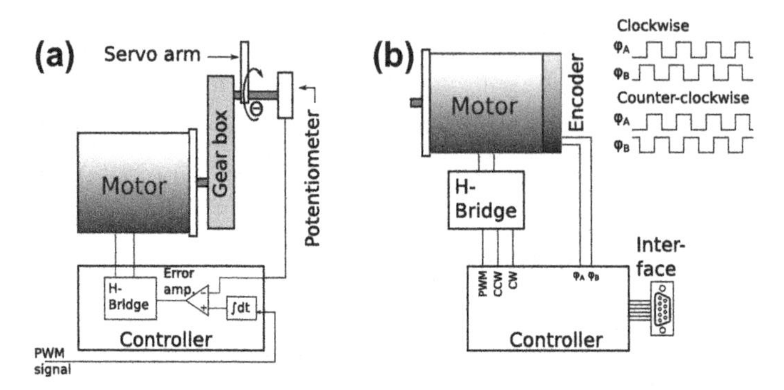

Figure 14.9 Schematic of a semi-analog servo system (a) and a closed-loop DC motor controller (b). Servo systems are often driven by digital PWM signals. The controller integrates this signal to obtain an analog signal that can be compared to the position-dependent potentiometer output. The controller can be realized with analog electronics, or a microcontroller can be used that converts both signals to digital values and processes those. The closed-loop system in (b) is based on a quadrature encoder. Two signals, φ_A and φ_B, are 90° out of phase, and their phase shift depends on the direction.

The two optointerrupters in this example are spaced apart by 40.5° (more generally, $k \cdot p + p/4$, where p is the pitch in degrees and k is an integer number). When one optointerrupter encounters the edge of a slot, the other is positioned over the center of another slot. In this fashion, the signals of the two optointerrupters are 90° out of phase. The highest possible resolution is achieved when both signals, φ_A and φ_B, cause an interrupt in the microcontroller. The interrupt service routine then samples the levels of φ_A and φ_B. When the interrupt was issued by φ_A and the levels are the same, the motor turns clockwise, whereas the motor turns counterclockwise when the levels are opposite (Figure 14.9b). The situation is reversed when the interrupt was issued by φ_B. By using all four edges from the optointerrupters, the angular resolution is increased fourfold.

The use of optical quadrature encoders in the fashion described here has one additional advantage. Not only do the pulse trains from the encoders allow closed-loop position and speed control, but any motor motion caused by an external torque or load would be tracked by the microcontroller. In a position control system, an external torque that turns the motor would cause a corrective action.

The simplicity of the hardware (one microcontroller and the H-bridge) is balanced by the need for fairly complex software. At least three interrupt service routines are required: one each for φ_A and φ_B, and the timed interrupt that triggers the computation of the corrective action. The outer software layer (that is, any software code outside of the interrupt service routine) is responsible for the interface communication, where the microcontoller can receive new parameters, such as speed or setpoint. Algorithm 14.1 demonstrates how the quadrature encoder interrupts can be serviced, and Algorithm 14.2 provides pseudo-code for computation of the motor speed and an example for a position control algorithm.

Algorithm 14.1. Interrupt service routines (ISR) to service the quadrature encoder.

1 *Interrupt-on-change from* φ_A;
2 $D \leftarrow \varphi_A$ xor φ_B;
3 **If** $D = 1$ **then**
4 $CCW \leftarrow 1$; `// Counter-clockwise`
5 $POS \leftarrow POS + 1$;
6 **else**
7 $CCW \leftarrow 0$; `// Clockwise`
8 $POS \leftarrow POS - 1$;
9 **end**
10 $SCNT \leftarrow SCNT + 1$;
11 return from interrupt;
12 *Interrupt-on-change from* φ_B;
13 $D \leftarrow \varphi_A$ xor φ_B;
14 **If** $D = 1$ **then**
15 $CCW \leftarrow 0$; `// Clockwise`
16 $POS \leftarrow POS - 1$;
17 **else**
18 $CCW \leftarrow 1$; `// Counter-clockwise`
19 $POS \leftarrow POS + 1$;
20 **end**
21 $SCNT \leftarrow SCNT + 1$;
22 return from interrupt;

Several registers are used. The CCW-register keeps track of the current motor direction and can be used, if necessary, outside of the ISR. The POS-register is the actual position: Each edge from the quadrature encoder increments or decrements this register, depending on the motor direction. To obtain a physical position, the POS-register needs to be scaled with a factor that depends on the quadrature encoder and any external gear or leadscrew elements. Moreover, POS needs to be reset after power-up when the motor is in a reference position: typically, the motor is driven into a limit switch and the POS-register reset. Lastly, the $SCNT$-register counts the absolute number of steps and is read and reset in regular intervals to determine the motor speed (Algorithm 14.2).

Algorithm 14.2 makes use of two independent timers. Timer 0 is set to a relatively long period to allow accumulation of quadrature encoder events. To stay with the previous example of an encoder disk with 20 slots, Algorithm 14.1 increases $SCNT$ for every 4.5° of rotation. Each full revolution increments $SCNT$ by 80. If Timer 0 has a period of 100 ms, the constant K_S could be set to 7.5 to calibrate $SPEED$ in revolutions per minute. Clearly, a balance between the lowest measurable speed and the lag after which $SPEED$ is available needs to be found. Furthermore, oscillatory motion of the motor faster than the period of Timer 0 leads to incorrectly high values of $SPEED$. One possible extension of Algorithm 14.1 is to reset $SCNT$ whenever the direction changes.

Algorithm 14.2. Interrupt service routines (ISR) for the speed determination and for the actual control algorithm.

1 *Interrupt from Timer 0*;
2 $SPEED \leftarrow SCNT \cdot K_S$;
3 $SCNT \leftarrow 0$;
4 return from interrupt;
5 *Interrupt from Timer 1*;
6 $\epsilon \leftarrow SET - POS$; `// Deviation from target position`
7 $Q \leftarrow \epsilon \cdot K_P$; `// Corrective action (with sign)`
8 **If** $Q < 0$ **then**
9 $P_{CCW} \leftarrow 0$;
10 $PWM \leftarrow |Q|$; `// Set the PWM register`
11 $P_{CW} \leftarrow 1$; `// Enable clockwise motion`
12 **else**
13 $P_{CW} \leftarrow 0$;
14 $PWM \leftarrow |Q|$; `// Set the PWM register`
15 $P_{CCW} \leftarrow 1$; `// Enable counterclockwise motion`
16 **end**
17 return from interrupt;

To keep this example relatively simple, we use *P* control for the position control algorithm. The setpoint is SET, and it needs to be provided by a function outside the ISR, for example, by a command from the interface. Extensions (see, for example, Algorithm 13.1) can be implemented in a straightforward fashion. The control algorithm is invoked by an interrupt from Timer 1, and the period of Timer 1 coincides with the sampling interval T of a time-discrete system. The period of T can be kept short to reduce the sampling lag. In this example, PWM is the register that controls the pulse width, and P_{CW} and P_{CCW} are the output port bits that control the direction of the H-bridge. The sequence of setting the port bits ensures that P_{CW} and P_{CCW} are never high at the same time.

Once the motor control hardware is completed, the control algorithm can be supplied. Methods to design the controller parameter for angular speed (first-order approximation) and position (second-order approximation with one pole in the origin) have been described in Chapter 9. *PID* control is a good starting point for both speed and position control.

14.4 Resonant Sine Oscillator

Among electrical engineers, the adage goes "amplifiers tend to oscillate, oscillators don't." Oscillators are a particularly good example of a feedback system that is tuned toward one specific purpose. Moreover, the principles of oscillator design provide a very illuminating example for the application of frequency-response methods. We have defined gain and phase margin (Chapter 11) such that a system becomes unstable when its gain is greater than unity when its phase shift reaches 180°.

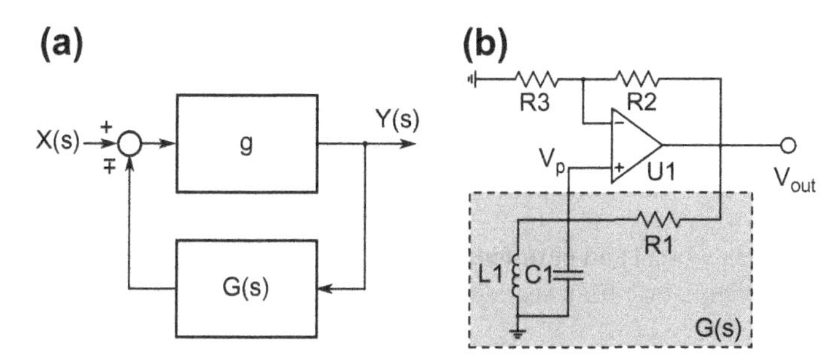

Figure 14.10 Canonical feedback form of an oscillator circuit (a) and a circuit realization (b). Any oscillator needs a gain element g and a feedback network $G(s)$. If the open-loop gain is greater than unity when the phase shift reaches $-180°$ (or 360° for positive feedback), the closed-loop system turns unstable and begins to oscillate. The operational amplifier has a gain $g = R_2/R_3 + 1$, and the resonant RLC circuit in this example has a second-order transfer function $G(s)$.

Oscillators are feedback circuits that are intentionally designed to be unstable, but where either the supply voltage or some form of gain control end the exponential increase of the amplitude at some point. Therefore, an oscillator has a start-up phase, after which it continues oscillating with constant amplitude. If the amplitude is limited by the supply voltage, the oscillator does not provide a sinusoidal output, but rather a clipped sinusoid or a square wave with considerable harmonic distortion.

The basic principle of an oscillator is shown in Figure 14.10a. The transfer function $H(s) = Y(s)/X(s)$ is

$$H(s) = \frac{g}{1 \pm g \cdot G(s)} \tag{14.2}$$

For any oscillator, the input signal $X(s)$ is irrelevant and usually kept at a constant reference voltage. Oscillation at a frequency ω_0 requires that $1 \pm g \cdot G(j\omega_0) = 0$. For negative feedback, i.e., the case that we covered more extensively in this book, the magnitude and phase criteria emerge as

$$\begin{aligned} |g \cdot G(j\omega_0)| &= 1 \\ \arg\left(g \cdot G(j\omega_0)\right) &= 180° \end{aligned} \tag{14.3}$$

For positive feedback, an oscillator would require $\arg\left(g \cdot G(j\omega_0)\right) = 0°$. Let us examine the circuit in Figure 14.10b as an example. This circuit provides positive feedback, and the feedback network acts as a complex impedance between the positive input of the op-amp and the circuit output:

$$G(s) = \frac{V_p(s)}{V_{out}(s)} = \frac{s}{RCs^2 + s + R/L} \tag{14.4}$$

The op-amp is configured as a noninverting amplifier with gain $g = 1 + R_2/R_3$. For the node at the noninverting input, Kirchoff's node rule yields the relationship

$$\frac{1}{L_1}\int V_p \mathrm{d}t + C_1\frac{\mathrm{d}V_p}{\mathrm{d}t} = \frac{1}{R_1}\left(V_{out} - V_p\right) \tag{14.5}$$

Rearranging Eq. (14.5) and eliminating V_{out} with $V_{out} = g \cdot V_p$ gives the homogeneous differential equation for a second-order system,

$$\frac{\mathrm{d}^2 V_p}{\mathrm{d}t^2} + \frac{1-g}{R_1 C_1}\frac{\mathrm{d}V_p}{\mathrm{d}t} + \frac{1}{L_1 C_1}V_p = 0 \tag{14.6}$$

with the natural frequency ω_0, the damping coefficient ζ

$$\omega_0 = \sqrt{\frac{1}{L_1 C_1}}; \quad \zeta = \frac{1-g}{2R_1}\sqrt{\frac{L_1}{C_1}} \tag{14.7}$$

and the solution

$$V_p(t) = \hat{V} e^{-\zeta\omega_0 t} \sin\left(\omega_0\sqrt{1-\zeta^2 t}\right) \tag{14.8}$$

The similarity to the spring-mass-damper system (*cf.* Section 2.2 and Table A.5) is obvious: for $g < 1$, we obtain an attenuated oscillation with $\zeta > 0$. When $g = 1$ and consequently $\zeta = 0$, Eq. (14.8) turns into a continuous sinusoid $V_p(t) = \hat{V}\sin(\omega_0 t)$. The active gain component actually allows $\zeta < 0$ for which the exponent turns positive, and the amplitude of the oscillation increases with time.

In practice, it is impossible to choose the components to precisely obtain $g = 1$. Rather, the gain is chosen to be slightly larger than unity, and additional measures, such as a nonlinear feedback component, are added to stabilize the amplitude of the oscillation. For the well-known radiofrequency LC oscillator circuits (e.g., Hartley or Colpitts oscillators), the gain is provided by a transistor. As the amplitude of the oscillation increases and the transistor approaches saturation, its gain is reduced, thereby stabilizing the amplitude. At high frequencies, the harmonics caused by nonlinear distortion or clipping can be easily removed by lowpass filtering.

The situation is different when lower-frequency sinusoids are required. The use of large inductors is impractical due to their size and susceptibility to external magnetic fields. Conversely, operational amplifiers advertise themselves as active elements, since their gain-bandwidth product is in the high kHz range.

A good starting point is to replace the RLC network[1] in Figure 14.10b by a Wien half-bridge that has a transfer function of

$$G(s) = \frac{V_p(s)}{V_{out}(s)} = \frac{RCs}{(RC)^2 s^2 + 3RCs + R + 1} \tag{14.9}$$

[1]Most electrical engineers *hate* inductors and avoid them whenever possible.

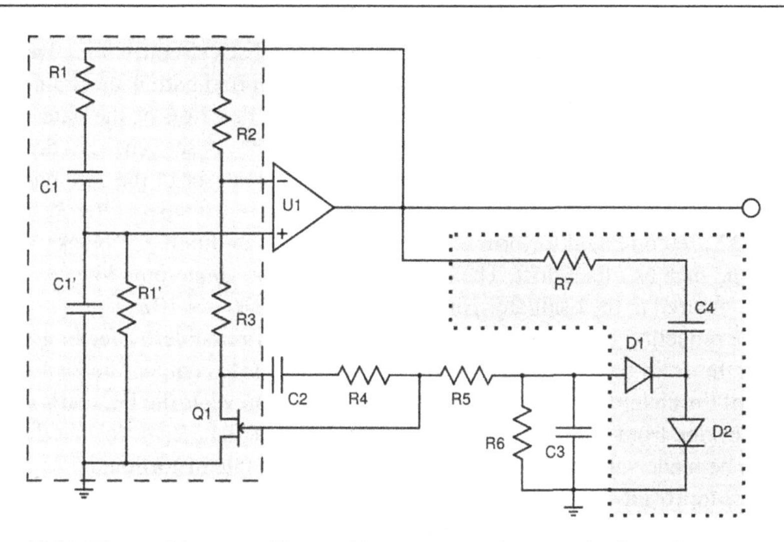

Figure 14.11 Wien-Robinson oscillator with automatic gain control. The active element (U1) together with R_2 and R_3 provides gain in exactly the same manner as the similarly labeled components in Figure 14.10b if transistor Q1 is thought to be shorted. The resonant feedback network $G(s)$ is composed of the Wien half-bridge R_1, C_1, R_1' and C_1' and replaces the RLC resonance circuit R_1, L_1, C_1 in Figure 14.10b. Gain control is provided through the voltage doubler/rectifier circuit (dotted area), the lowpass filter with R_6 and C_3, and Q1, which is used as voltage-controlled resistor.

Some similarity to Eq. (14.4) is evident, and the behavior of the Wien bridge oscillator corresponds closely to the behavior we examined above. The amplitude of the basic Wien bridge oscillator is limited by the power rails. Consequently, it has a fairly high harmonic distortion. A somewhat low-tech approach is to replace R_3 by a temperature-dependent resistor (positive temperature coefficient, PTC), such as a lamp. Increasing output amplitude increases the power dissipation in the PTC, thereby raising its temperature. The increased temperature causes an increased resistance, and the gain is reduced: a negative, gain-stabilizing amplitude feedback loop is established. This approach is unsuitable for precision applications, because the PTC element is highly dependent on environmental factors.

It is worth examining an oscillator circuit that features active gain control (Figure 14.11). This circuit is known as Wien-Robinson oscillator. Comparison to the basic schematic in Figure 14.10 reveals the gain element (U1, R_2, R_3) with the added field-effect transistor Q1 and the resonant feedback network R_1, C_1, R_1', C_1' (i.e., the Wien half-bridge). The other half of the bridge is the voltage divider R_2, R_3, and the complete Wien bridge is highlighted by the dashed area in Figure 14.10. When R_3 is thought to be grounded, and the entire feedback circuit composed of Q1, C_2 to C_4, R_4 to R_7, D1, and D2 omitted, the traditional Wien bridge oscillator emerges. Its feedback circuit (Eq. (14.9)) causes the output signal of U1 to obey a second-order differential equation similar to Eq. (14.6) and to oscillate at the frequency $\omega_0 = (R_1 C_1)^{-1}$.

The automatic gain control is realized by a second feedback circuit. Corrective action is provided by the field-effect transistor Q1. With good approximation, the drain-source channel resistance depends on the gate voltage, and a decrease of the gate voltage therefore increases the series resistance of Q1 with R_3. The consequence is a decreased gain factor g and a reduction of the amplitude $\hat{V}_{out}$. The sensor of the automatic gain control circuit is the rectifier with D1 and D2 followed by the lowpass filter R_6-C_3. C_4, D1, and D2 are combined to form a voltage doubler. In addition, C4 blocks any DC component, such as offset drift. The time constant of the single-pole lowpass should be long compared to the oscillator period, specifically, $R_6C_3 > 10R_1C_1$.

Further reduction of the total harmonic distortion is possible by feeding a small amount of the drain-source voltage through C_2 into the gate to compensate for nonlinear behavior of the channel resistance of the FET. Here, C_2 prevents the DC component of the Wien bridge from influencing the gain control. In practice, $R_4 \approx R_5$, but the value of R_4 can be made variable to adjust for minimum harmonic distortion.

Further improvement of this oscillator circuit is possible by replacing the passive lowpass filter (R_6, C_3) by an op-amp-based active filter. In fact, the filter could represent an analog *PI* controller (similar to the one presented in Figure 6.4) for even more precise amplitude control. In the oscillator application, however, the *PI* controller needs to have a second pole to reduce its high-frequency gain to prevent the gain control from acting over just a few oscillator periods. In practice, an additional capacitor would be placed in parallel to resistor R_2 in Figure 6.4, and the gain control transfer function becomes $(k/s) \cdot (s+z)/(s+p)$.

Two other, but related, approaches to realizing the second-order differential equation of a harmonic oscillation (Eq. (14.6)) are worth mentioning, because they directly link to feedback principles covered in this book. The first approach is to use two integrators in the feedback network, and the second approach is to chain low-order lag elements to obtain a phase shift larger than 180° in the feedback path.

Figure 14.12a is a direct circuit realization of a second-order differential equation built around two integrators. Inverting integrators are particularly easy to build, because they need only one op-amp, one capacitor, and one resistor. The transfer function of the integrators is $-a/s$ where $a = 1/(R_1C_1)$. By resolving the signals along the loop, we first obtain

$$\begin{aligned} v(t) &= -a \int u(t)\mathrm{d}t \\ w(t) &= -k \cdot v(t) - a \int v(t)\mathrm{d}t \end{aligned} \tag{14.10}$$

By substituting $v(t)$ and using $w(t) = -u(t)/g$, the differential equation emerges as

$$\frac{\mathrm{d}^2u(t)}{\mathrm{d}t^2} + agk\frac{\mathrm{d}u(t)}{\mathrm{d}t} + a^2gu(t) = 0 \tag{14.11}$$

Once again, the differential equation has the same form as Eq. (14.6), and the attenuated sinusoidal output has a natural frequency ω_0 and the damping factor ζ of

$$\omega_0 = a\sqrt{g} \;\; ; \;\; \zeta = \frac{k\sqrt{g}}{2} \tag{14.12}$$

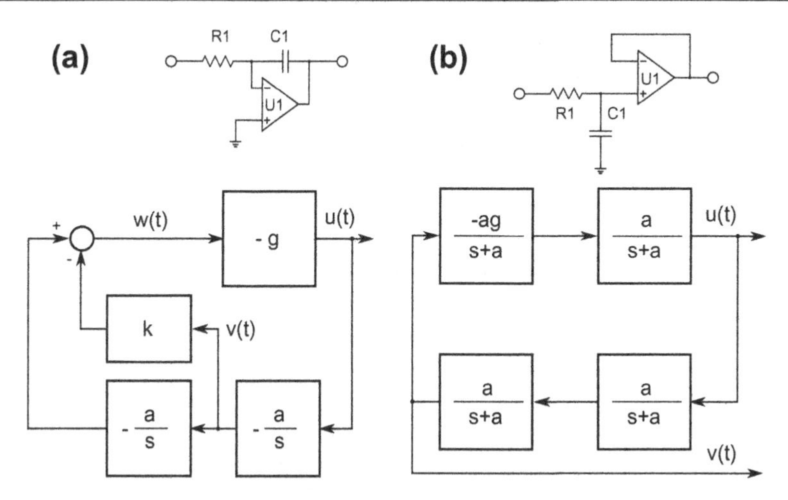

Figure 14.12 Two different oscillator principles to obtain sinusoidal output voltages. (a) Circuit implementation of a second-order differential equation built around two integrators (inset: op-amp-based integrator with the transfer function $-1/(R_1C_1s)$). (b) Phase shift oscillator to generate a quadrature signal (inset: op-amp realization of the transfer function $a/(s+a)$).

The undamped oscillation occurs when $k = 0$. As in the previous example, however, a slightly negative value for ζ is desirable. Adjustment of k is possible by feeding the signal $v(t)$ either into the positive or negative input of the gain stage, for example, with a potentiometer. However, even with highly accurate adjustment of k, the oscillation will either amplify or get attenuated over time. The situation is similar as in the previous example as some form of automatic gain control is needed. The technique for automatic gain control is similar to the previous example.

The circuit in Figure 14.12a can further be modified by using operational transconductance amplifiers (OTA), such as the NE5517. OTAs have become some of the more exotic components in analog circuit design, although a number of applications can be realized in a particularly elegant manner. Unlike a regular op-amp with a fixed, large gain g, the OTA has an additional input for a control current, traditionally labeled I_{ABC}, that determines the transconductance (i.e., the gain). The OTA's gain equation is therefore

$$I_{out} = g_m(V_p - V_m) \approx 19.4 I_{ABC}(V_p - V_m) \tag{14.13}$$

The output provides a current rather than a voltage. The proportionality constant of 19.4 V^{-1} is device-specific. The OTA is a voltage-controlled current source. Furthermore, its gain (more precisely, its transconductance) g_m is proportional to the control current. An integrator is therefore realized simply by feeding the OTA output current into a capacitor, and the integrator time constant depends on the control current I_{ABC}. It can be seen that the oscillation frequency (Eq. (14.12)) depends on the integration constant a and now becomes current-controlled. With a resistor as voltage-to-current converter, the oscillator can be turned into a voltage-controlled oscillator (VCO). With an additional rectifier, inverting lowpass filter and one more OTA, the automatic gain control is straightforward.

The second oscillator (Figure 14.12b) is based on first-order lag elements and most closely resembles the feedback systems examined in Chapter 11. Each lag element $a/(s+a)$ creates a phase shift of $-45°$ at the cutoff frequency $\omega_0 = 1/(R_1C_1)$. Total open-loop gain at the cutoff frequency is therefore 0.25 with a phase shift of $-180°$. The gain stage needs to provide an amplification of 4 or more to allow the circuit to oscillate. The circuit has two advantages over other circuits of comparable complexity. First, it provides four signals that are 45° out of phase, and the two signals labeled $u(t)$ and $v(t)$ in Figure 14.12b are a quadrature (sin/cos) pair. Second, each stage can be seen as a lowpass filter that suppresses harmonics. Even without gain control, this circuit provides sinusoidal signals with low harmonic distortion.

It is worth comparing the above purely analog circuits to digital waveform oscillators. A common technique is known as *direct digital synthesis* (DDS), which can be used to generate a sine wave or any other periodic waveform. The DDS oscillator circuit is shown in Figure 14.13. DDS oscillators are available as complete integrated circuits (e.g., the AD9858 or ISL5314). The digital part of the DDS oscillator can also conveniently be implemented in software, but a microcontroller implementation puts considerable limits on the output frequency. The advantage of the microcontroller implementation is the flexibility when it comes to changing the waveform.

A DDS oscillator consists of a numerically controlled oscillator, a digital-to-analog converter, and a lowpass filter. The central element of the numerically controlled oscillator is a lookup table (either a read-only memory table or a table in program memory) that converts a discrete input phase k into the corresponding amplitude. if k is a m-bit

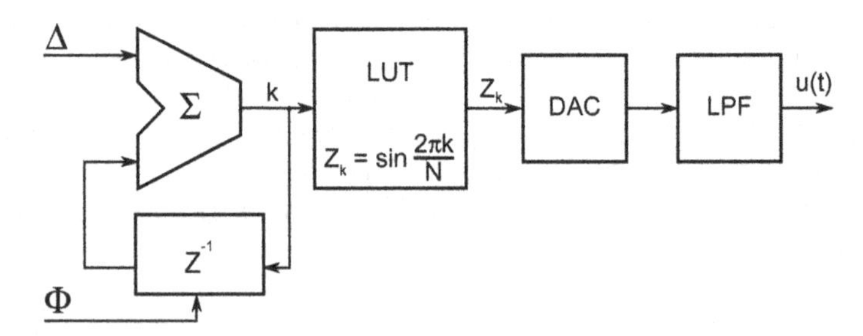

Figure 14.13 Direct digital synthesis oscillator. At the core is a lookup table (LUT) that contains $N = 2^m$ discrete digitized values of the desired waveform, in this case a sine function. When the digital value k is applied to the input, the lookup table outputs the corresponding function value $Z_k = \sin(2\pi k/N)$. The subsequent digital-to-analog converter (DAC) with lowpass filter (LPF) provides a proportional output voltage. The LUT input is provided by a phase accumulator that consists of a m-bit memory (indicated by z^{-1}) and a digital adder. At each clock cycle Φ, the phase k is incremented by Δ. Therefore, Δ determines (at constant clock speed) how fast the lookup table is traversed.

word, the lookup table contains $N = 2^m$ entries,

$$Z_k = f\left(\frac{2\pi k}{N}\right) = \sin\left(\frac{2\pi k}{N}\right) \tag{14.14}$$

where $f()$ represents a general 2π-periodic waveform function. As k is increased from 0 to $N - 1$, the function argument (i.e., the phase) increases from 0 to 2π. The flexibility of the numerically controlled oscillator comes from the ability to increment k by a constant phase step Δ at each clock cycle of the clock Φ. For example, when the clock is 1 MHz and $m = 10$, the lowest possible output frequency (when $\Delta = 1$) is 1 MHz/1024 = 976.6 Hz. The output frequency is proportional to Δ, since k traverses the range from 0 to N faster with larger Δ:

$$f_{out} = \Phi \cdot \frac{\Delta}{N} \tag{14.15}$$

The highest frequency is the Nyquist frequency $\Phi/2$, which at the same time limits $\Delta < N/2$. The lowpass filter that follows the DAC should have a cutoff frequency of $\Phi/2$, and a second-order filter would attenuate the clock frequency by 12 dB. The harmonic distortion of the output signal strongly depends on the quality of the output filter.

In a microcontroller, the clock could be represented by a timer that causes a regular interrupt. The numerically controlled oscillator is then contained in the interrupt service routine as shown in Algorithm 14.3.

Algorithm 14.3. Interrupt service routine (ISR) for the software implementation of a numerically controlled oscillator based on a periodic interrupt from a timer.

1 *Interrupt from Timer*;
2 $K \leftarrow K + DELTA$; `// Update phase accumulator`
3 **if** $K \geq N$ **then**
4 $K \leftarrow K - N$; `// Rollover of the phase accumulator`
5 **end**
6 $Z \leftarrow LUT[K]$; `// Obtain function value from LUT`
7 Output Z to DAC port;
8 return from interrupt;

A table of function values, LUT, needs to be provided, which contains N discrete values of the function at phase increments of $2\pi/N$. The accumulated phase K (note the rollover provision in Lines 3–5) is the index into the table, and the value Z found in the table is directly applied to the port that connects to the DAC. The phase increment $DELTA$ is supplied from outside the interrupt service routine. A typical basic 8 bit microcontroller (e.g., the Microchip PIC18F4520)—depending on the implementation—needs less than 5μs for the interrupt service routine, allowing for clock rates up to 100 kHz. It is noteworthy that a precision sine wave generator for the audio range can be implemented with a \$5 microcontroller and a few external components.

If minor jitter is acceptable, the lookup table can be shortened by making use of any symmetry of the function. For a sine wave, for example, only values from 0 to $\pi/2$ need to be stored. When $k \geq N/2$, a negative sign is set in memory and a new $k' = k - N/2$ is computed. If $k' > N/4$, we set $k' = N/2 - k'$ and finally use k' for the lookup table, and combine Z with the sign before applying Z to the DAC port. Many basic microcontrollers come with 16 k or more program memory and 1024 words of additional data storage. With this trick, a 1024 byte lookup table can service a 12 bit phase accumulation register. To remain with the example of an audio-frequency generator, a 41 kHz timer frequency allows to cover the frequency range approximately from 10 Hz to 20 kHz in 10 Hz steps.

Although direct digital synthesis is not directly related to feedback control systems, the example serves to complement the examples of time-discrete systems in this book, illustrates how digital systems can be used to replace or supplement analog systems, and specifically highlights how a design engineer can choose between the two alternatives of analog or digital systems. The software implementation is always attractive when a microcontroller is already included in the design and merely requires some additional program code. On the other hand, limits imposed by the discretization may not always be acceptable, and software solutions are often limited by the program execution time.

14.5 Low-Distortion (Hi-Fi) Amplifiers with Feedback

Since we covered oscillators, it is only fair to cover amplifiers as well. The design goal for an audio power amplifier is to amplify a moderately small input signal such that the amplifier can provide enough current to drive a low-impedance load, typically a 4 or 8 Ω loudspeaker. The Hi-Fi (short for high fidelity) sound reproduction standard dates back to the 1960s and in its essence prescribes a maximum harmonic distortion and a maximum noise level. Most of today's high-quality audio equipment easily exceeds the Hi-Fi standard, and today's audiophile listeners would probably consider equipment that merely *meets* the Hi-Fi standard to be an insult to the ears.

One of the primary design goals of amplifier design is to keep the harmonic distortion as low as possible over the entire audio frequency range (usually from 10 Hz to over 20 kHz). Moreover, the gain should be constant over the entire frequency range. Total gain is moderate, and amplifier gain values lie in the order of 15–30.

The main challenge in amplifier design is posed by its semiconductor components, which typically have highly nonlinear characteristic curves. Furthermore, transistors have several parasitic capacitances, some of them gain-dependent ("Miller effect"). Each transistor acts like a lowpass filter. Common to almost all amplifier designs is its similarity to an operational amplifier, usually built from discrete components. The use of discrete components allows to minimize the number of transistors and their associated lowpass effect and semiconductor noise. Although design of Hi-Fi audio amplifiers is as much an art as a science (with some black magic interspersed), some

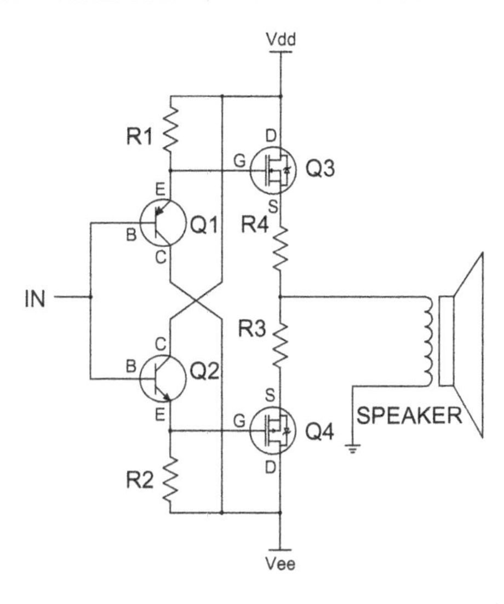

Figure 14.14 Simplified schematic of a complementary power amplifier output stage. Power transistors Q3 and Q4 are responsible for providing a push-pull current from a symmetric voltage supply. Q1 and Q2 act as driver transistors, and together with R_3 and R_4 they control the quiescent current through Q3 and Q4 when the input is zero. The quiescent current reduces the nonlinear behavior near the cross-over point and reduces total harmonic distortion.

fundamental guidelines can be established, and a low harmonic distortion can be related to feedback principles.

To drive a low-impedance load, such as the loudspeaker, a complementary output stage (Figure 14.14) is typically employed. The output power transistors are either bipolar or MOSFET types. In both cases, a pronounced nonlinear behavior of the output stage appears whenever the output signal changes its sign or is generally close to zero. At this point, the transistor gain is very low, and distortion occurs. One common solution is to allow a small continuous quiescent current to flow through both transistors even when the input voltage is zero. Amplifiers that build on this technique are referred to as class-AB amplifiers, as opposed to class-B amplifiers, which have no quiescent current. Clearly, a balance needs to be found between thermal losses caused by the quiescent current and the reduced harmonic distortion that comes with a higher current.

At this point, we could take the output stage from Figure 14.14 and complete it with an operational amplifier IC such that the positive input is the overall amplifier input, and the negative input is connected to the loudspeaker output through a voltage divider—the voltage divider creates a negative feedback that determines the overall gain and improves the linearity of the amplifier. Many low-end amplifiers follow this principle with acceptable results.[2]

[2]For lower-end Hi-Fi applications, a number of single-chip integrated circuit solutions exist that include the output transistors on the chip. Examples include the LM1875, TDA2050, and TDA7294 monolithic amplifiers. Integrated solutions are particularly attractive when space is a concern and when audio from poor sources (such as MP3-compressed audio) provides the actual quality bottleneck.

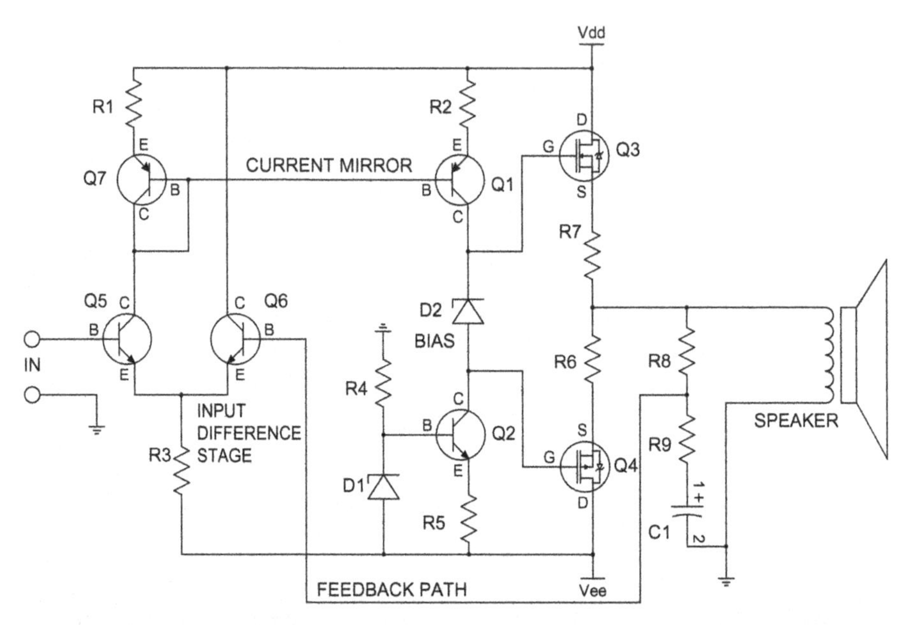

Figure 14.15 Basic schematic of a complete audio amplifier. The output stage from Figure 14.14 is widely recognizable, but Q1 now acts as a current source, which forms a current mirror with Q7. Q2 is configured as constant-current sink and is responsible for inverting the signal for the lower branch of the power driver, with D2 adding a constant bias voltage that controls the quiescent current through Q3 and Q4. Finally, Q5 and Q6 form a typical difference stage, with the audio signal entering the positive input and the feedback signal entering the negative input.

High-end audio amplifiers, on the other hand, are built with discrete transistors that constitute the input difference stage and the gain stage. A basic schematic of this configuration is shown in Figure 14.15. The input stage (Q5 and Q6) is a typical difference stage. The collector current of Q5 is fed into the output stage through the current mirror composed of Q7 and Q1. The current source around Q2 and the bias diode D2 provide the signal for the lower output branch and determine the quiescent current.

A negative feedback path is established through R_8, R_9, and C_1. AC gain is limited to $(1+R_8/R_9)$, but at frequencies below the pole created by R_9 and C_1, gain drops rapidly to unity. Reduction of the DC gain through the feedback path obviates the need for input and output capacitors and improves the frequency response for low frequencies. The main importance of the negative feedback is shown in Figure 14.16. The characteristic deadband-type nonlinearities of the output transistors lead to significant distortion of a sinusoidal input (Figure 14.16b). This distortion can be reduced by negative feedback, provided that the feedforward path provides more gain than needed for the closed-loop system.

The principles shown in Figure 14.15 can be found in many audio amplifier schematics. Often, additional gain is provided between the input difference stage and the power driver stage. Also, sometimes, phase-lead networks can be found in the gain stage that

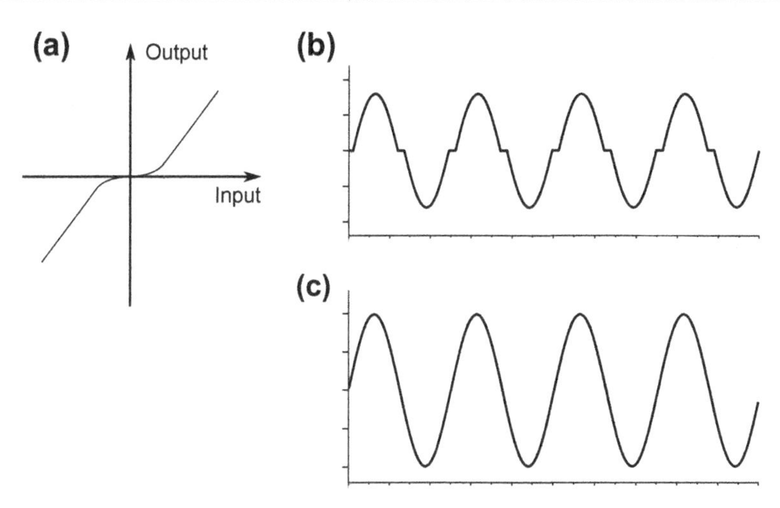

Figure 14.16 Effect of feedback on the nonlinearity of the output stage. (a) In class-B amplifiers and to some extent in class-AB amplifiers, the output transistors become nonlinear when the input signal is close to zero, and a form of deadband forms in the characteristic curve. (b) Output waveform of an output stage with deadband characteristic and without feedback when the input is a pure sine-wave signal. (c) When the gain of the forward path is increased and negative feedback established, the deadband distortion is reduced proportionally to the feedforward gain, and the signal distortion is markedly reduced.

compensate some of the lowpass effect from the transistors and improve the frequency response at high frequencies. With suitable current sources, the bias can be improved to provide better temperature stability. Lastly, the output often contains a passive network that may include a small inductor in series with the load and an RC series impedance to ground. The inverse time constants for the associated poles and zeros lie in the order of 150 kHz to 2 MHz, which is far above the audio range. It would be interesting to examine the actual physical effect of these networks on the audio signal.

As with almost all systems in this book, audio amplifiers, too, have begun to take the path toward digital systems. Digital amplifiers predominantly use pulse-width modulation (PWM) and—closely related to switching power supplies—use a two-pole LC lowpass to remove the switching frequency. The author has not yet had the opportunity to listen to a class-D (i.e., digital PWM) amplifier, but admits to a certain skepticism as to its audio quality. Without doubt, however, class-D amplifiers can provide extreme power levels to the loudspeakers, yet feature a small footprint and an extremely high degree of efficiency. Low-power monolithic integrated class-D amplifiers gain increasing popularity in many battery-driven devices, such as laptops, where low power demands are more important than high-end audio quality. Arguably the biggest advantage is the ability of class-D amplifiers to directly take digital signals, for example, from a CD player, and drive the loudspeakers without any intermediate analog conversion.

In its simplest form, the analog audio input in a class-D amplifier is compared to a triangle signal with a frequency far above the audio range (several hundred kHz are typical). The resulting digital PWM signal drives a complementary output stage, which

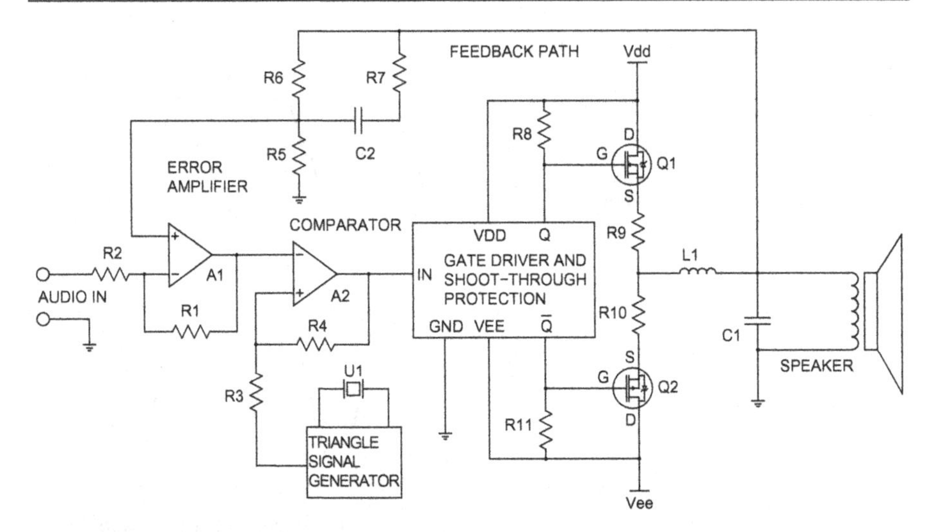

Figure 14.17 Simplified schematic of a class-D pulse-width modulated amplifier. Unlike in analog output stages, transistors Q1 and Q2 are either fully on or fully off. Consequently, thermal losses in Q1 and Q2 are at a minimum. Much like in a switch-mode power supply, the PWM signal is converted into an analog signal through L_1 and C_1. The digital switching signal is provided by the comparator A2 and a triangle signal generator. A negative feedback loop defines the overall gain and reduces nonlinear behavior. The additional phase-lead network C_2 and R_7 can reduce switching noise to some extent.

in turn generates a digital high-current output signal. A two-pole output filter removes the switching frequency. In Figure 14.17, the output filter consists of L_1, C_1, and the resistive load of the loudspeaker itself, which is required for damping. Typically, the three components would be combined into a Butterworth or Bessel filter.

Let us look at a dimensioning example for an 8 Ω loudspeaker. We place the cutoff frequency at $f_c = 19$ kHz to allow a minor attenuation at frequencies near the auditory limit. To obtain Bessel characteristic, we choose $\zeta = 0.85$ in the second-order filter function. Using Kirchoff's node rule, the filter transfer function quickly emerges as

$$H_F(s) = \frac{\frac{1}{LC}}{s^2 + \frac{1}{RC}s + \frac{1}{LC}} \tag{14.16}$$

where L and C are the values of L_1 and C_1, respectively, and $R = 8\ \Omega$. The first-order term provides the necessary capacitance:

$$C = \frac{1}{\zeta}\frac{1}{4\pi f_c R} = 620\ \text{nF} \tag{14.17}$$

The cutoff frequency provides the relationship between L and C as $\sqrt{LC} = 1/(2\pi f_c)$, from which we obtain $L = 4.5$ mH. A different loudspeaker impedance changes the damping behavior. Moreover, these considerations are an approximation, because loudspeakers (especially those with crossovers) are by no means a purely resistive load.

A switching frequency of 380 kHz, for example, would be attenuated by merely 52 dB in the two-pole output filter. One would hope that the loudspeakers themselves provide additional attenuation without intermodulation.

As in its analog counterpart, negative feedback from the output to the input determines the overall gain and reduces some of the nonlinearities. Overall gain is determined by the gain of the difference amplifier around A1. An additional phase-lead network (C_2 and R_7) can be introduced to the feedback path to feed back high frequencies (above 20 kHz) and therefore attenuate switching noise to some extent.

14.6 Phase-Locked Loop Systems

Phase-locked loops (PLL) are feedback-controlled oscillators that follow a reference oscillator to match both frequency and phase. Phase-locked loop circuits are frequently used as decoders in telecommunication devices or as frequency multipliers. PLL circuits are probably most widely used in dual-tone multifrequency (DTMF) decoders that identify the two frequencies from a touchtone phone. PLL circuits are usually limited to either sinusoidal waveforms or square waves (i.e., digital signals). The output can be either an oscillation (e.g., in a PLL-based frequency multiplier), a voltage (e.g., in a frequency-to-voltage conversion), or a digital signal (e.g., in a frequency decoder).

The control loop of a PLL system is shown in Figure 14.18. A voltage-controlled oscillator (VCO) generates the output signal with the frequency f_{osc}. The phase detector (or phase discriminator) generates an error signal that is proportional to the phase shift between the reference signal and the oscillator output. The error signal needs to be

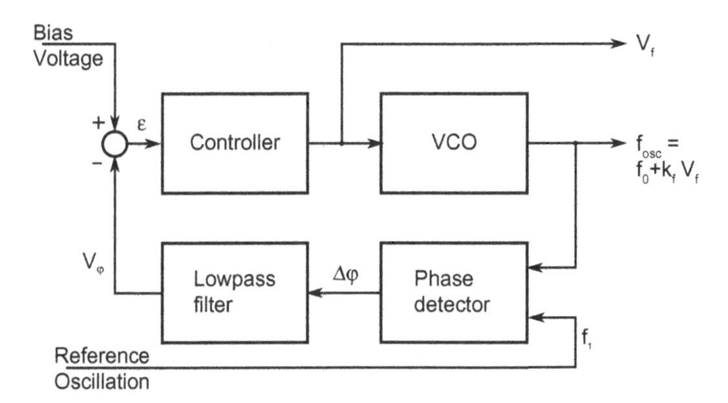

Figure 14.18 Basic schematic of a phase-locked loop. A phase detecter compares the phase of the reference signal and the oscillator signal and generates a voltage that is proportional to the phase error. An analog lowpass filter ensures that the oscillator frequency is suppressed. It is possible to supply a bias voltage to influence the capture range. A controller provides the corrective action, which is the input voltage to the VCO. The VCO, in turn, generates the oscillation that the PLL attempts to match with the reference signal.

filtered to remove the oscillator frequency components. After filtering, it can be used as input signal (biased, if needed) to the VCO.

An electromechanical analogy to the PLL can be imagined. A DC motor drives a disk with the rotary speed ω_{osc}. Just opposite this disk is another disk that is driven by an external drive with the reference speed ω_1. Both disks are linked by a potentiometer in such a way that the potentiometer output voltage increases when the motor disk advances with respect to the reference disk. The potentiometer acts as the phase detector. Negative feedback of the potentiometer voltage into the motor drive voltage provides the corrective action that minimizes the phase angle between the two disks. This analogy is valid both for the behavior of a PLL when the frequencies are grossly mismatched and for the typical phase rollover behavior.

When the oscillator frequency almost matches the frequency f_1 of the reference signal, and the oscillator lags behind, the phase-proportional voltage V_φ is negative. The negative feedback causes the VCO input voltage to rise and its frequency to increase. With increased frequency, the VCO accumulates phase and reduces the phase lag. It is important to realize that the frequency is the first derivative of the phase. This makes the PLL example similar in some respects to the motor/positioner example (Chapter 9) with one pole of its transfer function in the origin. The VCO output frequency follows its input voltage linearly,

$$f_{osc} = f_0 + k_f V_f \tag{14.18}$$

where f_0 is the center (or open-loop) frequency of the VCO, k_f its gain, and V_f the controlling input voltage. We define $\omega_{osc} = 2\pi f_{osc}$. Similar to a motor's angular position, the phase is

$$\varphi_{osc} = \int_0^t \omega_{osc}(\tau)\mathrm{d}\tau \tag{14.19}$$

If we define the phase difference between oscillator output and reference signal as $\Delta\varphi$, we obtain the relationship

$$\Delta\varphi = \varphi_{osc} - \varphi_1 = \int_0^t \omega_{osc}(\tau) - \omega_1(\tau)\mathrm{d}\tau \tag{14.20}$$

and we expect the phase discriminator to provide an output voltage $V_\varphi = k_\varphi \Delta\varphi$. If we disregard the constant components, such as the bias voltage and the center frequency f_0, and if we assume a simple P controller, we can close the loop:

$$\omega_{osc}(s) = \frac{g}{s+g}\omega_1(s) \tag{14.21}$$

where g is the product of the real-valued gain factors including the P controller gain: $g = k_\varphi k_f k_p$. Equation (14.21) also neglects any influence of the lowpass filter. Its purpose is merely to show that the oscillator frequency follows the reference frequency, and the steady-state tracking error for the frequency is zero. It can be shown, however, that the phase tracking error is inversely proportional to the loop gain g and a steady-state phase shift remains. Generally, the use of a *PI* controller forces the steady-state phase tracking error to zero.

Before we further examine the closed-loop behavior, it is important to introduce different phase discriminators and analyze their properties. A very common approach is the multiplication of the two signals. Assume that we can describe the normalized reference voltage with $V_1(t) = \sin(\omega_1 t)$ and the oscillator output (with its phase shift φ) with $V_{osc}(t) = \sin(\omega_{osc}t + \varphi)$, then the product is

$$[V_1 \cdot V_{osc}](t) = \frac{1}{2}\cos\big((\omega_1 - \omega_{osc})t - \varphi\big) - \frac{1}{2}\cos\big((\omega_1 + \omega_{osc})t + \varphi\big) \quad (14.22)$$

which can be simplified when the PLL is operating near its equilibrium point, that is, $\omega_1 \approx \omega_{osc}$:

$$\ldots \approx \frac{1}{2}\cos(\varphi) - \frac{1}{2}\cos(2\omega_1 t + \varphi) \quad (14.23)$$

The second cosine term can be removed with the lowpass filter, and the output of the multiplying phase discriminator is proportional to $\cos\varphi$. A number of disadvantages become immediately visible:

- Multiplying the two signals is only practical for sinusoidal signals, because other waveforms (such as square waves) would add numerous sums and differences of the harmonic content.
- The phase discriminator is nonlinear. Good operating points exist for phase shifts near $\pm\pi/2$ and $\pm 3\pi/2$ where the gain of the phase discriminator is $\mp 1/2$.
- Near $\varphi = 0$ and $\varphi = \pi$, the gain of the phase discriminator drops to zero, thereby opening the loop.
- The phase discriminator is 2π-periodic, and is therefore unsuitable when the PLL is completely unsynchronized. This is a common property of most phase detectors and brings up the question of the PLL's *capture range*.

An alternative is the use of a synchronous demodulator, such as the AD630. A synchronous demodulator has one analog and one digital input. Depending on the level of the second input, the demodulator multiplies the first signal either with $+1$ or with -1 (Figure 14.19). As a result, the time average of the output signal is maximal when $\varphi = 0°$ and becomes zero for $\varphi = 90°$. If the input signal is a sinusoid, the time-average value of the output signal is proportional to $\cos\varphi$. Once again, $\varphi = \pm\pi/2$ and $\varphi = \pm 3\pi/2$ are suitable operating points.

Most phase discriminators have a 2π-periodic transfer function, and for the unsynchronized PLL, the output is a zero-mean oscillation. Therefore, the capture range (that is, the frequency difference between ω_{osc} and ω_1 in which the PLL can reach a synchronized state) is very narrow. Once the PLL is locked, however, the PLL tends to remain synchronized in a much wider frequency range.

It is therefore desirable to use phase discriminators that provide additional information on the sign of the frequency difference to provide a refined error signal that drives the VCO toward the reference frequency. A brute-force approach would be the use of two digital counters and a subtraction stage. Every positive edge of the reference signal and the oscillator output increments its associated counter. The two-complement difference is then fed into a digital-to-analog converter. The resulting transfer function

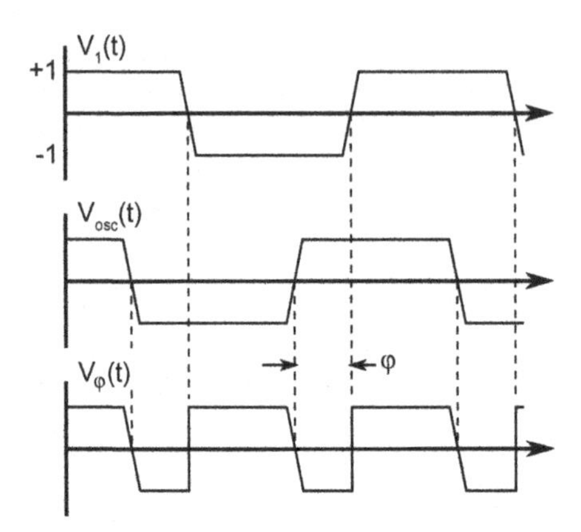

Figure 14.19 Input and output waveforms of a synchronous demodulator. The synchronous demodulator multiplies one signal with the sign of another. The average value of the output signal depends on the phase shift: for 0°, the output signal is maximal, and at 90°, zero.

shows a periodicity of $2^{m+1}\pi$, where m is the bit depth of the counter, which translates into an 2^mfold increased capture range.

Alternatively, a finite-state machine can be designed that keeps track of the sign of the phase difference. Such a phase comparator is provided in the popular CD4046 integrated PLL chip. The phase comparator is driven by the positive zero crossings (i.e., the positive edges) of the reference and VCO signal, respectively. The phase discriminator of the CD4046 has 11 states, and we introduce in Figure 14.20 a simplified version to explain its operation.

The disadvantage of the phase discriminator in Figure 14.20 is a lower sensitivity toward very low phase shifts due to its inherent delay. Practical applications would use both phase discriminators in the CD4046 and switch from the edge-driven phase detector to the level-driven alternative phase discriminator when the PLL enters the synchronized state.

We now return to the overall transfer function of the PLL. To eliminate the tracking error of the *phase*, an integrating controller (i.e., a *PI* controller) is typically used. We assume that the PLL is locked and tracks phase changes from the locked state. In the locked state, the phase discriminator can be linearized and acts as a gain element with the gain k_φ and the input signal $\Delta\varphi$. A first-order lowpass filter with the transfer function $\omega_f/(s+\omega_f)$ can be assumed. The VCO outputs a frequency, which, when multiplied with $1/s$, yields the phase. These elements lead to the open-loop transfer function

$$\varphi_{osc} = \frac{2\pi f_0}{s} - k_f \cdot \frac{k_P s + k_I}{s^2} \cdot \frac{\omega_f}{s+\omega_f} \cdot k_\varphi \cdot \Delta\varphi \tag{14.24}$$

The phase difference $\Delta\varphi$ is defined in Eq. (14.20), and its Laplace transform $\Delta\varphi = \varphi_{osc} - \omega_1/s$ can be substituted into Eq. (14.24). This leads to the closed-loop transfer

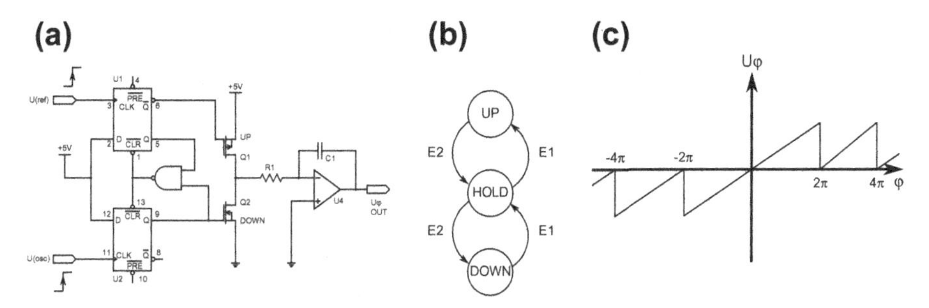

Figure 14.20 (a) Simplified schematic of the phase detector used in the CD4046 integrated PLL. The circuit is edge-sensitive, and a positive edge $E1$ from the oscillator sets flipflop U1 and charges the integrator capacitor C_1 through Q1. Conversely, a positive edge $E2$ from the reference sets flipflop U2 and discharges the integrator capacitor through Q2. The lagging edge from either input resets the flipflops. (b) State diagram of the circuit. There are three states, up (Q1 charges the integrator), down (Q2 discharges the integrator), and hold (integrator input is zero). A leading edge $E1$ from the reference source causes the integrator to charge C1 and increases its output voltage until the lagging edge $E2$ returns the circuit into the hold state. A leading edge $E2$ from the VCO enters the down state and decreases the output voltage until the lagging edge $E1$ returns to the hold state. (c) Transfer function of the phase detector. The function is still 2π-periodic, but the sign of the phase shift translates into an offset voltage that drives the oscillator toward $\varphi = 0$.

function for the phase

$$\varphi_{osc} = \frac{k_\varphi k_f \omega_f (k_p s + k_I)}{s^2(s+\omega_f) + k_\varphi k_f \omega_f (k_p s + k_I)} \cdot \varphi_1 + \frac{2\pi s}{s^2(s+\omega_f) + k_\varphi k_f \omega_f (k_p s + k_I)} \cdot f_0 \qquad (14.25)$$

Frequently, this equation is simplified by omitting the center frequency term and examining only changes near the center frequency. Furthermore, when the lowpass filter has a high cutoff frequency compared to the entire system dynamics, the pole at $-\omega_f$ can be considered nondominant and omitted. Finally, we can combine k_p and k_I with the other gain factors to obtain k'_p and k'_I. Equation (14.25) then simplifies to

$$\varphi_{osc} = \frac{k'_p s + k'_I}{s^2 + k'_p s + k'_I} \cdot \varphi_1 \qquad (14.26)$$

This is the transfer function of a second-order system, and k_p and k_I can now be optimized (under consideration of the other gain factors k_f and k_φ) to achieve the desired dynamic response. One possible choice is the ITAE-optimal response with $\zeta = k_p/(2\sqrt{k_I}) = 0.7$. Equation (14.26) appears relatively straightforward, but the actual PLL behavior may be more complex, primarily due to the nonlinear behavior of the phase discriminator and due to the presence of harmonics. In some cases, simulation or experimental optimization is necessary to improve on our linear-systems approximation.

Although many integrated PLL chips are available for various applications, the entire phase-locked loop can be implemented as a time-discrete system in software. The finite state machine in Figure 14.20 is particularly amenable to a software implementation. The rising edges of both oscillator and reference signal cause an interrupt and adjust the state accordingly. Depending on the precision requirements, an output pin can be set either high, low, or put into high-impedance mode. This output pin connects to an analog integrator. Alternatively, each interrupt that leads to the up- or down-state increments or decrements an accumulation register that holds the integrator value. Similarly, the lowpass filter is either an analog device connected to a DAC which receives the integrator value, or the lowpass filter is implemented as a digital filter. In that case, the digital filter can also include the *PI* control algorithm.

The software implementation of the VCO could potentially be based on a timer. The major drawback is the resolution of the timer: a 16 bit timer clocked at 1 MHz, for example, can provide time intervals in increments of 1 μs. A true steady state cannot be achieved in a purely digital implementation. Rather, the digital PLL will have significantly more phase jitter and phase noise than its analog counterpart.

14.7 Stabilizing an Unstable System

Arguably, the most well-known example for an unstable system that can be stabilized in a feedback control system is the inverted pendulum model. One example for this model is a rocket, where the thrust vector creates a torque when the vector does not pass exactly through the center of gravity. Another example are self-balancing vehicles, such as the Segway. Balancing the rocket is a two-dimensional problem, whereas the inverted pendulum with one axis of rotation is a one-dimensional problem.

The inverted pendulum has one pole in the right half-plane, and a linear approximation exists only for small angular deviations from the vertical (i.e., desired) position. The general geometry is shown in Figure 14.21. Let us define the orientation of the pendulum with respect to the horizontal as Θ. Therefore, the angular deviation from the vertical orientation is $\phi = 90° - \Theta$. The center of mass is shifted from the pivot point by $d = (L/2)\sin\phi$. Friction of the wheels and of the pivot point is assumed to be negligible. Gravity acts on the center of mass with $F = m \cdot g$. Therefore, a torque τ appears, which obeys

$$\tau = F \cdot d = F \cdot \frac{L}{2} \sin(\phi) \approx F \cdot \frac{L \cdot \phi}{2} \tag{14.27}$$

where the linear approximation is only valid for small values of ϕ and for ϕ given in radians. The torque causes an increase of the angular momentum of the mass,

$$\tau_m = m \cdot \frac{L^2}{4} \cdot \frac{\mathrm{d}^2\phi}{\mathrm{d}t^2} \tag{14.28}$$

where the mass is assumed to be concentrated in a point at the center of gravity. Let us introduce a corrective action where the wheels drive the pivot point in the x-direction.

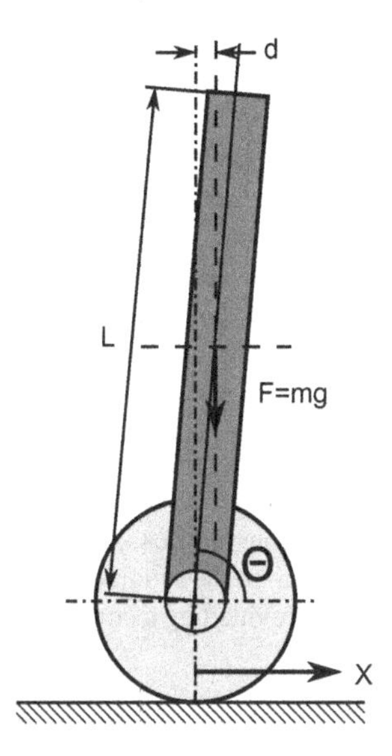

Figure 14.21 Inverted-pendulum behavior of a self-balancing vehicle. When the center of gravity is not vertically aligned with the point of rotation, a torque $\tau = d \cdot F$ pulls the mass further away from the vertical orientation. For $\Theta \approx 90°$, $d = (L/2)\sin(90° - \Theta)$ increases with increasing deviation from the vertical orientation, which in turn increases the torque. Positive feedback ensues. Feedback control would move the wheel to the right to correct for the angular deviation.

Acceleration in the x-direction translates directly into a torque τ_x acting on the mass,

$$\tau_x = m \cdot L\cos(\phi)\frac{\mathrm{d}^2 x}{\mathrm{d}t^2} \approx m \cdot L \cdot \frac{\mathrm{d}^2 x}{\mathrm{d}t^2} \tag{14.29}$$

The torque τ_x acts in the opposite direction of τ_m. Therefore, the balance-of-torques equation is for small angles ϕ

$$\frac{mL^2}{4} \cdot \frac{\mathrm{d}^2\phi(t)}{\mathrm{d}t^2} = \frac{mgL}{2}\phi(t) - mL\frac{\mathrm{d}^2 x(t)}{\mathrm{d}t^2} \tag{14.30}$$

In the Laplace domain and after dividing by $mgL/2$, Eq. (14.30) becomes

$$\frac{L}{2g} \cdot s^2\Phi(s) - \Phi(s) = -\frac{2}{g} \cdot s^2 X(s) \tag{14.31}$$

The corresponding transfer function

$$G(s) = \frac{\Phi(s)}{X(s)} = -\frac{\frac{2}{g}s^2}{\frac{L}{2g} \cdot s^2 - 1} \tag{14.32}$$

has two real-valued poles at $\pm\sqrt{2g/L}$ and a double zero in the origin. The unstable real-valued pole in the right half-plane causes the angular deviation to exponentially increase.

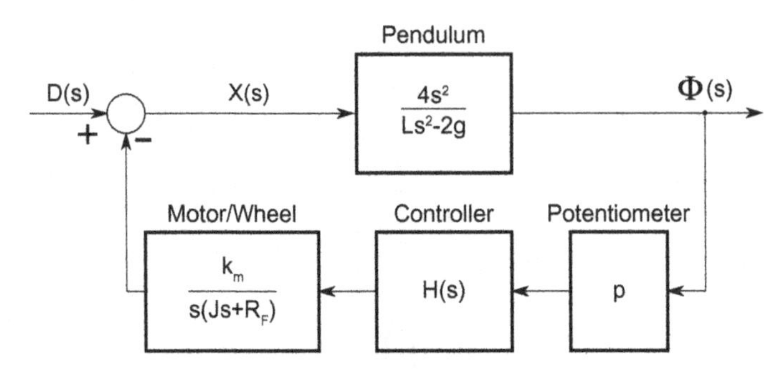

Figure 14.22 Proposed feedback control system for the inverted pendulum. A disturbance $D(s)$ can be modeled as acting on $X(s)$. The angular position $\Phi(s)$ is sensed by a symmetrical potentiometer with gain p, which also contains an implicit setpoint when its output is zero for $\phi = 0$. The negative sign has been carried forward to the summation point. Corrective action is provided by a DC motor whose drive voltage is provided by a controller with the transfer function $H(s)$.

We will now attempt to stabilize the inverted pendulum with the feedback control system shown in Figure 14.22. The angle ϕ is sensed by a potentiometer at the pivot point. The potentiometer can be conveniently adjusted so that its output voltage is zero when the pendulum is in the vertical position (e.g., with a symmetrical supply voltage). Its gain p can be integrated into the controller, and we can arbitrarily set $p = 1$.

A controller provides a motor drive voltage that is fed into a DC motor. The DC motor has the transfer function that was introduced in Chapter 9. The term k_m/s in the motor transfer function considers the motor constant, but also the wheel radius and any possible gear between motor and wheel. Similarly, J is the combined rotational inertia of motor and wheel. Lastly, a disturbance $D(s)$ can be modeled as a small shift of the wheels in the negative x-direction.

At this point, we can determine the transfer function between the disturbance $D(s)$ and the angle $\Phi(s)$:

$$\Phi(s) = \frac{4s^2(Js + R_F)}{LJs^3 + LR_F s^2 + (4k_m H(s) - 2gJ)s - 2gR_F} \cdot D(s) \tag{14.33}$$

Without even setting up a Routh-Hurwitz table, we can see that this transfer function always has at least one pole in the right half-plane due to the negative zero-order term $-2gR_F$, unless the controller has an integral component. We will therefore examine a *PI* controller with the transfer function

$$H(s) = k_p' + \frac{k_I'}{s} = \frac{k_p}{4k_m} + \frac{k_I}{4k_m s} \tag{14.34}$$

where k_p' and k_I' are the actual controller parameters, which have been scaled to include the term $4k_m$ to simplify Eq. (14.33). With this *PI* controller, Eq. (14.33) becomes

$$\Phi(s) = \frac{4s^2(Js + R_F)}{LJs^3 + LR_F s^2 + (k_p - 2gJ)s + k_I - 2gR_F} \cdot D(s) \tag{14.35}$$

Let us set up the Routh array explicitly to demonstrate how the stability requirements are obtained:

s^3	JL	$k_p - 2gJ$
s^2	$R_F L$	$k_I - 2gR_F$
s^1	β	0
s^0	$k_I - 2gR_F$	0

The coefficient β is determined through

$$\beta = \frac{R_F L(k_p - 2gJ) - JL(k_I - 2gR_F)}{R_F L} \tag{14.36}$$

For stability, k_p and k_I must meet the following criteria:

$$\begin{aligned} k_p &> 2gJ \\ k_I &> 2gR_F \\ k_p R_F &> k_I J \end{aligned} \tag{14.37}$$

To examine the effect of k_p and k_I on the location of the poles, let us assume the following values: $J = 0.1$ N m s^2; $k_m = 10$ N m^2/V; $R_F = 0.5$ N m s; $L = 0.2$ m; $g = 9.81$ m/s^2. With these values, Eq. (14.35) and the stability criteria in Eq. (14.37) become

$$\begin{aligned} \Phi(s) &= \frac{s^2(0.4s + 2)}{0.02s^3 + 0.1s^2 + (k_p - 1.96)s + k_I - 9.81} \cdot D(s) \\ k_p &> 1.96 \\ k_I &> 9.81 \\ k_p &> 0.2k_I \end{aligned} \tag{14.38}$$

The first condition, $k_p > 1.96$, emerges from the requirement to have a positive first-order coefficient in Eq. (14.35), and it is met when the other two conditions are met. A root locus for k_p has two unstable open-loop poles when only the condition $k_p > 1.96$ is met. As k_p increases, these poles emit two complex conjugate branches that rapidly move into the left half-plane and follow the two asymptotes toward higher oscillatory frequencies. Low values of k_I cause strong, underdamped oscillatory responses (Figure 14.23).

Root locus analysis can be performed for k_I as well. An initial value of k_p needs to be chosen to have the poles for $k_I = 0$ in the left half-plane. The open-loop function for k_I has three poles and no zero, and the two asymptotes at $\pm 60°$ rapidly attract the branches into the right half-plane, unless k_p is increased as well. Both scenarios increase the underdamped, oscillatory response. With a pure *PI* controller, k_I is therefore chosen as low as permissible. The lowest value of $k_I = 9.81$ would not provide any robustness against variations of any coefficient, and a certain safety margin needs to be applied for the selection of k_I.

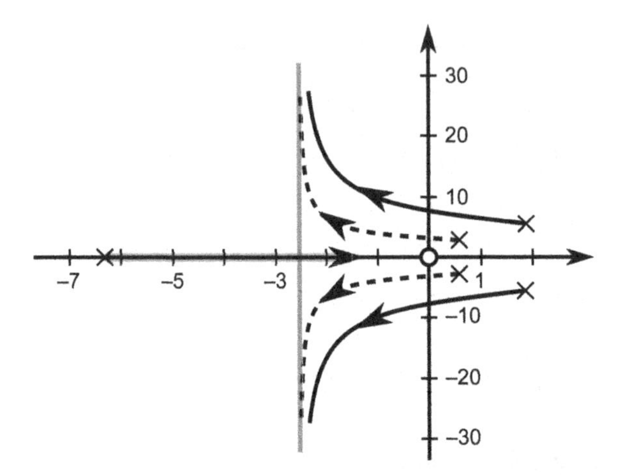

Figure 14.23 Root locus diagram for k_p for two selected values of k_I. When $k_I \approx 11$, the poles are closer to the origin (dashed curves). With a larger $k_I \approx 17$, the unstable poles move further into the right half-plane (solid curves). Ideally, k_I is chosen as low as permissible for stability.

Better performance can be achieved with a small phase-lead component. With a *PID* controller, the weight of the second-order coefficient can be increased (Eq. (14.39)):

$$\Phi(s) = \frac{s^2(0.4s + 2)}{0.02s^3 + (k_D + 0.1)s^2 + (k_p - 1.96)s + k_I - 9.81} \cdot D(s) \tag{14.39}$$

In this fashion, the stable pole can be moved further out in the left half-plane, and the complex conjugate branches shown in Figure 14.24 are attracted towards a double branchoff point at $s = -6.7$ which forms when $k_p = 2.7$ and $k_D = 0.3$. The additional phase-lead component can therefore dramatically improve the dynamic response. Moreover, the additional phase-lead component relaxes the stability criteria, allowing a larger k_I for a given k_p. The result is a more robust system.

The importance of a phase-lead component in a controller for an unstable system can further be demonstrated in a classroom experiment. The basis is the positioner system introduced in Chapter 9, but with a *PI* controller intentionally biased out of its stability range. The premise for this example is that the controller cannot be modified, and the only option for stabilizing the system is to introduce a compensator in the feedback path together with the sensor. Unlike the inverted pendulum, this second example has an unstable complex conjugate pole pair. Although this is a constructed example, it relates closely to weakly damped second-order systems that show a strong oscillatory response. The concept also relates to many electronic high-gain feedback circuits where a stable system with poles near the imaginary axis is made more robust with a phase-lead network in the feedback loop. Typical examples include switch-mode power supplies.

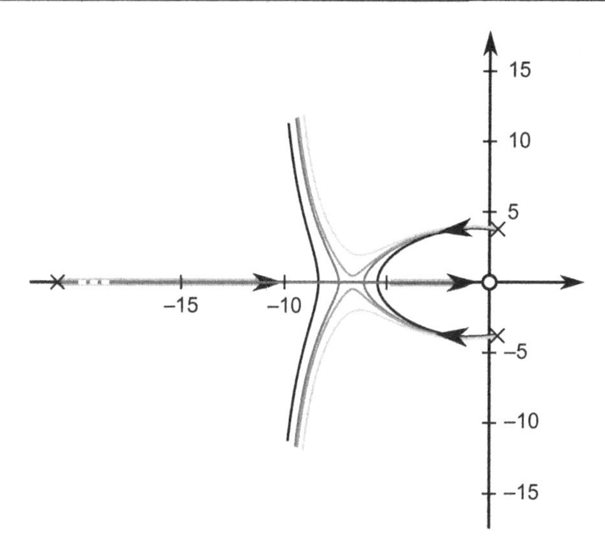

Figure 14.24 Root locus diagrams for k_p with a *PID* controller and increasing k_D, where higher k_D corresponds to a darker shade of the curve. The values for k_D are 0.29, 0.301, 0.305, and 0.32. The open-loop pole locations are only marginally influenced by k_D, and the stable, real-valued pole lies at about $s = -21$. The asymptote centroid lies at about $s = -10$ and slightly moves to the left with increasing k_D. Most importantly, a suitable choice of k_D allows to create a critically damped system or a slightly underdamped system.

The starting point is the system given in Figure 9.2, but with a *PI* controller $H(s) = k_p + k_I/s$. The transfer function (originally introduced as Eq. 9.5) now becomes

$$X(s) = \frac{\alpha(k_p s + k_I)}{s^3 + \frac{R_F}{J}s^2 + \alpha\, k_p s + \alpha k_I} X_{set}(s) \tag{14.40}$$

Equation (14.40) now describes a third-order system, and stability requires $k_I < k_p R_F/J$. If k_I is increased beyond this point, a complex conjugate pole pair moves into the right half-plane, and the system shows exponentially increasing oscillations. We can show that a zero, introduced by a phase-lead network in the sensor amplifier, can make the unstable system stable and can, in fact, improve the dynamic response of strongly underdamped systems.

By measuring several process constants, we can determine the zeros and poles of Eq. (14.40). When $R_F/J = 50$ ms, $\alpha = 2.4$ m/V, $k_p = 5$, and $k_I = 250$, the system has, as expected, a complex pole pair in the right half-plane, namely at $p_{1,2} = +0.02 \pm 3.8j$, a stable real-valued pole at $p_3 = -50$ and a zero at $z = -60$.

The first approach is to include a first-derivative operation in the sensor module. The zero in the feedback loop stabilizes the system. The transfer function is

$$X(s) = \frac{\alpha(k_p s + k_I)}{s^3 + (\frac{R_F}{J} + \alpha k_p)s^2 + \alpha k_I s} X_{set}(s) \tag{14.41}$$

We can see that the transfer function has three poles:

$$p_1 = 0; \quad p_{2,3} = -\frac{1}{2}\left(\frac{R_F}{J} + k_p\alpha \pm \sqrt{\left(\frac{R_F}{J} + k_p\alpha\right)^2 - 4k_I\alpha}\right) \tag{14.42}$$

By introducing the differentiator in the feedback loop, we have rearranged the poles that no combination of k_p and k_I can make the system unstable. However, we have also introduced an integrator pole in the origin, which leads to an unacceptable dynamic response.

The second approach is to add a phase-lead compensator to the sensor module (see Section 3.6.6). The phase-lead compensator has the transfer function $\tau_D s + 1$, and introducing the compensator in the feedback path changes the transfer function (Eq. (14.40)) to

$$X(s) = \frac{\alpha(k_p s + k_I)}{s^3 + (\frac{R_F}{J} + \alpha k_p \tau_D)s^2 + \alpha(k_I \tau_D + k_p)s + \alpha k_I} X_{set}(s) \tag{14.43}$$

Root locus analysis gives us an idea how τ_D influences the uncompensated poles. For very small τ_D, the root locus has two zeros (in the origin and at $z = -1$) and three poles; the poles coincide with those of Eq. (14.40). Notably, the pole at $p_3 = -50$ can be considered nondominant and thus ignored, particularly since this pole moves further to the left along the single asymptote. The branches from the unstable poles are pulled towards the zeros and meet at approximately $s = -0.5$ to turn into two real-valued branches. With a suitable (albeit large) τ_D, either a critically damped or a slightly underdamped configuration can be achieved under the premise that only a compensator in the feedback loop is possible.

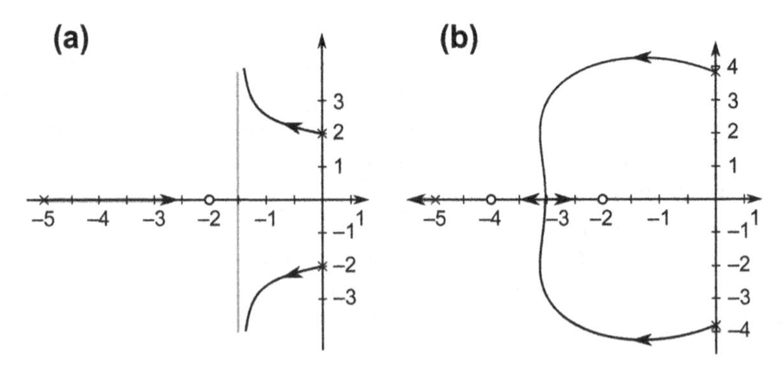

Figure 14.25 Root locus diagrams for strongly underdamped systems and phase-lead compensation. (a) A single zero between the real-valued pole and the origin attracts the root locus branches from the oscillatory pole pair and moves the poles to the left, with a more rapidly attenuated oscillation. (b) A second zero "bends" the root locus branches towards the real axis, and a branchoff point forms. This configuration allows high flexibility to create critically damped or slightly underdamped responses with a low oscillatory component.

We can generalize this example by assuming a third-order system with a complex conjugate pole pair near the imaginary axis (either stable or unstable) and a fast, real-valued third pole. With a phase-lead controller, we place a zero between the real-valued pole and the origin (Figure 14.25a). The zero attracts the root locus branches from the oscillatory poles and moves them toward the asymptotes. High loop gain creates an underdamped configuration at a higher frequency, but with intermediate gain values, oscillations decay rapidly.

Even more advantageous is a compensator that provides two zeros to the loop transfer function (Figure 14.25b). Depending on the location of the zeros and the loop gain, a nonoscillatory response can be obtained. The critically damped case is possible, as well as slightly underdamped configurations with faster step response and minor overshoot. This compensator is highly suited to reduce oscillations in electromechanical systems with a low-friction spring-mass system, for example, in an electromagnetic actuator. When technically possible, the second compensator pole should be placed to the left of the real-valued process pole. This configuration moves the branchoff point even further left, with an appropriately accelerated step response.

Appendix A
Laplace Correspondence Tables

Table A.1 Laplace transform pairs of some singularity functions.

Time-domain function $f(t)$, $t > 0$	**Laplace-domain correspondence $F(s)$**
$\delta(t)$ (unit impulse)	1
$u(t)$ (unit step)	$\frac{1}{s}$
t (unit ramp)	$\frac{1}{s^2}$
t^n	$\frac{n!}{s^{n+1}}$

Table A.2 Laplace transform pairs of some dead-time functions.

Time-domain function $f(t)$, $t > 0$	**Laplace-domain correspondence $F(s)$**
$\delta(t-\tau)$ (delayed unit impulse)	$e^{-s\tau}$
$u(t-\tau)$ (delayed unit step)	$\frac{1}{s}e^{-s\tau}$
$u(t)-u(t-\tau)$ (rectangular pulse)	$\frac{1}{s}\left(1-e^{-s\tau}\right)$

Table A.3 Laplace transform pairs of some first- and second-order functions with real-valued poles and exponential characteristic.

Time-domain function $f(t)$, $t > 0$	**Laplace-domain correspondence $F(s)$**
e^{-at}	$\frac{1}{s+a}$
$\frac{1}{a}\left(1-e^{-at}\right)$	$\frac{1}{s(s+a)}$
te^{-at}	$\frac{1}{(s+a)^2}$
$(1-at)e^{-at}$	$\frac{s}{(s+a)^2}$
$\frac{1}{b-a}\left(e^{-at}-e^{-bt}\right)$	$\frac{1}{(s+a)(s+b)}$
$\frac{1}{a-b}\left(ae^{-at}-be^{-bt}\right)$	$\frac{s}{(s+a)(s+b)}$

Linear Feedback Controls. http://dx.doi.org/10.1016/B978-0-12-405875-0.00022-X
© 2013 Elsevier Inc. All rights reserved.

Table A.4 Laplace transform pairs of some higher-order functions with real-valued roots and exponential characteristic.

Time-domain function $f(t)$, $t > 0$	Laplace-domain correspondence $F(s)$
$\frac{1}{a^2}\left(at - 1 + e^{-at}\right)$	$\frac{1}{s^2(s+a)}$
$\frac{1}{(n-1)!}t^{n-1}e^{-at}$	$\frac{1}{(s+a)^n}$
$\frac{1}{ab}\left(1 - \frac{be^{-at}}{b-a} + \frac{ae^{-bt}}{b-a}\right)$	$\frac{1}{s(s+a)(s+b)}$
$\frac{1}{a^2}\left(1 - e^{-at} - ate^{-at}\right)$	$\frac{1}{s(s+a)^2}$
$\frac{e^{-at}}{(b-a)(c-a)} + \frac{e^{-bt}}{(c-b)(a-b)} + \frac{e^{-ct}}{(a-c)(b-c)}$	$\frac{1}{(s+a)(s+b)(s+c)}$

Table A.5 Laplace transform pairs of some functions with at least one complex conjugate pole pair that describe periodic or attenuated oscillations.

Time-domain function $f(t)$, $t > 0$, $\zeta < 1$	Laplace-domain correspondence $F(s)$
$\sin \omega t$	$\frac{\omega}{s^2+\omega^2}$
$\cos \omega t$	$\frac{s}{s^2+\omega^2}$
$\sin(\omega t + \varphi)$	$\frac{s \sin\varphi + \omega\cos\varphi}{s^2+\omega^2}$
$e^{-at}\sin \omega t$	$\frac{\omega}{(s+a)^2+\omega^2}$
$e^{-at}\cos \omega t$	$\frac{s+a}{(s+a)^2+\omega^2}$
$1 - \cos \omega t$	$\frac{\omega^2}{s(s^2+\omega^2)}$
$e^{-at} - \cos\omega t + \frac{a}{\omega}\sin\omega t$	$\frac{a^2+\omega^2}{(s+a)(s^2+\omega^2)}$
$e^{-\zeta\omega_n t}\sin\omega_d t$	$\frac{\omega_d}{s^2+2\zeta\omega_n s+\omega_n^2}$
$\frac{1}{\omega_n^2} - \frac{1}{\omega_n\omega_d}e^{-\zeta\omega_n t}\sin(\omega_d t + \phi)$	$\frac{1}{s(s^2+2\zeta\omega_n s+\omega_n^2)}$
$\frac{a}{\omega_n^2}\left(1 - e^{-\zeta\omega_n t}\cos(\omega_d t)\right) + \left(\frac{1}{\omega_d} - \frac{a\zeta}{\omega_n\omega_d}\right)e^{-\zeta\omega_n t}\sin(\omega_d t)$	$\frac{s+a}{s(s^2+2\zeta\omega_n s+\omega_n^2)}$

where $\omega_d = \omega_n\sqrt{1-\zeta^2}$ and $\cos\phi = \zeta$.

Table A.6 Comprehensive correspondence table for systems with single or multiple real-valued poles in the origin and in the left half-plane.

Denom.	1	$(s+a)$	$(s+a)(s+b)$	$(s+a)(s+b)(s+c)$
1	$\delta(t)$	e^{-at}	$\frac{e^{-at}}{b-a}+\frac{e^{-bt}}{a-b}$	$\frac{e^{-at}}{(b-a)(c-a)}+\frac{e^{-bt}}{(a-b)(c-b)}+\frac{e^{-ct}}{(a-c)(b-c)}$
s	1	$\frac{1}{a}(1-e^{-at})$	$\frac{1}{ab}\left(1-\frac{b}{b-a}e^{-at}-\frac{a}{a-b}e^{-bt}\right)$	$\frac{1}{abc}\left(1-\frac{bc}{(b-a)(c-a)}e^{-at}-\frac{ac}{(a-b)(c-b)}e^{-bt}-\frac{ab}{(a-c)(b-c)}e^{-ct}\right)$
s^2	t	$-\frac{1}{a^2}(1-at-e^{-at})$	$-\frac{1}{(ab)^2}\left((a+b)-abt-\frac{b^2}{b-a}e^{-at}-\frac{a^2}{a-b}e^{-bt}\right)$	$-\frac{1}{(abc)^2}\left((ab+bc+ac)-(abc)t-\frac{(bc)^2}{(b-a)(c-a)}e^{-at}-\frac{(ac)^2}{(a-b)(c-b)}e^{-bt}-\frac{(ab)^2}{(a-c)(b-c)}e^{-ct}\right)$
s^3	$\frac{1}{2}t^2$	$\frac{1}{a^3}\left(1-at+\frac{1}{2}a^2t^2-e^{-at}\right)$	$\frac{1}{(ab)^3}\left((a^2+ab+b^2)-ab(a+b)t+(ab)^2\frac{t^2}{2}-\frac{b^3}{b-a}e^{-at}-\frac{a^3}{a-b}e^{-bt}\right)$	$\frac{1}{(abc)^3}\left((a^2b^2+b^2c^2+a^2+c^2+a^2bc+ab^2c+abc^2)-(a^2b^2c+ab^2c^2+a^2bc^2)t+(abc)^2\frac{t^2}{2}-\frac{(bc)^3}{(b-a)(c-a)}e^{-at}-\frac{(ac)^3}{(a-b)(c-b)}e^{-bt}-\frac{(ab)^3}{(a-c)(b-c)}e^{-ct}\right)$

Example how to use Table A.6. For a system with one pole in the origin and two real-valued poles, we seek the step response. The step response is given as $F(s)$ in Eq. A.1. The function $F(s)$ has two poles in the origin (third row in Table A.6) and two poles in the left half-plane in the form $(s+a)(s+b)$ (third column in Table A.6). We find the time-domain correspondence as $f(t)$:

$$F(s)=\frac{1}{s^2(s+a)(s+b)} \quad \bullet\!\!-\!\!\circ \quad f(t)=-\frac{1}{(ab)^2}\left((a+b)-abt-\frac{b^2}{b-a}e^{-at}-\frac{a^2}{a-b}e^{-bt}\right) \tag{A.1}$$

Table A.7 Some properties of the Laplace transform (a, b, τ are real-valued, positive constants).

Property	**Time domain**	**Laplace-domain correspondence**
Definition	$f(t)$	$F(s) = \int_{0+}^{\infty} f(t)e^{-st}\mathrm{d}t$
Linearity	$af_1(t) \pm bf_2(t)$	$aF_1(s) \pm bF_2(s)$
Convolution	$f_1(t) \otimes f_2(t)$	$F_1(s) \cdot F_2(s)$
Integral	$\int_0^t f(\tau)\mathrm{d}\tau$	$\frac{F(s)}{s}$
First derivative	$\frac{\mathrm{d}f(t)}{\mathrm{d}t}$	$sF(s) - f(t = 0^+)$
nth derivative	$\frac{\mathrm{d}^n f(t)}{\mathrm{d}t^n}$	$s^n F(s) - \sum_{i=1}^{n} s^{n-i} f^{<i-1>}(t = 0^+)$
Time delay	$f(t - \tau) \cdot u(t - \tau)$	$F(s) \cdot e^{-s\tau}$
Time scaling	$f(at)$	$\frac{1}{a}F(\frac{s}{a})$
Exponential attenuation	$f(t) \cdot e^{-at}$	$F(s + a)$

Appendix B
Z-Transform Correspondence Tables

Table B.1 Z-transform pairs of some singularity functions for sampling rate T. In all cases, $f_k = 0$ for $k < 0$.

Continuous function $f(t)$, $t > 0$		Discrete sequence f_k	Z-domain correspondence $F(z)$
$\delta(t)$	(unit impulse)	$f_k = \begin{cases} 1 & \text{for} \quad k = 0 \\ 0 & \text{for} \quad k \neq 0 \end{cases}$	1
$\delta(t - mT)$	(delayed unit impulse)	$f_k = \begin{cases} 1 & \text{for} \quad k = m \\ 0 & \text{for} \quad k \neq m \end{cases}$	z^{-m}
$u(t)$	(unit step)	$f_k = 1 \quad \text{for } k \geq 0$	$\frac{z}{z-1}$
t	(unit ramp)	$f_k = kT$	$T \frac{z}{(z-1)^2}$
t^2		$f_k = (kT)^2$	$T^2 \frac{z(z+1)}{(z-1)^3}$

Table B.2 Z-transform pairs of some functions with single- or double-exponential characteristic. In all cases, $f_k = 0$ for $k < 0$.

Continuous function $f(t)$,	Discrete sequence f_k	Z-domain correspondence $F(z)$
e^{-at}	$f_k = e^{-akT}$	$\frac{z}{z-e^{-aT}}$
$\left(1 - e^{-at}\right)$	$f_k = 1 - e^{-akT}$	$\frac{z(1-e^{-aT})}{(z-1)(z-e^{-aT})}$
te^{-at}	$f_k = (kT)e^{-akT}$	$\frac{Tze^{-aT}}{(z-e^{-aT})^2}$
$e^{-at} - e^{-bt}$	$f_k = e^{-akT} - e^{-bkT}$	$\frac{z(e^{-aT}-e^{-bT})}{(z-e^{-aT})(z-e^{-bT})}$
$1 - e^{-at} - ate^{-at}$	$f_k = 1 - e^{-akT} - akTe^{-akT}$	$\frac{z(1-(1+aT)e^{-aT})+e^{-aT}(aT-1)+e^{-2aT}}{(z-1)(z-e^{-aT})^2}$

Table B.3 Z-transform pairs of some discrete power series. In all cases, $f_k = 0$ for $k < 0$.

Continuous function $f(t)$,	Discrete sequence f_k	Z-domain correspondence $F(z)$
$a^{t/T}$	$f_k = a^k$	$\frac{z}{z-a}$
$1 - a^{t/T}$	$f_k = 1 - a^k$	$\frac{z(1-a)}{(z-1)(z-a)}$
$a^{t/T} - b^{t/T}$	$f_k = a^k - b^k$	$\frac{z(a-b)}{(z-a)(z-b)}$

Linear Feedback Controls. http://dx.doi.org/10.1016/B978-0-12-405875-0.00023-1
© 2013 Elsevier Inc. All rights reserved.

Table B.4 Z-transform pairs of some functions that describe periodic or attenuated oscillations. In all cases, $f_k = 0$ for $k < 0$.

Continuous function $f(t)$,	**Discrete sequence f_k**	**Z-domain correspondence $F(z)$**
$\sin(\omega t)$	$f_k = \sin \omega kT$	$\frac{z \sin(\omega T)}{z^2 - 2z\cos(\omega T) + 1}$
$\cos(\omega t)$	$f_k = \cos \omega kT$	$\frac{z^2 - z\cos(\omega T)}{z^2 - 2z\cos(\omega T) + 1}$
$e^{-at}\sin(\omega t)$	$f_k = e^{-akT}\sin \omega kT$	$\frac{ze^{-aT}\sin(\omega T)}{z^2 - 2ze^{-aT}\cos(\omega T) + e^{-2aT}}$
$e^{-at}\cos(\omega t)$	$f_k = e^{-akT}\cos \omega kT$	$\frac{z^2 - ze^{-aT}\cos(\omega T)}{z^2 - 2ze^{-aT}\cos(\omega T) + e^{-2aT}}$

Table B.5 Direct correspondence between transfer functions in the Laplace- and z-domains. The Laplace-domain transfer function $G(s)$ has a direct correspondence in $G(z)$, whereas $G_1(z) = G_0(z)G(z)$ includes the virtual zero-order hold (*cf.* Figure 4.6)

Laplace-domain function $G(s)$	**Direct z-domain correspondence $G(z)$**	**Z-domain correspondence with zero-order hold $G_1(z) = G_0(z)G(z)$**
$\frac{1}{s}$	$\frac{z}{z-1}$	$\frac{T}{z-1}$
$\frac{a}{s+a}$	$\frac{az}{z-e^{-aT}}$	$\frac{1-e^{-aT}}{z-e^{-aT}}$
$\frac{a}{s(s+a)}$	$\frac{z(1-e^{-aT})}{(z-1)(z-e^{-aT})}$	$\frac{1}{a}\frac{z(aT+e^{-aT}-1)+1-(1+aT)e^{-aT}}{(z-1)(z-e^{-aT})}$
$\frac{1}{(s+a)(s+b)}$	$\frac{1}{b-a}\frac{z(e^{-aT}-e^{-bT})}{(z-e^{-aT})(z-e^{-bT})}$	$\frac{z(a-b-ae^{-bT}+be^{-aT})-ae^{-aT}+be^{-bT}+(a-b)e^{-(a+b)T}}{ab(a-b)(z-e^{-aT})(z-e^{-bT})}$

Table B.6 Some properties of the z-transform (a, b are real-valued, positive constants, T is the sampling interval).

Property	**Time domain**	**Z-domain correspondence**
Definition	$f(t)$	$F(z) = \sum_{k=0}^{\infty} f(kT)z^{-k}$
Linearity	$af_1(t) \pm bf_2(t)$	$aF_1(z) \pm bF_2(z)$
Convolution	$f_k \otimes g_k$	$F(z) \cdot G(z)$
Initial value	$f(0)$	$\lim_{z\to\infty} F(z)$
Final value	$f(k \to \infty)$	$\lim_{z\to 1} (z-1)F(z)$
Time lag	$f(t-T)$	$z^{-1}F(z)$
Time lead	$f(t+T)$	$zF(z) - zf(0)$
Time multiplication	$t \cdot f(t)$	$-Tz\frac{\mathrm{d}F(z)}{\mathrm{d}z}$
Exponential attenuation	$f(t) \cdot e^{-at}$	$F(ze^{aT})$

Appendix C Introduction to Operational Amplifiers

Operational amplifiers (op-amps for short) are versatile electronic control elements that can be used in the practical realization of controllers and compensators, and as gain amplifiers for sensor signals. Appendix C provides a short introduction to its basic behavior and some application examples that help explain the op-amp-based circuit examples throughout this book.[1] Figure C.1 shows the schematic symbol of an op-amp and its block diagram that allows an interpretation of its function.

Operational amplifiers are relatively complex devices, consisting of an input difference stage, a gain stage, and the output driver. Various simplified mathematical models can be used to approximate their behavior. In the simplest model, the voltage difference between the inputs is multiplied by a large, real-valued gain factor g, and the op-amp acts as a voltage-controlled voltage source with a gain between 10^4 and 10^6. This gain is commonly referred to as *differential gain*, because it amplifies the voltage difference between the positive and negative inputs. Furthermore, input currents are considered to be zero in an ideal operational amplifier, that is, an infinite input impedance is assumed. Finally, any output offset voltage is considered to be negligible, and the output impedance is taken as zero. For low frequencies up to several kHz, the zero-order model of real-valued gain g is sufficiently accurate. However, the gain decreases with increasing frequencies, and for frequencies in the high kHz range and above, a higher-order transfer function needs to be used.

Electronic circuits with high gain tend to be unstable and oscillate. For this reason, most general-purpose op-amps come with a built-in compensation network to make them appear like a stable first-order time-lag system in a wide frequency range, and the op-amp pole is usually placed between 10 and 100 Hz. At much higher frequencies, typically far above 100 kHz, the gain drops rapidly due to internal delays. When the op-amp is used in a feedback configuration with a closed-loop gain much smaller than the native op-amp gain, the first-order model of the open-loop gain,

$$G(s) = \frac{g}{\omega_O^{-1} s + 1} \tag{C.1}$$

is valid for frequencies far above ω_O. The exact value of ω_O can be found in the manufacturer's data sheets. In a typical configuration, a feedback network is placed

[1] A detailed description of op-amp theory and practical applications can be found in *Op-Amps for Everyone* by Ron Manchini.

Linear Feedback Controls. http://dx.doi.org/10.1016/B978-0-12-405875-0.00024-3
© 2013 Elsevier Inc. All rights reserved.

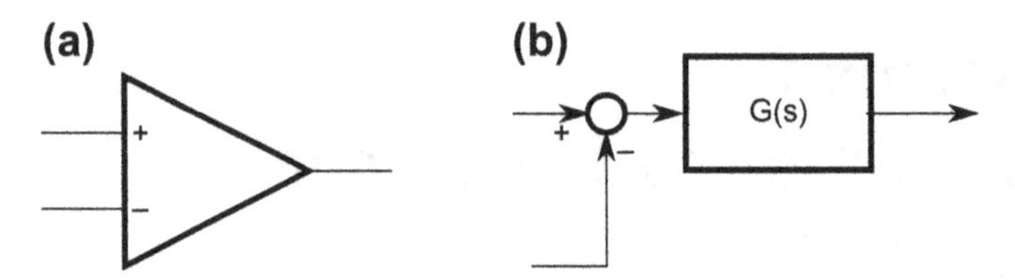

Figure C.1 (a) Schematic symbol of the operational amplifier. (b) Functional block diagram. The ideal operational amplifier multiplies the difference of the input signals by a large, real-valued gain factor $G(s) = g$, where g is typically between 10^4 and 10^6, depending on the type of operational amplifier. The input current is considered to be negligible.

between the output and the inverting input. When the feedback network has the transfer function $H(s)$, the closed-loop gain between non-inverting input and output becomes

$$\frac{V_{out}(s)}{V_{in}(s)} = \frac{G(s)}{1 + G(s)H(s)} \approx \frac{1}{H(s)} \tag{C.2}$$

where the approximation $1/H(s)$ is valid as long as $G(s)H(s) \gg 1$, that is, for relatively low frequencies up to several kHz. In many typical control applications, this frequency range is sufficient, and the zero-order approximation $G(s) = g$ can be used.

In the simplest case, the output of the op-amp is connected to the negative input ($H(s) = 1$). Because the output voltage is identical to the input voltage, this configuration is called a *voltage follower*, and it can be used to decouple a first-order lowpass (Figure 2.2) or highpass and ensure that no current flows through the output branch.

With one op-amp and a few discrete components for the feedback network, several first-order systems can be realized. Two examples are given in Figure C.2. The first

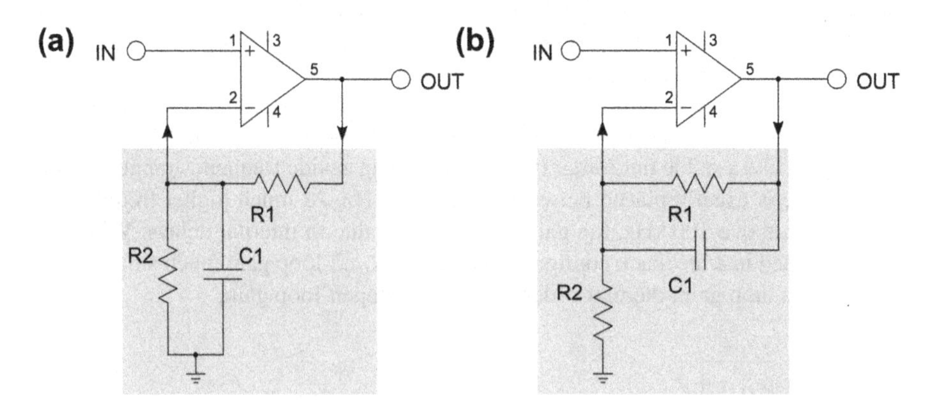

Figure C.2 Two example first-order systems realized with op-amps. The passive feedback network with the transfer function $H(s)$ is shaded gray, and arrows indicate the signal flow into and out of the feedback networks. (a) *PD* compensator with one zero in the transfer function $k_p(\tau_D s + 1)$. (b) Lead-lag compensator with one zero and one pole in the transfer function $k_p(\tau_D s + 1)/(\tau_P s + 1)$.

example in Figure C.2a represents a noninverting *PD* compensator (*cf.* Section 3.6.6) with the overall transfer function (under the assumption of a very large g):

$$\frac{V_{out}(s)}{V_{in}(s)} = \frac{C_1 R_1 R_2 s + R_1 + R_2}{R_2} = k_p\,(\tau_D s + 1) \tag{C.3}$$

The circuit has a DC gain of $k_p = (1 + R_1/R_2)$ and a zero at $1/\tau_D = -(R_1 + R_2)/(C_1 R_1 R_2)$. A simplification is possible by omitting the resistor R2 ($R_2 \rightarrow \infty$), which produces a circuit with unit DC gain and a zero at $-1/C_1 R_1$.

The circuit in Figure C.2b creates a pole-zero pair. The overall transfer function of the circuit is

$$\frac{V_{out}(s)}{V_{in}(s)} = \frac{C_1 R_1 R_2 s + R_1 + R_2}{C_1 R_1 R_2 s + R_2} \tag{C.4}$$

and it has a DC gain of $(1 + R_1/R_2)$, a pole at $-1/C_1 R_1$, and a zero at $-(R_1 + R_2)/(C_1 R_1 R_2)$. This circuit can be used as noninverting lead-lag compensator, and as an alternative to the inverting compensator in Figure 13.9.

Frequently, the op-amp is used in an inverting configuration where the positive input is grounded and the negative input receives the input signal through an impedance Z_1. Feedback is established from the output to the negative input through a different impedance Z_2. Since the current into the negative input is assumed to be negligible, the current i through both impedances is identical. Therefore,

$$i = \frac{V_{minus} - V_{in}}{Z_1} = \frac{V_{out} - V_{minus}}{Z_2} \tag{C.5}$$

holds, where V_{minus} is the voltage at the negative input. With the simplified equation for the differential gain, $V_{out} = g \cdot (V_{plus} - V_{minus})$ and the grounded positive input ($V_{plus} = 0$), we can eliminate V_{minus} from Eq. C.5 and obtain the transfer function for the circuit as

$$\frac{V_{out}(s)}{V_{in}(s)} = -\frac{g\,Z_2}{Z_1(g+1) + Z_2} \approx -\frac{Z_2}{Z_1} \tag{C.6}$$

where the approximation $-Z_2/Z_1$ is valid for $g \gg 1$. For example, by using a resistor R for Z_1 and a capacitor C for Z_2, the circuit assumes the transfer function $-1/RCs$ and acts as an integrator. By using a capacitor C for Z_1 and a resistor R for Z_2, the circuit becomes a differentiator with the transfer function $-RCs$. By using the same approach, the *PI* controller in Figure 6.4, the *PID* controller in Figure 13.8, and the lead-lag compensator in Figure 13.9 are obtained.

Op-amp circuits are important for digital control systems as well. Frequently, analog signals need to be converted to a digital value, and analog-to-digital converters (ADC) require a specific input voltage range. Typically, ADC that are integrated with a microcontroller are limited to input voltages between ground and the microcontroller's supply (often +5 V). A difference amplifier circuit can be used to adjust the voltage levels. Furthermore, bandwidth limitation *must* be performed with an analog filter before ADC to prevent aliasing. A combined filter/gain/level adjustment circuit can be realized with a single op-amp.

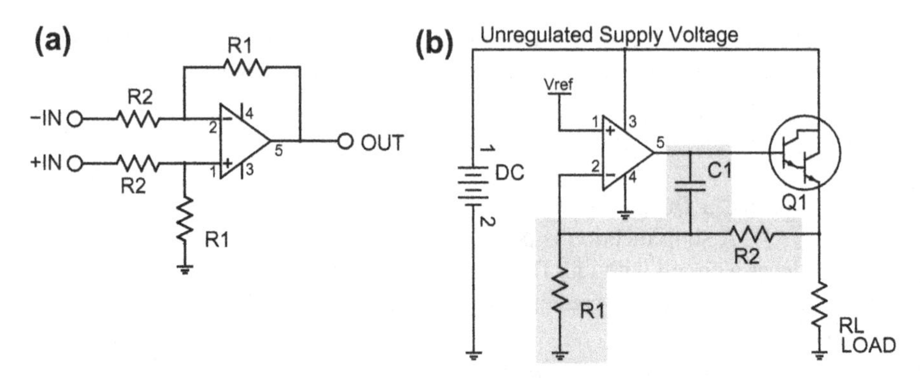

Figure C.3 Control applications of op-amps. (a) A difference amplifier with the overall gain $k_p = R_1/R_2$. (b) Op-amp as combined difference stage and error amplifier in a power supply application. The voltage feedback path is shaded.

Because of its capability to compute the difference of two signals, the op-amp can be used in the difference stage of a feedback control system. With four external resistors (Figure C.3a), an op-amp can be configured to provide the output voltage $V_{out} = k_p \cdot (V_{plus} - V_{minus})$. The gain k_p is determined by the ratio of resistors with $k_p = R_1/R_2$.

An example application where the large differential gain g of an op-amp is used directly to combine difference stage and error amplifier is provided in Figure C.3b. The circuit is representative for linear regulated power supplies, and the purpose is to keep the voltage across the load RL constant, even if the load changes (disturbance). An additional element of interest is the nonlinear behavior of the transistor Q1, which serves as current amplifier. The base voltage needs to be about 1.4 V higher than the emitter voltage, but the base current is about two to three orders of magnitude less than the emitter current (i.e., the current available for the load). In power supplies, an additional disturbance exists when the primary supply voltage shows variability. A reference voltage (Vref) is provided at the positive input of the op-amp. The voltage across the load is divided by a factor $R_1/(R_1+R_2)$ before being applied to the negative input of the op-amp. Because of its large differential gain, the op-amp adjusts its output voltage to minimize the voltage difference between the two inputs and therefore provides enough base current to transistor Q1 so that its emitter voltage is Vref$(R_1 + R_2)/R_1$. A small capacitor C1 in the feedback loop acts as phase-lead element and improves the relative stability of the circuit.

Appendix D
Relevant Scilab Commands

To make use of the many Scilab commands related to control theory, the complex variable s needs to be defined first:

```
s  = poly (0,'s')
```

To obtain the roots of a polynomial $p(s)$, first define $p(s)$, then use the `roots` function. Note that all coefficients need to be known numerically.

```
p  = s^3  + 5 * s^2  + 16
roots (p)
```

Higher-level functions require to define a linear system from its transfer function. This can be done with the `syslin` function. We first define a transfer function $H(s)$ similar to the polynomial $p(s)$ in the previous example, then we create a time-continuous linear system `Hsys` from `H`:

```
 H = (s + 1)/(s^2 + 2 * s + 4)
Hsys  = syslin ('c', H)
```

The linear system is a prerequisite for a number of convenient functions, including Bode, Nyquist, and root locus plots. Key functions are:

- `plzr(Hsys)`—Draw a pole-zero plot for the linear system
- `pfss(Hsys)`—Perform a partial fraction expansion of the linear system `Hsys` and return the additive components as a vector of linear systems
- `bode(Hsys)`—Draw a bode diagram for the linear system
- `show_margins(Hsys)`—Draw a Bode diagram and indicate the phase and gain margins in the Bode plot
- `g_margin(Hsys)`—Returns the gain margin of the linear system `Hsys`
- `p_margin(Hsys)`—Returns the phase margin of the linear system `Hsys`
- `nyquist(Hsys)`—Draw a Nyquist plot for the linear system
- `show_margins (Hsys, 'nyquist')`—Draw a Nyquist plot for the linear system and indicate the phase and gain margins
- `evans(Fsys)`—Draw a root locus plot for the root locus polynomial `Fsys`

For the root locus plot, Scilab requires the root locus polynomial as defined in Eq. (12.3), and *not* the system transfer function. Specifically, if you have a system with the transfer function $H(s) = p(s)/q(s)$ and you want to examine its root locus for one specific parameter κ that occurs in the coefficients of the denominator polynomial

Linear Feedback Controls. http://dx.doi.org/10.1016/B978-0-12-405875-0.00025-5
© 2013 Elsevier Inc. All rights reserved.

$q(s)$, you need to extract κ from $q(s)$ in the form

$$1 + \kappa \cdot F(s) = 0 \tag{D.1}$$

and create a linear system from $F(s)$ before using it in the `evans` function. Although Scilab will accept `Hsys` for your system $H(s)$, the root locus plot will be incorrect if you omit this step.

Scilab can also help you simulate the dynamic response of a system. Once again, you need to have a linear system, such as `Hsys`, created with the `syslin` command. We then need a time vector (we create this with the `linspace` command, in this case from 0 to 10 s with 500 steps) and pass both to `csim`, which then simulates and returns the dynamic response, which in this example is the step response. We then use `plot` to plot the response and add a few more commands to embellish the plot:

```
t  = linspace (0, 10, 500);
stprsp  = csim ('step', t, Hsys);
plot (t, stprsp);
xgrid ();
xtitle ("Step Response From 0 To 10 Seconds");
```

Instead of `'step'`, we can use `'imp'` for the impulse response or `'t'` for the unit ramp response.

For the simulation of the response of a time-discrete system, `csim` cannot be used, because it is limited to continuous systems. The equivalent function for time-discrete systems is `flts`. A time-discrete system receives a sequence of values as input. In the following example, the sequence is created as a vector `xk` on discrete time points kT.

```
z = %z                              // define z variable
t = linspace (0, 1, 101);           // Time vector with T=10ms
xk = linspace (1, 1, 101);          // Step input sequence
Zsys = syslin ('d', 0.5/(z-0.5))    // the time-discrete system
stprsp = flts (xk, Zsys);           // simulate its response
plot (t, stprsp);                   // and plot it over matching time
```

Lastly, `xcos` starts the interactive simulation tool.

References and Further Reading

1. Åström KJ, Wittenmark B: *Adaptive control*, 2008, Dover Publications.
2. Åström KJ, Hägglund T: *Advanced Pid control*, 2006, ISA-The Instrumentation, Systems, and Automation Society.
3. Baudin M: *Introduction to Scilab*, 2010, Consortium Scilab.
4. Bennett S: *A history of control engineering, 1930–1955*, 1993, Peter Peregrinus Ltd.
5. Bennett S: A brief history of automatic control, *IEEE Trans Control Syst* 16(3):17–25, 1996.
6. Bequette BW: *Process control: modeling, design, and simulation*, 2003, Prentice Hall.
7. Bishop RH, Dorf Richard C: *Modern control systems*, 2004, Prentice Hall College Division.
8. Black HS: Stabilized feed-back amplifiers, *Trans Am Inst Elec Eng* 53(1):114–120, 1934.
9. Bode HW: Feedback—the history of an idea. *Proceedings of the symposium on active networks and feedback systems*, Polytechnic Press, 1960.
10. Bracewell R: *The fourier transform & its applications*, ed 3, 2000, McGraw-Hill Science/Engineering/Math.
11. LePage WR: *Complex variables and the Laplace transform for engineers*, 1980, Dover Publications, Inc.
12. Levine WS, et al.: *The control handbook*, ed 2, 2011, CRC Press.
13. Lyons RG: *Understanding digital signal processing*, 2010, Pearson Education.
14. Mancini R: *Op amps for everyone*, ed 2, 2003, Newnes/Elsevier.
15. Maxwell JC: On governors, *Proc Roy Soc Lond* 16:270–283, 1867.
16. Mayr O: *The origins of feedback control*, 1970, MIT Press.
17. Narendra KS, Annaswamy AM: Stable Adaptive Systems, ed 2, 2005, Dover Publications.
18. Minorsky N: Directional stability of automatically steered bodies, *Nav Eng J* 32(2), 1922.
19. Nyquist H: Regeneration theory, *Bell Syst Tech J* 11(3):126–147, 1932.
20. Ogata K: *Modern control engineering*, 2010, Prentice Hall.
21. Tietze U, Schenk C: *Electronic circuits—handbook for design and applications*, ed 2, 2008, Springer.
22. Van De Vegte J: *Feedback control systems*, ed 3, 1993, Prentice Hall.

Linear Feedback Controls. http://dx.doi.org/10.1016/B978-0-12-405875-0.00026-7
© 2013 Elsevier Inc. All rights reserved.

Glossary

ADC	The ADC (analog-to-digital converter) is a device that takes samples from a continuous signal $f(t)$ and converts the momentary amplitude into a proportional digital value Z
Block diagram	One possible formal representation of a linear system in which functional units are represented by blocks, with signals represented by arrows pointing to, and away, from the block. Blocks typically contain the transfer function of the system they represent
Bode plot	The Bode plot is a frequency-domain plot of a system response to sinusoidal input signals of varying frequency. In a Bode plot, magnitude and phase shift of the transfer function are plotted over the frequency in two separate diagrams
Characteristic polynomial	In the Laplace domain, the characteristic polynomial is the denominator polynomial of a transfer function
Closed-loop configuration	The closed-loop configuration is achieved when the path of a feedback loop is completed, notably, when the sensor signal is subtracted from the setpoint signal, and the resulting control deviation leads to a corresponding control action. Often, it is easier to determine system behavior in an open-loop configuration, but the feedback control system is only engaged when the loop is closed
Compensator	A compensator is a part of a control system that is introduced to improve stability or dynamic response
Control action	This is the output signal of the controller and serves to actuate the process and therefore to move the controlled variable toward the desired value
Control deviation	The control deviation $\epsilon(t)$ is the time-dependent difference between setpoint and sensor output. The control deviation is often also referred to as the *error variable* or *error signal*
Controlled variable	This is the output of the process. The design engineer specifies the controlled variable in the initial stages of the design
Controller	The controller is a device that evaluates the control deviation and computes an appropriate control action. In many cases, the controller is the only part of the feedback control system that can be freely *designed* by the design engineer to meet the design goals of the closed-loop system

Linear Feedback Controls. http://dx.doi.org/10.1016/B978-0-12-405875-0.00027-9
© 2013 Elsevier Inc. All rights reserved.

DAC	The DAC (digital-to-analog converter) converts a digital value to a proportional analog current or voltage. After momentary application of the digital value, the DAC usually holds the output constant
Dead-time system	A dead-time system is a system where the response to an input function $x(t)$ follows after a delay. Examples include transport delays in conveyor belts and extruders. The delay introduces an irrational term e^{-sT} into the transfer function, and polynomial-based methods can no longer be applied
Dependent variable	Any signal that is influenced by other signals anywhere in the entire system (such as the control deviation) is a dependent variable
Differentiator	A differentiator is an elemental system whose output is proportional to the first derivative of the input
Disturbance	Any influence other than the control action that influences the controlled variable. Examples are variable load conditions in mechanical systems, electromagnetic interference, shear winds for airplanes, etc. A disturbance $d(t)$ can often be modeled as an additive input to the process
Feedback path	The feedback path carries the information about the controlled variable back to the controller. The definition of the feedforward and feedback paths is not rigid and merely serves as orientation
Feedforward path	The feedforward path includes all elements that directly influence the controlled variable
Fourier transform	An integral transform that decomposes a time-dependent periodic signal into its harmonic content, that is, the strength of the constituent harmonic oscillations
Gain margin	A metric of relative stability. The gain margin is the amount of additional gain that can be introduced to a feedback system before the closed-loop system becomes unstable
H-bridge	The H-bridge is an electronic circuit composed of four transistors that allows to control the current through a component by magnitude and sign
Independent variable	Any external signal that is not influenced by other signals is an independent variable. Examples are the setpoint R and the disturbance D
Integral windup	Integral windup occurs in *PI* and *PID* controllers when, during the transient phase, large control deviations are accumulated in the integral component of the controller. Integral windup leads to considerable overshoot in the system response
Integrator	An integrator is an elemental system whose output is proportional to the time integral of the input. In a control context, an integrator can reach a steady-state only when its input is zero

ITAE	ITAE stands for integrated, time-averaged absolute error. The ITAE criterion is one of several time-integrated metrics to quantify the performance of a system during the transient phase. These criteria complement other metrics, such as the steady-state control deviation
Laplace transform	An integral transform that relates a time-dependent system or function to a function of the complex variable s that contains the periodic and aperiodic (i.e., exponential) components
Loop gain	The loop gain $L(s)$ of a feedback system is the product of all transfer functions along a feedback loop
LTI system	A linear, time-invariant system (LTI system) is any system whose signals obey the superposition and scaling principles, and that responds to a delayed input signal with an equally delayed output signal
MIMO	MIMO stands for *multiple input, multiple output* and describes a system with more than one input signal and more than one output signal. A linear MIMO system can be decomposed into multiple SISO systems
Model	A model is an abstract description, usually in mathematical terms, of a system. The model allows us to predict the response of a system to a given input signal. Models are often based on simplifying assumptions (such as linearity) that limits applicability of the model to the real-world system
Noise	Noise is a broadband random signal that may be introduced in many places of a system. Amplifiers typically create noise; dirt or rust in a mechanical system would also create random influences. It is often possible to model noise as a single signal $n(t)$ as an additive input to the sensor
Nyquist frequency	The Nyquist frequency is upper frequency limit in a digital system for the signal to be exactly reconstructed. Shannon's sampling theorem states that the Nyquist frequency is exactly one half of the sampling frequency
Nyquist plot	The Nyquist plot is a frequency-domain plot of a system response to sinusoidal input signals of varying frequency. In a Nyquist plot, real and imaginary parts of the Fourier-domain transfer function are plotted as a function of the input frequency on orthogonal axes
Open loop configuration	An open-loop configuration is achieved when any part of the feedback loop is interrupted, and feedback control cannot be achieved. The open-loop configuration is useful to determine system coefficients and to draw conclusions on closed-loop stability without engaging the feedback control system

Operational amplifier The operational amplifier is a linear element whose output signal is proportional to the difference of its two input signals. The operational amplifier can be used to realize an analog difference operation and a large number of other linear operations, including integral, derivative, time-lag, and time-lead functions

Oscillator Oscillators are feedback systems that are deliberately designed to be unstable to create an oscillatory output signal. Special design provisions need to be made to maintain a constant frequency (resonance) and amplitude

Partial fraction expansion Partial fraction expansion is a method to find a set of transfer functions with low order (specifically, first and second order) whose sum is equal to a higher-order transfer function

Phase margin A metric of relative stability. The phase margin is the amount of phase delay that can be introduced to a feedback system before the closed-loop system becomes unstable

***PI* controller** A type of linear system whose output is a weighted additive combination of the input signal and its time integral. In a feedback loop, the *PI* controller forces the control deviation to zero, because it cannot reach a steady state otherwise

***PID* controller** A controller whose output is a weighted additive combination of the input signal, its time integral, and its first derivative. The *PID* controller is one of the most widely used control elements

PLL *PLL* stands for phase-locked loop. A *PLL* is a feedback control system that controls an oscillator such that its output matches a reference signal in both frequency and phase

Process The process (also referred to as plant) is the linear system to be controlled. The process has the controlled variable as its property and provides some means to influence the controlled variable

PWM *PWM* stands for pulse-width modulation. *PWM* is a method to use two digital states (on and off) to create pseudo-analog control. *PWM* requires that the controlled process is slow enough to integrate (average) the switching frequency

Relative stability The notion of relative stability reflects how close system poles are to the imaginary axis, or, in other words, how long it takes for transient components of the response to decay. The notion of relative stability is usually applied to oscillatory responses. Relative stability relates to the degree of robustness or tolerance of a system toward changes that drive a system instable

Resonance	Resonance is a form of energy accumulation in underdamped second-order systems that leads to an oscillatory output. Resonance is related to a complex conjugate pole pair close to the imaginary axis
Root locus	The root locus is the set of all locations of the roots of a polynomial in the s-plane as one coefficient is varied from zero to infinity
Saturation	Saturation occurs when the output signal of a component is limited by physical constraints (such as linear travel stops or power supply limits). In saturation, the system is no longer linear and has a differential gain of zero
Sensitivity	The sensitivity of a system is defined from the loop gain $L(s)$ as $S(s) = 1/(1 + L(s))$. Low sensitivity (high loop gain) improves the dynamic and steady-state response of a system and improves its disturbance rejection
Sensor	The sensor is an apparatus to measure the controlled variable and make the measurement result available to the controller
Setpoint	This signal determines the operating point of the system and influences the controlled variable through the controller
Signal	A signal is any *observable* or *measurable* time-varying property of a system. Signals can be, for example, voltages or currents in electronic systems, but also a force, liquid level, pressure, speed, or position
Signal flow graph	One possible formal representation of a linear system in which transfer function is represented by links (arrows) in a graph
SISO	stands for *single input, single output* and describes a system with exactly one output signal and exactly one independent input signal
Spring-mass-damper system	The spring-mass-damper system is a model for many second-order mechanical and electronic systems that involve energy storage both as kinetic energy (mass) and as potential energy (spring) simultaneously
Stability	A system is absolutely stable when all transient components decay over time. A strict definition is BIBO stability (bounded input-bounded output) where a system is stable when its response to a bounded input signal is also bounded. Under a slightly different definition, a system is stable if its impulse response is absolutely integrable
State-space model	A mathematical representation of control systems where input and output variables are related by sets of first-order differential equations. Intermediate variables that link the first-order differential equations are called state variables. The state-space model advertises itself for an elegant and compact matrix representation of the system.

Steady state	The steady state of a system is achieved when all transients have decayed and time-dependent signals have reached an equilibrium. Steady state implies that all time derivatives of all signals become zero
Step response	The step response of a system is the time-varying output signal when a step function (i.e., a sudden change) is applied to the input. Often, step response refers to the more specific *unit step* response, in which a sudden change from zero to 1 is applied to the input
System	A very broad term to describe an assembly of interacting components (e.g., electrical or mechanical parts) to serve a specific goal or to meet a specific design purpose
Transfer function	The transfer function of a system is a mathematical model of a system, usually in an integral-transformed space (Laplace transform, Fourier transform or z-transform) in such a fashion that the equally transformed input signal, multiplied by the transfer function, yields the output signal
Transient response	The transient response of a system is the time-dependent output signal (or its Laplace transform) in response to a specified input signal immediately after application of the input signal, that is, before any equilibration can take place
Two-point controller	A control element whose output can assume only two states (e.g., on and off). The trip points of the input signal where the two-point controller changes its output state are different (hysteresis) to avoid undefined switching behavior
VCO	A VCO (voltage-controlled oscillator) is an oscillator whose output frequency linearly follows an input voltage. The VCO is a key element in a phase-locked loop
Zero-order hold	The zero-order hold is a mathematical model for a sampling system that takes samples $f(kT)$ at equal time intervals T and holds the value constant to the next sampling interval.

Printed in the United States
By Bookmasters